T0313720

Business Chemistry

Business Chemistry

How to Build and Sustain Thriving Businesses in the Chemical Industry

Edited by

Jens Leker
University of Münster, Department of Chemistry and Pharmacy, Germany

Carsten Gelhard
University of Twente, Faculty of Engineering Technology, The Netherlands

Stephan von Delft
University of Glasgow, Adam Smith Business School, United Kingdom

This edition first published 2018

Registered Office(s)
John Wiley & Sons, Inc., 111 River Street, Hoboken, NJ 07030, USA

Editorial Office
9600 Garsington Road, Oxford, OX4 2DQ, UK

For details of our global editorial offices, customer services, and more information about Wiley products visit us at www.wiley.com.

Wiley also publishes its books in a variety of electronic formats and by print-on-demand. Some content that appears in standard print versions of this book may not be available in other formats.

Library of Congress Cataloging-in-Publication data applied for

Hardback: 9781118858493

Cover image: Evonik Industries production plant
Photograph by Stephan Kube; Courtesy of the editors
The cover photo shows Evonik's new production plant for C-4 based products in Marl (Germany). Thanks to a globally new process, the company can now use FCC C4 (FCC = Fluid Catalytic Cracking) material flows for the production of a broad portfolio of chemicals.
Cover design: Wiley

Set in 10/12pt Warnock by SPi Global, Pondicherry, India

10 9 8 7 6 5 4 3 2 1

Contents

List of Contributors

Manuel Bauer is the Global Head of Innovation Management at LEDVANCE, a global leader in the general lighting market. Previously, he was a Manager at Innosight Consulting, a boutique consultancy founded by Professor Clayton Christensen, which focuses solely on helping clients develop and implement innovation and organic growth strategies. Previously he had worked as Corporate Innovation Manager at Clariant, a global specialty chemical company headquartered in Switzerland. There, he co-developed the new innovation management approach and implemented it in two of Clariant's largest global business units, including the project management process and tools, as well as the innovation portfolio management process and performance management concept. Manuel started his career at McKinsey & Company, in the Munich office, where he was a member of the global chemicals practice. He holds a PhD from the University of Münster, Germany, and a master of science in Chemistry from the University of Würzburg.

Carsten Gelhard is an Assistant Professor of Product-Market Relations at the Faculty of Engineering Technology of the University of Twente, The Netherlands. Prior to this, he was a postdoctoral researcher at the Amsterdam Business School, The Netherlands. Carsten received his PhD from the University of Münster, Germany, where he also studied Business Chemistry. His research is at the intersection of marketing, innovation management, and operations management. In particular, Carsten examines how firms can achieve sustained competitive advantage: (i) by managing trade-off situations that are associated with value creation activities and (ii) by collaborating with external partners across the value chain. His work has been published in the *Journal of Business Research* and the *Journal of Operations Management*.

Gerald Kirchner is Head of the Department of Corporate Environment, Health, and Safety at ALTANA, Wesel, Germany. Previously he headed the Innovation Management Group and the Global Regulatory Affairs Department within BYK, a division of ALTANA. Gerald started his professional career at

Chemie Linz/DSM (Austria) as a research chemist in the area of organic intermediates. He spent his postdoctoral year at the Massachusetts Institute of Technology, USA, within the Department of Chemical Engineering/Bio Catalysis. Gerald studied Technical Chemistry at the Technical University of Graz, Austria. In his thesis, he elaborated different synthesis routes to plant growth hormones.

Jens Leker is a Professor at the University of Münster, Germany, and Director of the Institute of Business Administration at the Department of Chemistry and Pharmacy. His research focuses on forecasting techniques, open innovation and knowledge sharing in R&D collaboration. Jens has published more than 30 articles on technology and innovation management in journals such as the *International Journal of Innovation Management, R&D Management, Technological Forecasting and Social Change* and *Technovation.* He studied business administration at the University of Kiel, Germany, where he also received his PhD. Jens is a member of the German Chemical Society (GDCh), Head of the Advisory Board of the International Society of Professional Innovation Management (ISPIM) and an Editor-in-Chief of the *Journal of Business Chemistry.*

Thibaut Lenormant is an Innovation Manager at Gebauer & Griller in Vienna, Austria, and a PhD candidate at the University of Münster, Germany. He studied chemical engineering at the ESCPE Lyon, France, and innovation management at EMLyon, France. Thibaut also worked for several years as a Business Developer in the chemical industry where he contributed to the development of major innovations. His research focuses on organizational learning, complexity theory, and new product development processes.

Tobias Lewe is Partner and Managing Director at A.T. Kearney Management Consulting GmbH, Germany. He has more than 17 years of experience in top management consulting, leads A.T. Kearney's EMEA practice for Energy & Process Industries, and is a member of the A.T. Kearney's EMEA senior leadership team. Since starting his career he has consulted for a broad range of multi-national clients, concentrating on downstream oil and petrochemicals as well as on chemical and coatings industry chains. His consulting expertize includes exogenous and organic growth, operating model design, innovation management, and digitization strategies as well as operations and performance transformation. Prior to joining A.T. Kearney, Tobias worked in the downstream oil industry, where he had management functions in supply and distribution and manufacturing/refining for ExxonMobil/Esso in Central Europe. He received his graduate and doctoral degrees in Chemistry at the University of Cologne.

Eric Meyer is Researcher at the Institute of Cooperative Systems at the University of Münster, Germany. After studying mathematics and economics at the University of Oldenburg he received his PhD from the University of Münster. His research focuses on the cooperation of companies. He is especially interested in how existing management methods have to be extended and adapted in order to make them applicable in business cooperation.

Theresia Theurl is Professor at the University of Münster, Germany, and Director of the Institute of Cooperative Systems. She studied economics at the University of Innsbruck, Austria, where she received her diploma. After three years as a Lecturer at the University of Munich, she returned to Innsbruck and received her PhD in economics. Theresia is member of numerous committees and councils related to cooperatives and cooperating companies. Since 2014 she has been Dean of the Münster School of Economics and Business. Her research focuses on the governance mechanisms that can be observed in cooperatives and cooperative groups and on the mechanisms that govern the cooperation of companies.

Irina Tiemann is a Senior Researcher at the Department of Business Administration, Economics and Law of the University of Oldenburg, Germany, and a member of the Oldenburg Center for Sustainability Economics and Management (CENTOS). Irina received her PhD from the University of Münster, Germany, where she also studied Business Chemistry. She subsequently worked as Business Development Manager at aleo solar GmbH in Oldenburg, Germany. Her academic interests are in the field of innovation management and sustainable entrepreneurship. Her work has been published in the *International Journal of Innovation Management* and *International Journal of Business Venturing.*

Hannes Utikal is a Professor for Strategic Management and Sustainability at Provadis School of International Management and Technology, Frankfurt, Germany, where he leads the Center for Industry and Sustainability. His research interests focus on strategic management and the design of transition processes. He initiated the "rhein-main-cluster chemie and pharma" where successor companies of the former Hoechst AG exchange knowledge about good management practices in the chemical and pharmaceutical industries. Hannes co-edited the book *Future. Chemistry. Glimpses at the World of Tomorrow* and has published on the industries' managerial challenges in the *Journal of Business Chemistry* and *CHEManager*. In addition, Hannes is involved in innovation and education projects at Climate-KIC, the world's largest public–private partnership working in the fields of climate adaptation and mitigation. Hannes studied business administration at the University of Cologne, where he also received his PhD. He is a member of the German Chemical Society (GDCh) and an Editor-in-Chief of the *Journal of Business Chemistry.*

Stephan von Delft is a Lecturer in Strategy at the Adam Smith Business School, University of Glasgow, UK. His research focuses on business model design, business model innovation, and organizational capabilities. Stephan has published in *CHEManager, Nachrichten aus der Chemie*, the *Journal of Business Chemistry* and the *Journal of Business Research*. He studied Business Chemistry at the University of Münster, Germany, where he also obtained his PhD. Stephan has been a visiting researcher at the University of San Diego, USA, and a postdoctoral researcher at the Amsterdam Business School, the Netherlands. He is a member of the German Chemical Society (GDCh), the Strategic Management Society (SMS), and a member of the Scientific Panel of the International Society of Professional Innovation Management (ISPIM).

Daniel Witthaut is Head of Corporate Innovation Strategy at Evonik Industries AG, a leader in specialty chemicals. Prior to his current role, he worked for six years as a Vice President in building up and leading the New Business Development Department for the former Advanced Intermediates Business Unit of Evonik. Daniel has more than 18 years of professional experience in innovation, strategic controlling, and strategy functions in different areas of the chemical industry and has lived in diverse cultures (USA, Germany, China, Singapore). He is a lecturer and a frequent (key note) speaker at conferences on Innovation Management. Furthermore, Daniel is a member of the Board of the European Industrial Research Management Association (EIRMA). He holds a PhD in Organic Chemistry from the University of Münster, Germany, and an MBA from the University of Chicago, USA.

Preface

Marketing-, R&D-, and production-related activities need to be orchestrated with the architecture of a firm's business model to build, grow, and sustain a chemical company in today's competitive environment. This orchestration inevitably requires that chemists, chemical engineers, and other R&D experts acquire business skills from the fields of strategy and innovation to jointly create value with marketers, business developers, and executives with backgrounds in business administration. Consequently, there is a growing demand in the chemical industry for trained specialists who not only have a solid chemical knowledge but also a good understanding of the underlying management processes. What is needed are experts in both chemistry *and* business – business chemists.

Business Chemistry is a practitioner-oriented book that grew from this demand. It takes the characteristics of the chemical industry (e.g., research intensity, business-to-business relationships) into consideration while introducing experts with backgrounds in science and engineering to the most relevant and latest managerial topics for the chemical industry and related sectors, such as biotechnology, consumer products, and pharmaceuticals. The book is structured into two parts. The first part deals with key topics from the field of strategy, such as industry-specific challenges impacting strategy formulation and execution, as well as analytical methods and concepts of strategic analysis applied by chemical companies. The second part covers key topics from the field of innovation, such as concepts and tools for new product and new business development in the chemical industry as well as collaborative activities with customers and suppliers. All chapters within these two parts of the book are written by experienced practitioners from companies such as ALTANA, A.T. Kearney, and Evonik Industries, and leading academics from the field of business chemistry.

We would not have been able to edit this book without the support from several individuals. Firstly, we would like to thank all co-authors for their valuable contributions and great commitment to offering insights into the chemical industry. Secondly, Birte Golembiewski, Gerrit Knispel, and Nicole vom

Stein are thanked for their feedback and for facilitating the editing process. We also thank Walter W. Zywottek for a vital discussion about various topics covered in the book. Of all the great contributors behind the scenes, we finally wish to thank our publisher Wiley, especially Shagun Chaudhary, our project editor, Sarah Higginbotham, our lead manuscript editor, and Rebecca Ralf, Managing Editor Life Sciences Books. Without their support, this book would not have been possible.

We hope that this book will be informative, useful, and enjoyable for you, and that it will enable you to build and sustain thriving businesses in one of the most exciting and versatile industries of all.

Jens Leker, Carsten Gelhard, and Stephan von Delft
Münster, Enschede and Glasgow 2018

Part I

Strategy

1

Management Challenges in the Chemical and Pharmaceutical Industry

Jens Leker[1] *and Hannes Utikal*[2]

[1] *University of Münster, Department of Chemistry and Pharmacy*
[2] *Provadis School of International Management and Technology AG*

For time and the world do not stand still. Change is the law of life. And those who look only to the past or the present are certain to miss the future.

John F. Kennedy (1917–1963), 35th President of the United States of America

The first chapter of this book outlines the specific characteristics of the chemical and pharmaceutical industry regarding, for example, products, site locations, competition, and research efforts. Additionally, the chapter summarizes results of a survey in the German chemical and pharmaceutical industry on business transformation processes and drivers of change that affect the industry. From these findings, management challenges and solutions to these problems will be derived.

1.1 Introducing the Chemical Industry as a Source of Innovation and Prosperity

The chemical industry is one of the major global industries affecting all parts of human life. Advances in chemicals and pharmaceuticals have contributed to improving living conditions and particularly nutrition and health levels worldwide. Enhancements in the field of automobiles as well as new developments concerning battery electric or fuel cell vehicles have resulted, not least because of new materials and new formulations originating from the chemical industry. New electronic devices such as smartphones have only been possible due to a change of pace in the development of electronic materials and an increase in

Business Chemistry: How to Build and Sustain Thriving Businesses in the Chemical Industry, First Edition. Edited by Jens Leker, Carsten Gelhard, and Stephan von Delft.

their purity. Continuous research for and production of active pharmaceutical ingredients (APIs) are of central importance for fighting (new) diseases and improving therapeutic methods.

The chemical and pharmaceutical industry alters modern life through the transformation of scientific findings into marketable products. The invention and industrialization of production pathways such as the Haber(–Bosch) process for ammonia synthesis, the Fischer–Tropsch process to produce liquid hydrocarbons, or the contact process for producing sulfuric acid laid the foundations of the chemical industry. These processes acted as prerequisites for overall industry growth, technological change and wealth creation, whereby the underlying reaction pathways still apply today. Enormous advancements in technology in recent years have additionally enabled the sector to have an economically, ecologically, and socially positive impact on society in the future as well as today. In order to continue to achieve this goal, the chemical industry is reconsidering its modes of operation and finds itself in a phase of transformation [1].

From an economic perspective, the crucial role of the chemical industry for different customer value chains and the connection to nearly every end-consumer market is reflected by the impressive size of world chemicals sales in 2013 of €3156 billion and the average global growth rate slightly above the global gross domestic product (GDP) [2]. It has to be considered that this overall development is mainly driven by high growth rates in the Asian–Pacific region, eventually compensating for lower growth rates in Western countries. Asia has already become the largest market for chemicals, with now more than 50% of the global market. This share is very likely to increase even more due to the growing population in Asia and the declining demand in the West, especially in Europe.

All these facets and volatilities make the chemical industry one of the most fascinating industries, not only from a scientific, technological or societal perspective, but also from a business point of view. In the following, we first characterize this highly interesting industry with regard to its specific characteristics and then subsequently highlight current managerial challenges relevant to the industry. In order to do so, we combine results from a literature review with findings from one of our own empirical studies on the management challenges in the chemical industry.

1.2 Characteristics of the Chemical and Pharmaceutical Industry

The chemical industry today is one of the largest industries in the world, with an impressive history (see, e.g., [1, 3]). This is reflected by the variety of products, processes, and market characteristics.

1.2.1 Product and Process Characteristics

The chemical industry is a **process industry** where firms "add value to materials by mixing, separating, forming, or chemical reactions" [4:28]. Process industries differ from so-called discrete industries with regard to the production process. In discrete industries, for example the automotive or engineering industry, production pathways converge as final products are assembled by using multiple discrete input components [5]. In contrast, a product in the chemical industry can simultaneously act as an intermediate, be processed further to synthesize other products, or serve as a finished, salable product. Production processes can therefore be convergent and divergent at the same time, resulting in an increased complexity for the planning and optimizing of such processes. In each process, components are mixed and react under well-defined physical conditions. In order to obtain high reaction yields, chemical companies rely on experience and knowledge from different fields, especially chemistry and engineering, and in some cases biology and biotechnology. Hence, the special nature of the highly complex processes sets the framework for all managerial decisions in the chemical industry.

By adapting a value chain perspective, the chemical industry appears to convert organic and inorganic raw materials into value added products (see Figure 1.1). The upstream stages are closely linked to the petrochemical and exploration industry and are only manufacturing a few products, such as fertilizers or basic plastics originating from the Naphtha fraction of crude oil, and, in the case of inorganic materials, deriving from chlorine and salts. In the downstream steps, products of the upstream operations are further processed into a variety of products, which then enter various end markets. The customers of chemical companies are usually other firms who process the materials into end products, so that most relations are business-to-business (B2B) in nature.

Within the chemicals value chain, the **production processes** vary. One can distinguish continuous, campaign, and batch production processes. Each process requires specific production assets, which tie up capital:

- Continuous processes run on single-purpose resources, steadily producing one product and not requiring regular changeover decisions. This type of process avoids downtime and scrap. However, flexibility in applying a different feedstock and input is limited as the production line is specialized for a certain product or process. Continuous processes can typically be found at the beginning of the chemicals value chain, involving petrochemicals, basis chemicals, and bulk polymers.
- Campaign production is related to multi-purpose assets, so that different processes and products can run on the same production resource.
- Batch production is also related to multi-purpose resources and, in addition, is suitable for steps implying a well-defined start, throughput, and end production time as well as the ability to customize the huge amounts of the desired product. This is typically the case in the specialty chemicals segment [5].

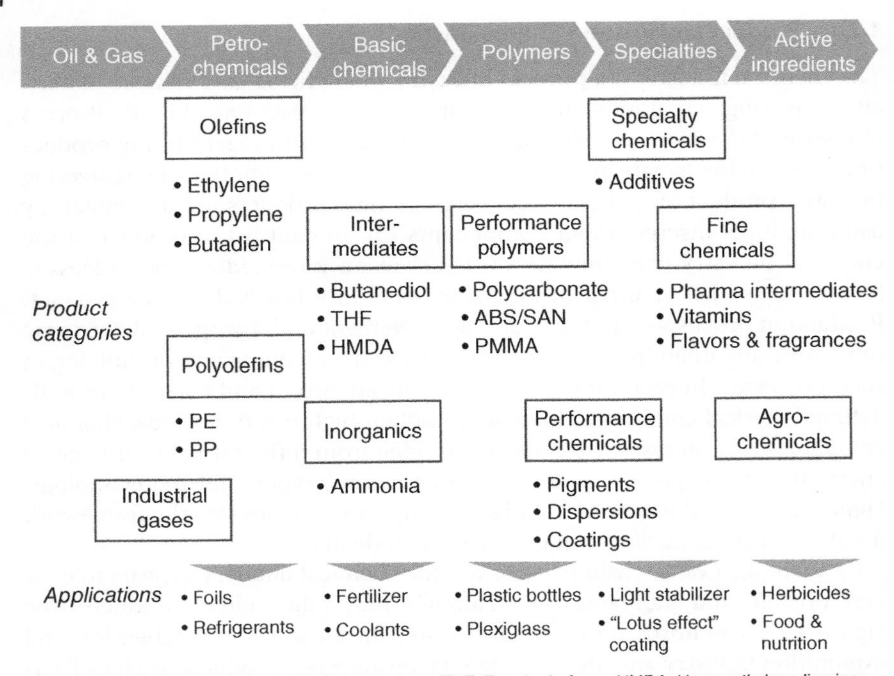

Abbreviations: PE: Polyethylene, PP: Polypropylene, THF: Tetrahydrofuran, HMDA: Hexamethylenediamine, ABS: Acrylonitrile butadiene styrene, SAN: Styrene acrylonitrile, PMMA: Polymethy methacrylate

Figure 1.1 Chemical products in the industry value chain

There are different approaches to classify products of the chemical industry. Kline (1976) [6] distinguishes between **commodities** and **specialty/fine chemicals.** Following this categorization, commodities demonstrate a low degree of differentiation and a high production volume. They can be found in the early stages of the value chain and are manufactured by means of continuous production processes. These standard, high-volume products with few variants are typically characterized by a low unit value and a low unit margin. Thus, the main buying criterion is the price. On the contrary, the specialty segments show a high degree of differentiation and a small production volume. Specialty chemicals are typically located at later stages in the value chain and are produced in batches. These products are often available in many variations and generate fairly high unit values and margins. Customers buy specialty chemicals due to their specific and unique product properties.

The different segments of the chemical sector can furthermore be described by looking at the relative importance of additional key success factors, such as the intensity and contribution of research and development (R&D) to success, the relevance of distinctive knowledge about specific markets and customer insights, and the importance of highly qualified personnel for the success of a business. These aspects are particularly decisive for the pharmaceutical and specialty chemicals segment, whereas the extent of investment in production facilities, the energy intensity of manufacturing, and an immediate access to raw materials significantly affect the success of commodity businesses.

1.2.2 Market Characteristics

The chemical industry has been growing since its emergence in the 1860s. By encompassing all parts of modern life and creating new materials or new active ingredients for pharmaceuticals, the chemical industry has always been a trigger for innovation in its customer industries. The chemical industry has a share of 3 to 4% of the global GDP. The main markets are the European Union, the United States, and Asia, with Japan and China as central markets. While the growth rates of chemical consumption in mature economies such as Germany and the United States are similar to the rates of the respective national GDP, emerging economies, especially China, are demonstrating significant growth.[1]

Geographically, the chemical industry acts within at least three different markets. For a very limited number of products, companies produce the entire quantity of a product for the **global market** at one location. In this case, transportation costs must be negligible in view of the total cost position of a product and economies of scale. As a consequence, consolidation of the production in one plant is preferred over a global duplication of production activities. This is particularly valuable for producing APIs, where production processes typically have to be accredited. Nevertheless, **regional production** for the European, North American, and Asian markets is pursued for the majority of products. While there are limited trade flows between these main manufacturing regions, trading within the regions, for example within the European Union, is more intense. In addition to the global and regional markets, **local markets** can be identified, where products are only delivered around or even within one specific production facility. This can be observed in a so-called *Verbund* production system, which is an integrated production where products are delivered directly, via pipes, to customers that are based on the same chemical park. Overall, the chemical industry occupies a multiregional role.

With its different segments, the chemical industry provides significant **profit earning potential**. In rankings comparing the profitability of different industries, the pharmaceutical industry is often found among the top industries with an EBIT (earnings before interest and taxes) of about 20%. Other

---

1 Current figures on the industry's development might be found, for example, at the websites of the European Chemical Industry Council (cefic), the American Chemical Association (ACS), or the German Chemical Industry Association (VCI).

profitable industries included within this class are petroleum, tobaccos, or consumer foods. The chemical industry (without pharmaceuticals) ranks in the middle of the list of 16 sectors [7]. Other industries with a much higher visibility in the business news, such as electronics, telecommunications, or aviation, have much lower results. In the following, we will discuss reasons for this favorable profit position.

First of all, companies active in the chemicals value chain **provide value to their customers**. Pharmaceutical firms produce highly differentiated products bought by price-insensitive consumers and new products often benefit from their monopoly-like position due to patent protection. The producers of specialty chemicals manufacture highly differentiated products as well, and can often charge high prices, as customers need these specific products and might even be able to generate a competitive advantage for their firm by buying them. On the other side, the price pressure is much higher for commodities where products are highly standardized, so that companies can only differentiate themselves from their competitors through product prices. However, it would be wrong to assume that only highly price-differentiated companies can be profitable. In the chemical industry, those companies with an access to low-cost raw materials, low-cost energy or a highly effective interlinked production can realize above-average profits in the field of commodities as well [8].

The **intensity of competition**, another main driver influencing industry profits, varies by segment and region. While the whole chemical industry is somewhat less consolidated than other industries, a higher degree of concentration can be observed at the segment level. This is, for instance, reflected by the top six manufacturers of crop protection products, who account for around 80% of this market [9]. In regional terms, the North American market shows the highest concentration, implying a rather oligopolistic market structure (with few players of similar size and power). Although Asian markets show lower degrees of concentration, profit-destroying price wars can be impeded due to the strong growth [10]. In addition to the number of players active in the sector, risks for the profitability of the chemical industry stem from its capital intensity and high barriers to exit the market. For instance, in times of an economic downturn, firms are not able to reduce their production volume gradually due to process requirements, especially in the case of continuous production processes. The resulting overcapacities eventually lead to deteriorating prices and, in turn, to deteriorating profits. Moreover, exit barriers such as lay-off protection and environmental regulations, which primarily apply to European production sites, constitute high exit barriers. High market entry barriers, notably the major investments in production plants, R&D, and marketing, have mostly prevented new companies from entering the chemical industry in the past. Consequently, the industry is characterized by a specific set of companies, where mergers and acquisitions occur frequently but new players are rare [3, 11, 12].

The evolution of the chemicals industry can be explained by means of **its underlying basic sciences** [13]. Business historian Alfred Chandler finds that the success of companies in the chemical and pharmaceutical industry results from transferring findings from basic research into marketable products and using the profits and experience gained from each new generation of products to commercialize the next generation [3]. Such companies have yet to be aware of a future where science and technology essential to the continuing growth of high-technology companies might stop being the engine for innovation and growth. The chemical industry, with its periods of research-based growth between the 1880s and 1920s and again during the 1940s and 1950s, has to cope with the fact that since the 1950s, only a few major new developments have been created by chemical sciences or engineering [1, 14]. Incremental product and process developments have thus gained more importance for successful companies in the chemical industry than basic research (which is rather aimed at radical inventions). Also, the successful model of pharmaceutical companies developing new products based on basic research findings (blockbuster products) has stumbled lately. However, in spite of this, in the 1960s and 1970s, biology, as well as the related disciplines of microbiology, enzymology, and the beginnings of molecular biology, contributed to the generation of new pharma products. Since the 1980s, advances in the field of biotechnology have fueled the development of innovative products from basic research findings.

To sum up, the chemical industry is actually a process industry encompassing thousands of products used in different applications and enabling innovations in their customers' industries. The industry is capital intensive and consists of various segments, each having specific success factors and typically showing a multiregional character. The industry has a long tradition, with the initial industrial chemistry dating back to the 1860s in Great Britain. Applying insights from industry lifecycle theory [15], the industry can be classified to be in a maturity phase where the basic technological know-how is well diffused and the focus is – except for patent-heavy pharmaceuticals and some specialty chemicals – moreover set on technological improvements rather than on breakthrough innovations.

1.3 Business Transformation in the Chemical Industry

How can we then explain companies' success in this industry? And how can companies prepare for future success? While the perspectives and methods differ, these core questions are of importance for management practitioners and scholars alike [16].

1.3.1 Business Transformation and Organizational Change Processes

Business historians have analyzed the successful companies in the chemical and pharmaceutical industries by (mainly) focusing on past events. As mentioned earlier, Chandler (2005) identified a company's ability to create learning processes from one product generation to the next as being key to success. He found that companies with a focused strategy, limited in complexity in terms of different markets and products, are often more successful than firms pursuing strategies of unrelated diversification [3]. Another striking finding addresses the capability of successful companies to manage relationships within a value network. A chemicals firm's position in industry networks, encompassing other chemical and pharmaceutical firms, and a supporting nexus of specialized suppliers of products and services, serves as a market entry barrier. While these networks were basically established for the chemical industry between the 1880s and 1920s, developments in biotechnology might open up a new field where positions for new as well as established players are not yet fixed.

In recent years, the questions of whether and how companies can proactively adapt to upcoming changes have gained a lot of attention. Approaches have touched various aspects at all levels within a company, from path-dependent strategic behavior over continuous innovation cultures to the presence of (certain) dynamic capabilities that enable firms to adapt to changing environments. On the one hand, exogenous developments, such as globalization, demographics, and technological changes, have had a profound impact on the way companies do business. On the other hand, endogenous dynamics, such as product and process innovation or the re-invention of business models, may also lead to large-scale organizational change. Organizational change is defined as a shift in form, quality or state of an organizational entity over time [17]. Change processes can be observed for multiple entities (e.g., a whole industry) or for a single entity (e.g., a single company). One influential field analyzing change at the level of multiple entities is the so-called population ecology school, stating that the ability of a single entity to change is very limited. This school proposes a Darwinian view, describing change processes as a result of variation, selection, and retention to be adequate in order to understand change processes (e.g., [18]). The opposite position is taken by the school of planned change. This model in turn views developments at the level of the individual organization as a result of an active organizational design process, where decision makers formulate goals, implement measures, and evaluate the impact on the defined goals (cf. [19] for the different models). In the following, we discuss organizational change from a single company perspective and base our reasoning on the assumption that companies have some discretionary power in actively designing change processes.

Organizational change processes can differ according to their intensity. Incremental changes encompass minor modifications of the status quo, whereas radical changes have a profound impact on different fields of an organization [20]. Transformation processes can additionally be distinguished in terms of the question of whether the organization anticipates an upcoming need to change or whether it reacts passively as a response to external influences [21]. Even though the term proactive transformation appears to have a positive connotation in the practice-oriented management literature, proactive behavior might not be a successful concept *per se*. On the contrary, it is challenging for managers to balance the need for stability and exploitation of today's resource base, on the one hand, with the prospects of exploring new paths, on the other.

Summing up, we use the term business transformation to describe processes of intended organizational change. We conceptualize managers as change agents who proactively or reactively try to develop and shape their fields of responsibility (company, business unit, or department) in order to achieve prior defined organizational objectives.

In a recent study, we analyzed the need for business transformation in the German chemical and pharmaceutical industry by means of a large-scale online survey, conducted in 2014. In this survey, we addressed upcoming trends, potentially creating a need for transformation, and also asked participants about the relevant management activities to cope with these trends. In total, 270 people participated in the online survey: 141 managers possessing relevant experience in the industry completed the questionnaire; 34% of the respondents considered themselves as being experts in the segment of specialty chemicals, 16% in the field of polymers, 22% in pharmaceuticals, 10% in basis chemicals, 8% in agrochemicals, and 10% in other fields; 50% of the participants were top-managers (board level), 20% were experts in R&D and innovation, 25% had other leading positions in chemical and pharmaceutical companies, and 5% held other positions. The sample covered different company sizes: 13% of the participants were affiliated with companies of up to 100 employees, 14% were in firms with 101–1000 employees, 21% were in companies with 1001–10 000 employees, 42% worked in large companies with 10 001–100 000 employees and 10% in companies with more than 100 000 employees.

In the following, we present findings from this survey. In doing so, we distinguish between "successful" and "less successful" companies based on participants' self-evaluation.[2] Following this distinction, 31% of the respondents classified their companies as being very successful, while 57% designated their companies as being on average successful, and 12% as not successful.

2 The participants were asked to answer the following question: "Overall, we are more successful than our strongest competitors" using a scale from 1 = "I strongly disagree" to 7 = "I strongly agree." Participants answering with 1 or 2 were classified as "not successful," those answering 3, 4, 5 = "average successful," and those answering 6, 7 = "very successful."

1.3.2 Drivers for Change

The future of the chemical industry, and particularly the impact of so-called global megatrends, is actively debated in the literature [1, 22, 23]. **Global megatrends** are long-term trends that may have a global reach lasting for more than 20 years and are defined as drivers of change that affect all parts of society, business, and politics. On the basis of these megatrends and their complex interplay, political institutions, industry associations, and companies create different scenarios for the future. Industry associations employ these pictures to communicate potential opportunities and risks for an industry to politicians, while companies utilize these scenarios to identify relevant fields for action, for example the need for cost cutting in one division and for investment in another [13].

There might also be a critical side to taking megatrends as a starting point for industry scenarios. Megatrends are often vague, for example the megatrend of urbanization. Information about how many people are moving from rural communities is just given in a span. The selection of relevant trends is always subjective and their interaction does additionally hinder the determination of precise scenarios. On the other hand, taking the trends into account may increase companies' understanding of forces influencing their market and technological environment as well as their current business model. In our study, we focused on megatrends since they are one of the prevailing topics in the chemical management literature in the 2010s. Megatrends may thus serve as a common frame of reference when analyzing the necessity of transforming business in the chemical industry.

There are different ways to group relevant trends for the chemical and pharmaceutical industry [22, 24, 25]. For our study, we distinguished between 12 trends and asked the participants to rate their importance for their business activities in the years 2014 and 2024 (Figure 1.2).

Across all segments, the most important trends for the chemical industry in 2014 are the ongoing globalization, including the increasing importance of the Asian market, the need for interdisciplinary innovation, for example in the field of bio- or nanotechnology, and the growing significance of a higher employee qualification. The German chemical industry is thus becoming more international, opening up to adjacent scientific disciplines, and assigning significant importance to a highly skilled workforce in order to attain its goals. It is striking that for the year 2014, so-called "green issues," for example sustainable products, the shift to alternative energy sources, and the use of renewable resources, are considered to have the least relevance of all potential megatrends. At the same time, participants assume that these aspects will increase in significance until the year 2024. Successful companies – as defined earlier – attribute higher importance to these trends than less successful ones.

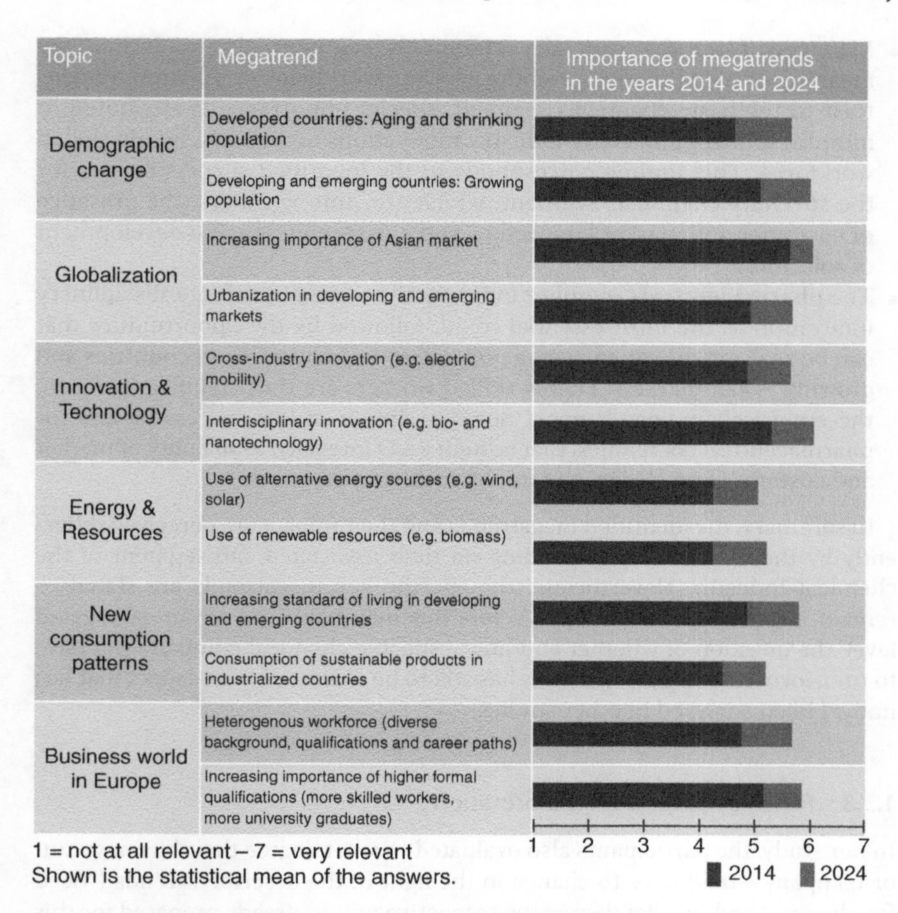

Topic
Megatrend
Importance of megatrends in the years 2014 and 2024
Demographic change
Developed countries: Aging and shrinking population
Developing and emerging countries: Growing population
Globalization
Increasing importance of Asian market
Urbanization in developing and emerging markets
Innovation & Technology
Cross-industry innovation (e.g. electric mobility)
Interdisciplinary innovation (e.g. bio- and nanotechnology)
Energy & Resources
Use of alternative energy sources (e.g. wind, solar)
Use of renewable resources (e.g. biomass)
New consumption patterns
Increasing standard of living in developing and emerging countries
Consumption of sustainable products in industrialized countries
Business world in Europe
Heterogenous workforce (diverse background, qualifications and career paths)
Increasing importance of higher formal qualifications (more skilled workers, more university graduates)
1 2 3 4 5 6 7
1 = not at all relevant – 7 = very relevant
Shown is the statistical mean of the answers.
2014
2024

Figure 1.2 Importance of different trends for the chemical industry in the years 2014 and 2024

The variation in importance is also observable when comparing different industry segments:

- For managers from the **basic chemicals segment**, the most relevant trends are the increasing significance of the Asian market, rising living standards in developing and emerging countries, and urbanization. They perceive cross-industry and interdisciplinary innovations in addition to the shrinking and more diverse workforce in Europe to be of less relevance. The results reflect the aforementioned characterization of the basic chemicals segment as being highly automated, capital intensive, and based on established product and process know-how.

- Managers from the **specialty chemicals segment** also underline the meaning of Asian markets and the growing worldwide population. In contrast to the basic chemicals segment, special importance is attributed to interdisciplinary and cross-industry innovations as well as a highly skilled workforce. This finding corresponds to the identified success factors for the specialty chemicals segment, which are, among others, the presence of customer and market knowledge, and a customer-specific development of solutions.
- The **pharmaceuticals segment** indicates the realization of interdisciplinary innovation as the most essential trend, followed by the opportunities that can be realized due to an ageing population in industrialized countries and growing Asian markets. Highly skilled workers are thus significant. Again, the empirical findings support our description of key success factors for pharmaceutical companies, that is, high R&D intensity, availability of market and customer knowledge, and use of qualified personnel.

In summary, the identified megatrends and their impact are perceived differently by the respondents depending on their associated sub-segment of the chemical industry. Nevertheless, the specific key megatrends are stated to remain important in the future. While this may hold true on an aggregated level, the question of whether and how a specific chemical company will have to transform its business activities has still to be examined – an aspect that has not yet been analyzed in other studies.

1.3.3 Fields of Business Transformation

In our study, the participants also evaluated to what degree their business unit or company would have to change in the light of the described trends ("need for change") and to what degree the respective unit is already prepared for this upcoming change ("degree of preparedness"). They identified a medium need for change for all three segments. Regarding this aspect, a significant difference between the chemical and the pharmaceutical industry, on the one hand, and other industries such as electronics, newspaper or financial industries, on the other, can be observed. While the chemical industry actually shows an evolutionary change pattern, the other mentioned industries are characterized by a more radical or "disruptive" change. Thus, radical innovations might not be expected in the chemical industry in the future.

The degree of preparedness in the chemical industry corresponds to the required change when considering the means of the answers. Differences can however be identified across the relevant fields of change. The degree of preparedness coincides with the existing need in the areas "strategy and business model" and "business processes," whereas a relevant discrepancy can be identified in the fields "workforce qualification" and "company culture."

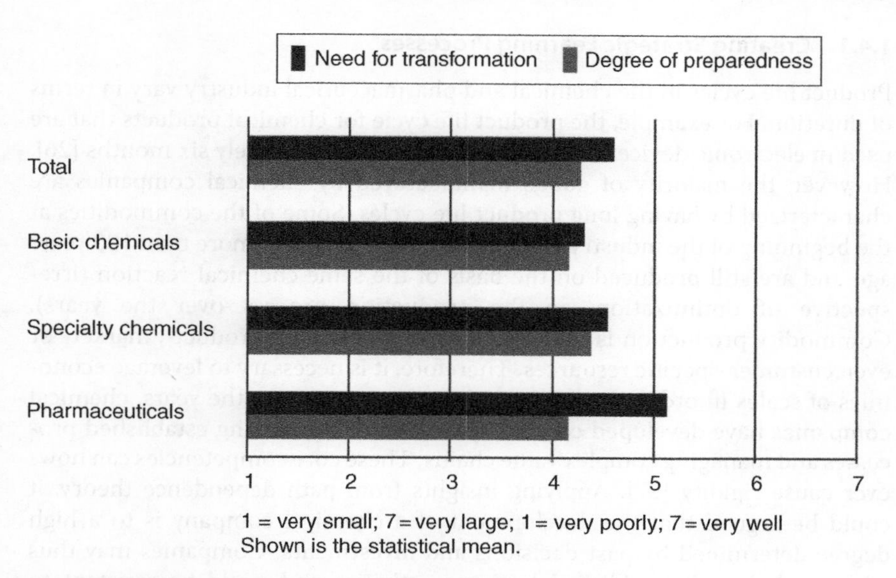

Figure 1.3 Business transformation in the chemical industry

Dividing the sample by industry segments as shown in Figure 1.3 reveals additional insights. The field of basic chemicals seems to be very well prepared, thus facing a rather small need for change. In the field of specialty chemicals, participants indicate a higher need for change. They assume that expected shifts within the key field of cross-industry and interdisciplinary innovation will imply changes in the workforce qualification and the company's values.

With regards to the pharmaceutical segment, the highest levels of required change encounter the lowest degree of preparedness. A great need for transformation is seen within the fields "corporate culture," "employee qualification," "strategy/business model," and "business processes." Compared with the other segments, the pharmaceuticals segment shows – with the exception of the topic strategy/business model – higher gaps, not only regarding so-called "soft issues" of "corporate culture" and "workforce qualification" but also concerning specific business processes.

1.4 Managerial Challenges in the Chemical Industry

After identifying major trends and fields of business transformation, the following section will present the findings from our study on current managerial challenges and elaborate on how to cope with the upcoming changes in the chemical industry.

1.4.1 Creating Strategic Learning Processes

Product life cycles in the chemical and pharmaceutical industry vary in terms of duration. For example, the product life cycle for chemical products that are used in electronic devices is often very short – lasting merely six months [26]. However, the majority of goods manufactured by chemical companies are characterized by having long product life cycles. Some of the commodities at the beginning of the industry's value chain were invented more than 100 years ago and are still produced on the basis of the same chemical reaction (irrespective of optimizations in the production process over the years). Commodity production is capital-intensive and ties up product-, market- or even customer-specific resources. Therefore, it is necessary to leverage economies of scales in order to achieve a cost advantage. Over the years, chemical companies have developed core competencies in optimizing established processes and managing complex value chains. These core competencies can however cause rigidity [27]. Applying insights from path dependence theory, it could be argued that the development of a chemical company is to a high degree determined by past decisions and investments. Companies may thus stick to their well-established business activities and could be resistant to change. As a consequence, such a high continuity of relevant product, process, and market know-how may prevent companies from looking outside the company, identifying future trends, and accepting the need for transformation [28, 29]. At the same time, routines and subsequent capabilities have been found to be developed in path-dependent learning mechanisms.

Strategic learning capability is defined as a company's ability to derive knowledge from past strategic actions and to use this knowledge to adjust strategy [29, 30]. As illustrated in Figure 1.4, we asked participants in our study to assess the strategic learning capability of their company or business unit (according to the measure used in [30]).

It turns out that successful companies stand out due to their strong strategic learning capability. More precisely, they are superior at assessing failures in strategic approaches and recognizing alternative strategies. Hence, these firms learn from their mistakes and are more flexible in adapting their current strategy and business practices. These firms significantly surpass other companies that are, according to their self-assessment, not as successful.

Strategic learning capability is crucial for chemical companies in light of the discussed megatrends. It is, for instance, a perquisite for adopting influences from bio- or nanotechnology and, accordingly, redirecting firms' research and/or production efforts. In addition, it enables companies to reconsider whether traditional success parameters on which their business model is assessed still apply. Recognizing non-sustainable pathways and quickly adjusting strategies is facilitated when companies have such a capability – only then can a company take advantage of innovation and growth opportunities.

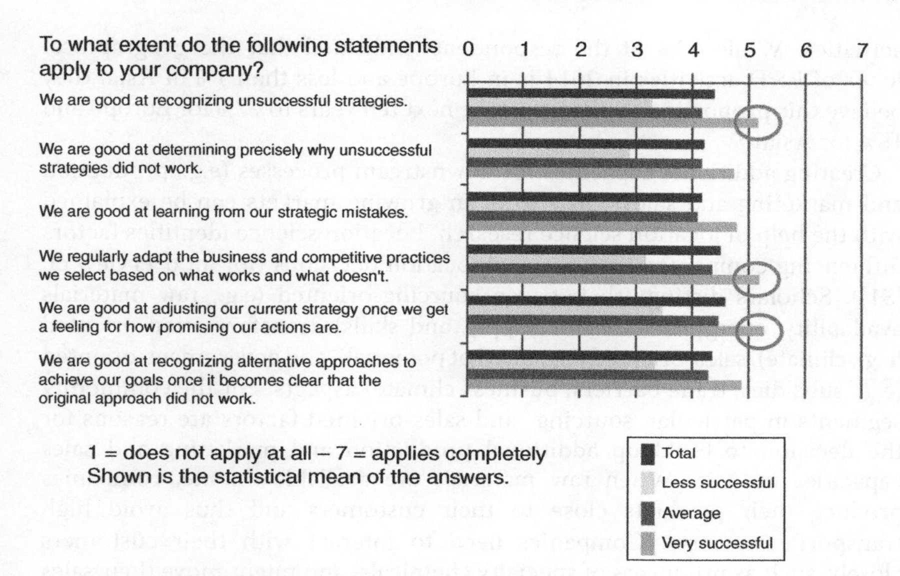

Figure 1.4 Strategic learning capability: successful versus less successful companies

1.4.2 Managing Value Chains Across the Globe

A growth of 4.5% per year up to 2030 is predicted for the global chemical industry [22]. The extent of growth will presumably vary by region and industry sector. A modest increase with a growth rate of 1.8% p.a. (per annum) is forecasted for mature chemical markets such as Germany, while Asian markets are expected to grow above average. The rising demand in Asia is explained by the increased prosperity in the region, resulting in a greater number of people buying chemical-intensive products. While the current share of Asian countries in the worldwide chemical production accounts for 40%, forecasts believe it will accumulate to 55% in 2030. By taking a company perspective, the key questions are how to participate in this growth and how to organize the value chain accordingly. In particular, companies have to decide about the geographic location of their value chain activities and about the way they handle interfaces across their globally dispersed activities.

Our study thus included a question asking companies about the geographic center of their business activities. Across all business functions and business segments, respondents answered that the relative importance of Europe as a location will decrease as Asia's importance will increase. For 15% of the companies, their current geographical production focus is located in Asia. This share is estimated to rise to 44% in the year 2024. Participants assume that they will additionally shift their marketing and sales activities to Asia: this number rises from 11% for 2014 to 41% for 2024. A shift is also expected for R&D

activities. While 89% of the respondents indicated that the geographical focus of R&D activities in 2014 is in Europe and less than 1% in Asia, they believe this proportion to change in the next ten years to 77% for Europe and 15% for Asia.

Creating additional capacities for downstream processes (e.g., production and marketing and sales) close to or in growing markets can be explained with the help of location science research. Location science identifies factors influencing companies' international location decisions (for an overview cf. [31]). Scholars distinguish between sourcing-oriented (e.g., raw materials availability, energy costs, labor supply, and skills), transformation-oriented (e.g., climate), sales-oriented (e.g., market potential), and government-oriented (e.g., subsidies, trade barriers, business climate) aspects. For the commodity segments in particular, sourcing- and sales-oriented factors are reasons for the decision to build up additional production and marketing and sales capacities in Asia. When raw materials are available on-site, companies produce their products close to their customers and thus avoid high transportation costs. Companies need to interact with their customers closely, such as producers of specialty chemicals, and might move their sales and application engineering employees to the target markets as well. They thereby create rich communication channels [32] that might be more appropriate for discussing innovative topics.

The still existing advantages for conducting R&D in North America and Europe explain why respondents only observe a low tendency to move R&D activities to Asia. Beneficial attributes are the well-established academic systems and a highly skilled workforce. An additional advantage is the presence of strong networks between chemical companies, their customers in lead markets such as the automotive industry, and specialized innovation partners in related industries such as machinery. For instance, many leading chemical producers in principal customer markets are still carrying out their research and production activities in Germany. The physical proximity and a comparable level of professionalism thus facilitate organizing cross-industry and cross-disciplinary collaborative projects. These agglomeration effects (e.g., in-sourcing, transformation, and sales) described by location theory still favor R&D in Europe and North America. Though – according to business associations and managers in the chemical industry – the limited innovation climate and openness might be detrimental to allocating R&D activities to European countries. This is, for instance, reflected in the fields of green biotechnology and fracking, where R&D is concentrated in North America, due to the less favorable legislation and the business climate in Europe.

In summary, it could be expected that the multiregional character of the chemical industry, with manufacturing activities in the major markets of North America, Europe, and Asia, each of which mainly serve their regional markets with little trading between the regions, will probably prevail. Only for active ingredients in pharmaceuticals is a more global production pattern anticipated. Furthermore, it is important to consider the international interfaces of multinational companies where sourcing, production planning, product innovation activities, quality assurance, and marketing activities need to be coordinated across the globe [33, 34]. This additional complexity is difficult to cope with from a leadership and human resource management perspective even though it can be reduced by powerful IT systems.

1.4.3 Optimizing Processes

The chemical industry has reached a mature stage in its life cycle. This can be illustrated by the results from our study concerning the focus of organizational change during the last five years. Companies in the chemical industry create profits by putting significant emphasis on optimizing the cost structure of their operative processes and adapting their production strategy accordingly, for example through projects for production excellence such as Six Sigma. Successful companies have concentrated even more on these aspects than less successful ones (Figure 1.5).

Changes in market-related strategies such as the pricing and sales or the commercialization strategy, for example, from sales to leasing or licensing models, have received little attention over the last five years. Thus, process

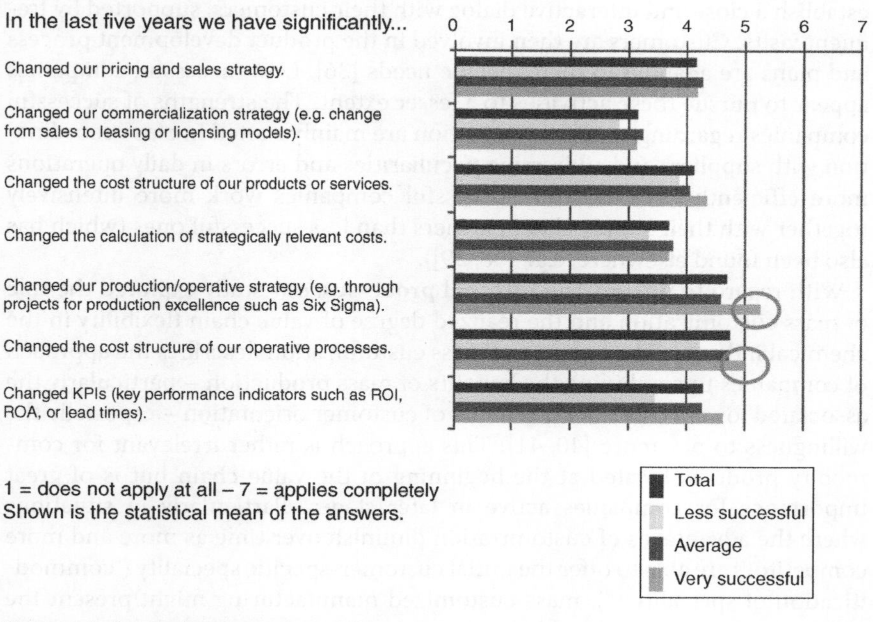

Figure 1.5 Focus of optimization activities

optimization aspects have dominated companies' activities to increase profits – a behavior that is typically associated with an industry in the maturity phase of the industry life cycle.

In order to grasp the findings in more detail, specific results for external and internal process optimization are described in the following [35], again by simultaneously distinguishing between successful and less successful companies.

With regard to **external process optimization,** the responses of the participants show that integrating customers into value creation is much more prominent than integrating suppliers. This seems to be particularly true for companies merely buying commodities from their suppliers in a standardized process, with price being the main buying criterion. In this type of supplier–buyer relationship, the supplier's role for value creation is limited, thus, expanding the relationship from the buyer's perspective does not provide a particular advantage. Managing the relations to suppliers should attract much more attention, whereby the extent should be evaluated based on the scale of each supplier's influence on the company's competitive position through their individual offering and role for innovation. This can be observed in the automotive industry where suppliers are segmented and managed depending on their strategic importance.

We also find that successful companies put more emphasis on managing external relations than less successful companies. Successful companies often establish a close and interactive dialog with their customers, supported by frequent visits. Customers are then involved in the product development process and plans are adapted to their specific needs [36]. Less successful companies appear to pursue these activities to a lesser extent. The strengths of successful companies regarding supplier integration are mainly rooted in close coordination with suppliers and addressing peculiarities and errors in daily operations more efficiently [37]. Overall, successful companies work more intensively together with their supply chain partners than less successful ones (which has also been found elsewhere, e.g., [38, 39]).

With regard to **optimizing internal processes**, our study explored the role of mass customization and the realized degree of value chain flexibility in the chemical industry. The concept of mass customization describes the approach of companies to combining the benefits of mass production – particularly the associated lower costs – with benefits of customer orientation – especially the willingness to pay more [40, 41]. This approach is rather irrelevant for commodity products located at the beginning of the value chain but is of great importance for companies active in later stages. Particularly in situations where the advantages of customization diminish over time as more and more competitors are able to offer the initial customer-specific speciality ("commoditization of specialities"), mass-customized manufacturing might present the only paths to remain profitable.

Successful companies appear to have a better understanding of and capacity to implement mass customization than other companies. They produce large quantities and simultaneously provide a great variety of products without sacrificing quality. These companies are able to react to customer-specific requirements more flexibly, can change their production processes quickly, and have fairly low (additional) set-up expenses. Less successful companies responding to the survey did not report comparable experiences.

This wide gap between successful and less successful companies also exists with regard to value chain **flexibility** (Figure 1.6). This flexibility implies processing and providing better individual services, sharing information with customers and suppliers, distributing customer-related information internally, and, therefore, an overall stronger performance in boundary-spanning activities [42, 43].

Value chain flexibility is not equally relevant for the three chemical segments. Companies involved in continuous production face less need in this regard than, for instance, volume-oriented producers of specialty chemicals. In the production of APIs, once processes have been accredited, a high level of flexibility is not a goal in itself. As production chains are becoming diversified and flexible due to an increasing modularization of processes and additional manufacturing processes, value chain flexibility may also gain more importance for companies in the chemical and pharmaceutical industry.

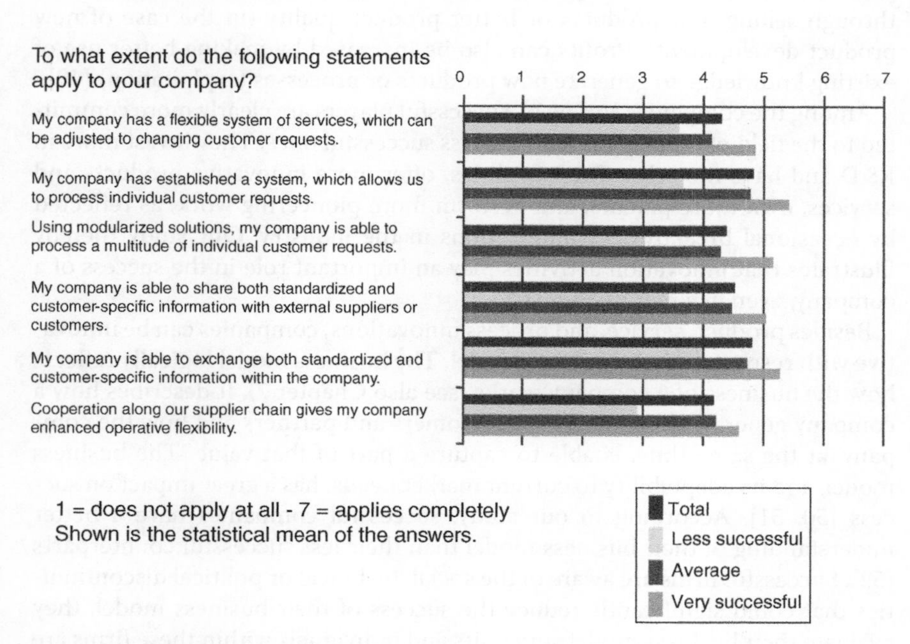

Figure 1.6 Value chain flexibility

1.4.4 Creating Product, Process, and Business Model Innovations

Innovation has always been a key driver of success in the chemical industry. While the incentive for innovation traditionally came from the core discipline of a company and thus stemmed from basic chemical research, stimuli for further developments are nowadays expected to be derived from related disciplines such as biology, bio- and nanotechnology [44]. In addition, business opportunities are not necessarily found within the "defined" boundaries of an established industry. At some interfaces, for example, in the field of electric mobility, the beginnings of converging processes between different industries can be observed [45, 46]. Such a development is accompanied by firms acquiring new knowledge that was not part of their traditional expertise and by adapting technology bases distinct from former core competencies [47]. Thus, companies in the chemical industry have to open up their innovation activities to emerging academic disciplines and other industries, as has also expressed in the open innovation concept by Chesbrough [48].

One further challenge for companies is finding the right balance between product and process innovation as well as between explorative and exploitative innovation activities. Companies in the chemical industry can improve their profit position by generating and applying new knowledge for innovation, that is, exploration, and thus either improve their cost position through enhanced productivity (in the case of process innovations) or increasing their revenues through selling new products or better product quality (in the case of new product development). Profits can also be increased by making better use of existing knowledge to generate new products or processes (exploitation) [49].

Among the companies surveyed, successful players are clearly more committed to the field of innovation than the less successful ones. They invest more in R&D and have first-class R&D facilities, offer more innovative products and services, own more patents, and perform more pioneering work, as reflected by occasional breakthrough innovations in the industry. Our study thereby illustrates that innovation activities play an important role in the success of a company, even in a mature industry.

Besides product, service, and process innovations, companies can be innovative with respect to their business model. The business model logically reflects how the business of a company works (see also Chapter 7). It describes how a company generates benefits for its customers and partners and how the company, at the same time, is able to capture a part of that value. The business model, and its adaptability to current market needs, has a great impact on success [50, 51]. According to our study, successful companies have a better understanding of their business model than their less successful counterparts [52]. Successful firms are aware of the social, technical or political discontinuities that could significantly reduce the success of their business model, they evaluate their business model, and units and individuals within these firms are aware of how they contribute to the firm's business model (Figure 1.7).

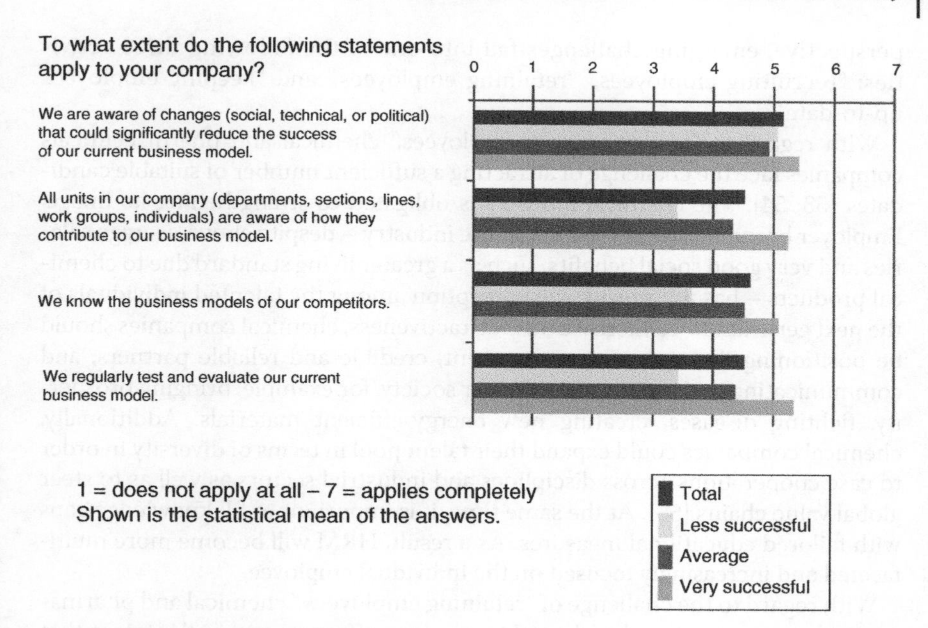
To what extent do the following statements apply to your company?
0 1 2 3 4 5 6 7
We are aware of changes (social, technical, or political) that could significantly reduce the success of our current business model.
All units in our company (departments, sections, lines, work groups, individuals) are aware of how they contribute to our business model.
We know the business models of our competitors.
We regularly test and evaluate our current business model.
1 = does not apply at all – 7 = applies completely
Shown is the statistical mean of the answers.
Total
Less successful
Average
Very successful

Figure 1.7 Business model sensing

It is important to note that business models evolve over time. A company's creativity with regard to the development of innovative business models can thus be a key driver for company success. Particularly within a mature industry where standardized and interchangeable products prevail, unusual solutions gain importance. Companies active in other sectors might yield different perspectives and serve as good examples for developing new business models for chemical companies, for example, telecommunication providers offering value-added services such as data streaming or cloud computing in order to sell more data volume. Pursuing sustainability objectives might open up future business opportunities for chemical companies, and the trend towards personalized medicine might facilitate the creation of new business models for pharmaceutical companies.

1.4.5 Developing Human Resources

In our study, participants ranked the growing importance of higher qualified personnel as the third most important megatrend influencing business activities in the chemical and pharmaceutical industry. This is not surprising as – of course – all of the other trends outlined (such as the growing importance of the Asian markets, the need for cross-industry and cross-disciplinary collaboration to foster innovation) have a direct impact on human resources (HR) and human resource management (HRM) within companies. From an internal HR perspective, emerging challenges fall into one of the following three activities: "recruiting employees," "retaining employees," and "keeping employees up-to-date."

With regard to "recruiting new employees," chemical and pharmaceuticals companies face the challenge of attracting a sufficient number of suitable candidates [53, 54]. The chemical industry is obligated to strengthening its image. Employer branding studies show that the industry – despite above-average salaries and very good social benefits, such as a greater living standard due to chemical products – has to improve the perception among the talented individuals of the next generation. To increase their attractiveness, chemical companies should be positioning themselves as transparent, credible and reliable partners, and communicating their generated value for society, for example, bringing prosperity, fighting diseases, creating new energy-efficient materials. Additionally, chemical companies could expand their talent pool in terms of diversity in order to ease cooperations across disciplines and industrial sectors as well as to steer global value chains [55]. At the same time, it is important to fill knowledge gaps with tailored educational measures. As a result, HRM will become more multifaceted and increasingly focused on the individual employee.

With regard to the challenge of "retaining employees," chemical and pharmaceutical companies are already making various efforts to retain the talent that they have attracted [56]. Flexible working-hour models allow for better work–life balance ("dynamic work place") and are an important factor in employee retention. The development of a variety of career paths that value leadership qualities as well as functional expertise can help ensure higher engagement and continued employment. A company culture that embraces diversity in the company and sees sustainability as a business driver, will support necessary change processes to gain global competitiveness [57]. These different measures have to be part of a global HR strategy, ensuring common values and standards across business units and personnel exchange as well as allowing regional adaptations in order to acknowledge differences between mature chemical markets in Europe and the fast growing Asian markets.

The third challenge in HRM focuses on the question of how to keep employees up-to-date with regard to their technical expertise and working skills: the time employees will be working in chemical and pharmaceuticals companies will be extended as the average retirement age increases and some employees are entering the labor market at a younger age due to shortened school periods and the Bologna's bachelor and master system. At the same time, market changes and new emerging business models necessitate employees to provide even more learning capabilities to facilitate companies' long-term success. In addition to keeping up with state-of-the-art knowledge in the fields of chemistry and engineering, it is important to develop interdisciplinary competencies and the ability to establish collaboration with other industries [1]. In the context of continuing globalization, emerging interfaces need to be managed when leading international teams are spread over two or three continents. A new focus will thus be placed on decentralized management and will directly affect qualification requirements for managers. They not only have to master different languages but have to provide intercultural competences and the ability to manage employees in different time and cultural zones [58]. As traditional lectures and workshops might be not be sufficient to acquire these leadership qualifications, new training concepts, including coaching elements, must be created [13]. Educational pathways within the chemical field evolve from a pre-Bologna paradigm (including one full-time study phase before entering professional life, which is complemented by further occasional training periods) to more differentiated and individualized careers in the future (shaped by a short initial study followed by gaining professional experience and continuing professional education in modular (master) programs) that will continually keep employees up-to-date.

1.5 Summary

The characteristics of the chemical industry, the global megatrends and associated changes, and the adaptation of business strategies and activities the industry is facing, along with the resulting challenges for managing chemical companies, can be summarized as follows:

- **The chemical industry is a capital-intensive industry characterized by high diversity in terms of products, segments, and end markets.** The portfolio of products in the chemical industry ranges from energy- and cost-intensive production of bulk chemicals produced and sold in large volume to highly individualized specialty chemicals manufactured and marketed in small quantities. Companies that are working with basic chemicals, specialty chemicals or pharmaceuticals require different capabilities to succeed in the given segment.
- **The industry finds itself at a mature stage shaped by a lower intensity of innovation of the (traditionally) underlying scientific disciplines.** New influences are anticipated to be derived from other areas such as biotechnology or nanotechnology. Thus, companies have to update their knowledge base through cross-industry or cross-disciplinary activities.
- **The industry has a multiregional character, whereby the majority of products are produced in the respective regions where they are sold.** North America, Europe, and Asia represent the main markets. Above GDP growth is expected in particular for Asia. As globalization continues, chemical companies will build up additional resources in the growing Asian markets. Consequently, managing global value chain activities will present a constant challenge for multinational companies.

- **Companies, active in the chemical industry, recognize the need to align their business activities according to the continuing globalization, see the opportunity to benefit from cross-disciplinary innovations, and anticipate the increasing importance of a highly skilled workforce.**
- **Innovation continues to be a key driver for the success of the chemical industry.** Prospering companies invest more effectively in R&D and implement business model innovations earlier than less successful firms. In addition, optimizing processes is crucial for chemical companies: successful players establish closer relationships with external partners than less successful businesses, and internal process optimization, via mass customization and high value chain flexibility, is adopted more seriously by successful companies than their less successful competitors.
- **In light of the identified megatrends, one core management challenge for chemical companies lies in the adequate development of human resources** – to have a highly skilled workforce, that is able to steer global value chains, cooperate with different industries and disciplines, and which embraces the principle of lifelong learning.

References

1 Whitesides GM. 2015. Reinventing chemistry. *Angewandte Chemie International Edition*, **54**(11): 3196–3209.
2 CEFIC. 2015. *The European Chemical Industry Facts & Figures 2014*. European Chemical Industry Council: Brussels.
3 Chandler AD. 2005. *Shaping the Industrial Century: The Remarkable Story of the Evolution of the Modern Chemical and Pharmaceutical Industries*. Harvard Studies in Business History. Vol. 46. Harvard University Press: Cambridge, MA.
4 Wallace TF. 1984. *APICS Dictionary: The Official Dictionary of Production and Inventory Management Terminology and Phrases*. 5th edn. American Production and Inventory Control Society: Falls Church, VA.
5 Kannegiesser M, Günther HO, Van Beek P, Grunow M, and Habla C. 2008. *Value Chain Management in the Chemical Industry – Global Value Chain Planning of Commodities*. Physica: Heidelberg.
6 Kline C. 1976. Maximizing profits in chemicals. *Chemtech*, **6**(2): 110–117.
7 Ernst &Young. 2014. Die jeweils 300 umsatzstärksten Unternehmen Europas und der USA im Vergleich. http://www.ey.com/echannel/publications.nsf/0/82D01992232B2B7985257CE000478BBB/$file/EY-Top-300-Europa-USA-2014.pdf?OpenElement (accessed 7 June 2015).
8 Bartels E, Augat T, and Budde F. 2006. Structural drivers of value creation in the chemical industry. *Value Creation: Strategies for the Chemical Industry*. 2nd edn. Wiley Online Library: 27–39.

9 IVA. 2015. Die Pflanzenschutzindustrie: Mit Kompetenz an die Spitze. http://www.iva.de/verband/die–pflanzenschutzindustrie-mit-kompetenz-die-spitze (accessed 12 February 2016).
10 Hofmann K and Budde F. 2006. Today's chemical industry: Which way is up? in *Value Creation: Strategies for the Chemical Industry* (eds F. Budde, U.-H. Felcht, and H. Frankemölle). Wiley-VCH Verlag: Weinheim, pp. 1–10.
11 Cesaroni F, Gambardella A, and Mariani M. 2007. The evolution of networks in the chemical industry, in *The Global Chemical Industry in the Age of the Petrochemical Revolution* (eds L. Galambos, T. Hikino, and V. Zamagni). Cambridge University Press: Cambridge, pp. 21–51.
12 Giannetti R and Romei V. 2007. The chemical industry after World War II, in *The Global Chemical Industry in the Age of the Petrochemical Revolution* (eds L. Galambos, T. Hikino, and V. Zamagni). Cambridge University Press: Cambridge, pp. 407–452.
13 Utikal H and Woth J. 2015. From megatrends to business excellence: Managing change in the German chemical and pharmaceutical industry. *Journal of Business Chemistry*, **12**(2): 41.
14 Schröter H. 2007. Competitive strategies of the world's largest chemical companies, 1970–2000, in *The Global Chemical Industry in the Age of the Petrochemical Revolution* (eds L. Galambos, T. Hikino, and V. Zambagni). Cambridge University Press: Cambridge, pp. 53–80.
15 Utterback JM and Abernathy WJ. 1975. A dynamic model of process and product innovation. *Omega*, **3**(6): 639–656.
16 Van De Ven A and Andrew H. 2011. Building a European community of engaged scholars. *European Management Review*, **8**(4): 189–195.
17 Van de Ven AH and Poole MS. 1995. Explaining development and change in organizations. *Academy of Management Review*, **20**(3): 510–540.
18 Hannan M and Freeman J. 1989. *Organizational Ecology*. Harvard University Press: Cambridge, MA.
19 Van de Ven AH and Sun K. 2011. Breakdowns in implementing models of organization change. *The Academy of Management Perspectives*, **25**(3): 58–74.
20 Levy A and Merry U. 1986. *Organizational Transformation: Approaches, Strategies, Theories*. Praeger: New York.
21 Nadler DA and Tushman ML. 1990. Beyond the charismatic leader: Leadership and organizational change. *California Management Review*, **32**(2): 77–97.
22 VCI. 2013. *Die deutsche chemische Industrie 2030*. Verband der chemischen Industrie: Frankfurt.
23 VNCI and Deloitte. 2012. *The Chemical Industry in the Netherlands: World-leading Today and in 2030–2050*. Vereniging van de Nederlandse Chemische Industrie: Den Haag.
24 Matlin S and Abegaz B. 2011. *The Chemical Element – Chemistry's Contribution to our Global Future* (eds J. Garcia-Martinez and E. Serrano-Torregrosa). John Wiley & Sons Ltd: Chichester, pp. 1–69.

25 Johansson A, Guillemette Y, and Murtin F. 2012. *Looking to 2060: Long-term Global Growth Prospects*. OECD Publishing: Paris.
26 Gocke A, Willers Y-P, Friese J, Gehrlein S, Schönberger H, and Farag H. 2014. *How 20 Years Have Transformed the Chemical Industry – The 2013 Chemical Industry Value Creators Report*. Boston Consulting Group: Boston.
27 Leonard-Barton D. 1992. Core capabilities and core rigidities: A paradox in managing new product development. *Strategic Management Journal*, **13**(2): 111–125.
28 Teece DJ, Pisano G, and Shuen A. 1997. Dynamic capabilities and strategic management. *Strategic Management Journal*, **18**(7): 509–533.
29 Eisenhardt KM and Martin JA. 2000. Dynamic capabilities: What are they? *Strategic Management Journal*, **21**(10–11): 1105–1121.
30 Anderson BS, Covin JG, and Slevin DP. 2009. Understanding the relationship between entrepreneurial orientation and strategic learning capability: An empirical investigation. *Strategic Entrepreneurship Journal*, **3**(3): 218–240.
31 Hübner R. 2007. *Strategic Supply Chain Management in Process Industries: An Application to Specialty Chemicals Production Network Design*. Springer: Berlin.
32 Daft RL and Lengel RH. 1986. Organizational information requirements, media richness and structural design. *Management Science*, **32**(5): 554–571.
33 Bartlett CA and Ghoshal S. 1999. *Managing Across Borders: The Transnational Solution*. Vol. 2. Taylor & Francis: Abingdon.
34 Ghoshal S and Bartlett CA. 1990. The multinational corporation as an interorganizational network. *Academy of Management Review*, **15**(4): 603–626.
35 Gelhard C. 2015. Kunden- und Lieferantenintegration entlang der Wertschöpfungskette, in *Von den Megatrends zum Geschäftserfolg* (eds Provadis School of International Management and Technology). Wiley-VCH: Weinheim, pp. 30–32.
36 Brown SL and Eisenhardt KM. 1995. Product development: Past research, present findings, and future directions. *Academy of Management Review*, **20**(2): 343–378.
37 Koufteros X, Vonderembse M, and Jayaram J. 2005. Internal and external integration for product development: The contingency effects of uncertainty, equivocality, and platform strategy. *Decision Sciences*, **36**(1): 97–133.
38 Frohlich MT and Westbrook R. 2001. Arcs of integration: An international study of supply chain strategies. *Journal of Operations Management*, **19**(2): 185–200.
39 Flynn BB, Huo B, and Zhao X. 2010. The impact of supply chain integration on performance: A contingency and configuration approach. *Journal of Operations Management*, **28**(1): 58–71.
40 Pine BJ. 1992. *Mass Customization: The New Frontier in Business Competition*. Harvard Business Press: Boston.

41 Da Silveira G, Borenstein D, and Fogliatto FS. 2001. Mass customization: Literature review and research directions. *International Journal of Production Economics,* **72**(1): 1–13.

42 Nair A. 2005. Linking manufacturing postponement, centralized distribution and value chain flexibility with performance. *International Journal of Production Research,* **43**(3): 447–463.

43 Gelhard C and Von Delft S. 2016. The role of organizational capabilities in achieving superior sustainability performance. *Journal of Business Research,* **69**(10): 4632–4642.

44 Leker J and Golembiewski B. 2015. *Disziplinübergreifende Innovationen in der chemischen Industrie,* in *Von den Megatrends zum Geschäftserfolg* (eds Provadis School of International Management and Technology). Wiley-VCH Verlag: Weinheim, pp. 22–23.

45 von Delft S. 2013. Inter-industry innovations in terms of electric mobility: Should firms take a look outside their industry? *Journal of Business Chemistry,* **10**(2).

46 Golembiewski B, vom Stein N, Sick N, and Wiemhöfer H-D. 2015. Identifying trends in battery technologies with regard to electric mobility: Evidence from patenting activities along and across the battery value chain. *Journal of Cleaner Production,* **87**: 800–810.

47 Curran CS and Leker J. 2011. Patent indicators for monitoring convergence – examples from NFF and ICT. *Technological Forecasting and Social Change,* **78**(2): 256–273.

48 Chesbrough HW. 2003. The era of open innovation. *MIT Sloan Management Review,* **44**(3): 35–41.

49 Bauer M and Leker J. 2013. Exploration and exploitation in product and process innovation in the chemical industry. *R&D Management,* **43**(3): 196–212.

50 Johnson MW, Christensen CM, and Kagermann H. 2008. Reinventing your business model. *Harvard Business Review,* **86**(12): 57–68.

51 Chesbrough H. 2007. Business model innovation: It's not just about technology anymore. *Strategy & Leadership,* **35**(6): 12–17.

52 von Delft S. 2015. Wachstumschance Geschäftsmodellinnovation, in *Von den Megatrends zum Geschäftserfolg* (eds Provadis School of International Management and Technology). Wiley-VCH Verlag: Weinheim, pp. 28–29.

53 Collins CJ and Kehoe RR. 2009. Recruitment and selection (electronic version), in *The Routledge Companion to Strategic Human Resource Management* (eds J. Storey, P.M. Wright, and D. Ulrich). Routledge: New York, pp. 209–223.

54 Posthumus J. 2015. *Use of Market Data in the Recruitment of High Potentials: Segmentation and Targeting in Human Resources in the Pharmaceutical Industry.* Springer: New York.

55 Milliken FJ and Martins LL. 1996. Searching for common threads: Understanding the multiple effects of diversity in organizational groups. *Academy of Management Review,* **21**(2): 402–433.

56 Affairs FMoWaS. 2015. Green Paper Work 4.0. http://www.bmas.de/EN/Services/Publications/arbeiten-4-0-greenpaper-work-4-0.html;jsessionid=A389C3A4C685CC490161405D83B1CDCB (accessed 6 June 2017).
57 Henderson R, Gulati R, and Tushman M. 2015. *Leading Sustainable Change: An Organizational Perspective.* Oxford University Press: Oxford.
58 Steers RM, Nardon L, and Sanchez-Runde CJ. 2013. *Management Across Cultures: Developing Global Competencies.* Cambridge University Press: Cambridge.

2

Principles of Strategy: How to Develop Strategy

Jens Leker[1] *and Tobias Lewe*[2]

[1] *University of Münster, Department of Chemistry and Pharmacy*
[2] *A.T. Kearney Management Consulting GmbH, Energy & Process Industries Practice*

> *Strategy is about making choices, trade-offs; it's about deliberately choosing to be different.*
>
> Michael Porter, Professor at Harvard Business School

This chapter introduces the fundamentals and basic principles of strategy development. We first describe a fictitious situation of an executive manager in the chemical industry to illustrate the market environments and choices managers face when developing strategies. This example ends with a couple of questions that should periodically be asked by executives to challenge existing strategies and to guide strategy development. The main issues with these questions are discussed in depth later as well as in Chapter 3, which particularly focuses on strategic analysis. In Section 2.2 of the present chapter, the main definitions of strategic management are presented, which are further illustrated by examples from the chemical industry. In Section 2.3 we present an overview of strategy trends and in Section 2.4 you will learn about the strategy development process. Finally, the general conditions and trends actually guiding strategy development in the chemical industry are depicted in Section 2.5, to emphasize the dynamic and complexity of industrial practice. But let's start with Walter Brown.

2.1 The First Day for CEO Walter Brown

Walter Brown leaned back in his office chair. Beyond the corner window, he could see the production site spread almost to the horizon, and he felt a quiet sense of achievement. He had always assumed that one day he would take over

Business Chemistry: How to Build and Sustain Thriving Businesses in the Chemical Industry, First Edition. Edited by Jens Leker, Carsten Gelhard, and Stephan von Delft.

as Chief Executive Officer (CEO) of the company, and that day had come. Nonetheless, the challenges he confronted were daunting.

The company had been struggling for some time. The costs for raw materials and energy were high relative to the prices its competitors paid and the company faced a dilemma: Should it relocate parts of the production abroad or should it make significant capital investments in the site? The workforce was aware of the company's struggle and was concerned; people had the feeling that their jobs were at stake. Meanwhile, the company was also aging, and reluctant to give up hard-won benefits. Although production continued to run smoothly and the industry outlook was bright, the stock market had reacted to the struggle of the company: its share prices had declined significantly within a relatively short period of time and on top of that the future of the company had been questioned in leading newspapers.

Over the weekend, the Board of Directors had asked Walter if he was ready to take the reins. He assured them he was. But there was no time to celebrate – the stakeholders had high expectations and he would need to deliver results fast. To restore the company's profitability and to make that profitability sustainable he needed to tackle the core problems immediately.

It was obvious to close observers that the company had lost its identity both internally and externally. The reasons that employees identified with the company and that customers enthused about its products had become increasingly less compelling in recent years. In addition, coordination problems among the different businesses, and even within the business divisions, were proliferating, leading to quality and delivery issues. Key emerging trends and changes in the competitive environment had been addressed too late, if at all. A recent development in raw material supply was causing a strong decline of prices for molecules sourced in regions where Walter Brown's company had not yet been active. Thus, competitors that were more cost competitive with respect to sourcing these raw materials had a competitive edge. As a result, the company's rivals gained market share while Walter's company was in some way adrift.

Walter Brown knew he could turn things around, but he also knew he could not do it alone. He picked up the phone to schedule the first board meeting under his aegis. Further meetings with key investors, managers, and staff should follow fast.

This case is fictional, but scenarios like this can play out at any time in a chemical company. Granted, the circumstances that Walter Brown faces are extreme, but in times of radical change and reconfiguration of chemical value chains they seem to be more the norm than being unrealistic. The development of many chemical players over the last 10–25 years is characterized by a constant need to adapt a company's strategy and strategic position in light of globalization, changing customer needs, the reconfiguration

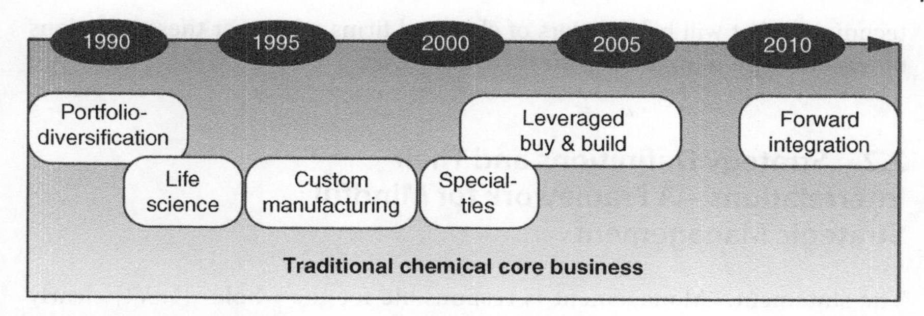

Figure 2.1 Strategic moves of chemical players

of value chains and new technologies that change the entire competitive landscape. Although the traditional chemical core business has not dramatically changed in terms of molecules to be produced and supplied to a broad variety of customers, the need for chemical players to formulate a strategic response to market challenges can be observed in a number of principal strategic moves (see Figure 2.1).

The case of Walter Brown has already illustrated the key tasks and considerations central to the responsibility of leading a business in the chemical industry in an increasingly complex and dynamic environment. The case also suggests a series of questions about strategic management that the leadership teams of actual chemical companies should ask themselves periodically, whether they face a version of Walter Brown's dilemma or – even more propitiously – during economically successful times:

- What are the global business-, industry-, and technology-related trends and how are they evolving? How are chemical industry value chains changing?
- Which customer value-drivers are relevant and how will they change in the future?
- What are the core competitors and how are they addressing the market? What are their future strategies?
- What are the core competencies and how can they be developed further to increase distinctiveness?
- What is the right level of ambition to be set? How does it translate into a mission and vision statement?
- What are the implications for the business strategy, the business portfolio as well as the underlying business model?
- What does the strategic roadmap look like to implement the strategy and what major changes will be implemented and until when?

In the following two sections, we discuss basic strategy definitions and terms, their relationships and interdependencies, and strategic management techniques that will help leaders of chemical firms to answer these questions in ways that are relevant to their own businesses.

2.2 Strategy Definitions and Their Interrelations – A Framework for Mindful Strategic Management

The statement, "Management is responsible for the whole" [1: VI] clearly illustrates that corporate governance is not just about fulfilling shareholders' expectations in the sense of increasing the share price. Management is also responsible for balancing the interests of shareholders and other stakeholders, with the ultimate goal of developing a sustainable and profitable business. Before we clarify the role of the management regarding strategic choices, the term **whole** needs more clarification. Clearly defining the company and its borders and, thus, its area of responsibility and activities it performs is a first step. Here, factual, legal, organizational, and systemic aspects can be described generally, to allow for quite different and broad scopes for the management of the business [2]. After defining the company (e.g., the organizational boundaries, the corporate entity, and the company's legally independent subsidiaries) the next step is to clarify what is distinctive about the company. Finally, demanding (but not unachievable) strategic objectives and goals must be set.

In summary, the question is: what does the company overall stand for? The company's mission statement, as it is commonly called, says, in a few memorable words, what the company excels at and what justifies its *raison d'être*. In recent history we are observing a trend to differentiate between a company's **mission** statement and a company's **purpose**. Mainly in order to create a positive perception with customers, the employees of the company, as well as other stakeholders, a more emotional formulation of the mission statement is aimed for [3].

Procter & Gamble (P&G), for example, defines its purpose as follows: "We will provide branded products and services of superior quality and value that improve the lives of the world's consumers, now and for generations to come. As a result, consumers will reward us with leadership sales, profit and value creation, allowing our people, our shareholders and the communities in which we live and work to prosper" [4]. Compared with P&G, BASF describes its purpose quite briefly as "We create chemistry for a sustainable future" [5]. For both, P&G and BASF, although one can find remarks about values and strategic principles to be quite prominent on the respective company webpages, a formulation about the mission is missing. This supports the impression that formulating a purpose rather than a mission statement has become more popular in recent years. Other chemical companies still formulate a mission statement. It is intended to be lasting and communicated to all stakeholders and the wording is typically kept general. The Dow Chemical Company's mission statement marks a good example in that sense: "To passionately create innovation for our stakeholders at the intersection of chemistry, biology, and physics" [6]. Another one – and somewhat different – can be found at Clariant: "Our mission is to create value by appreciating the needs of: Our customers – by providing competitive and innovative solutions; our employees – by adhering and embracing our corporate values; our environment – by acting sustainably at all levels" [7]. While somewhat generic, a mission statement can have direct consequences for corporate governance. So, for example, a company that establishes its mission as holding only brands with significant global presence, will re-evaluate its regional brands, and, regardless of their profitability, potentially separate from them. Or, a chemical player highlighting **sustainability** in its mission will challenge its entire business portfolio, as well as the operating model, to ensure adherence to this.

Strategic management is, of course, concerned with the future, so establishing a vision statement, as it is commonly called, that encapsulates the direction in which the company should develop itself in the future, has to be addressed. As Kaspar Rorstedt, Chairman of the Management Board of Henkel, points out: "Our new vision gives us a sense of direction and destination. It captures our aspiration of being the best in everything we do. It is the basis for what we all stand as 'One Henkel.' The corresponding vision is: "A global leader in brands and technologies" [8].

A vision statement should address a sufficiently long period of time that can be assessed by forecasting tools. In general, timeframes for vision statements vary between 5 and 10 years, depending on the company's dynamics and the environment in which it functions. Both factors can change, of course, and subsequently a chemical firm's vision and where necessary its mission or purpose statement can be adapted over time.

While the mission and vision statements are practical for establishing a framework for corporate governance and serve as a guideline for further strategy development, they are not specific enough for result-oriented strategic management. Therefore, formulating more detailed corporate goals is the next step. Although such goals can be formulated for different levels or areas, they have the following in common:

1) They are consistent with the standards set in the mission statement and further develop those set in the vision statement.
2) They are sufficiently precise and defined in a verifiable way. Each goal contains clear targets, which are specified in content and scope, along with an indication of progress to be achieved toward the goal in a given time period.
3) They should be understood by customers, the employees as well as other stakeholders to the same extent.

Goals can be developed top-down by corporate leaders as well as bottom-up by employees. In the classic understanding of management, top management determines the overall objectives, which lower levels and their managers subsequently adopt and clarify as needed. This process ensures that the direction that corporate leaders decide upon for the company is understood and followed by all divisions.

As an example for a top-down approach to develop goals, consider the case of the chemical company Clariant. To achieve its mission and, thus, in the long run its vision ("Our vision is to be the leading company for specialty chemicals" [7]), Clariant has formulated what it calls a "five-pillar-strategy" to: increase profitability, reposition portfolio, add value with sustainability, foster innovation and R&D, and intensify growth. For each of the five pillars, Clariant has developed a number of corresponding company goals, defining the firm's ambitions and allowing the firm to measure the successful implementation of its strategy as well as the degree of realization. Two of the five strategic pillars and their corresponding company objectives are:

- **Pillar 1: Increase Profitability**
  Target: Improve Clariant's EBITDA[1] margin by 1–2 percentage points through concentrated performance management, by the continuous improvement program Clariant Excellence, and by improving businesses' cost position by tackling areas with lower profitability.
- **Pillar 2: Foster Innovation and R&D**
  Target: Improve innovation success rate and generate additional sales growth at 1–2 percentage points by launching innovative products [7].

From here, appropriate sub-strategies, and additional measures to track progress, can be developed so that goals are achieved in the way that best suits the company. In other words, this is how to play in a chosen game [9]. In its assessment of the company, competitors, and the business environment, and in the setting of subsequent goals and strategies, executives can decide if they are going to engage in the so-called game that is already in play or if they determine the very game itself.

Traditionally, strategy scholars have assumed that strategies are developed and implemented on the basis of analysis and formal rational planning, and that these strategies are derived from long-term, valid objectives [10]. For this understanding of strategy to work and by using the analogy of gaming again, it has to be noted that the game must be predictable in its fundamentals and the player must be able to significantly influence it. However, if the game strongly depends on the maneuvers of other players or external conditions, its outcome

1 EBITDA (= earnings before interest, taxes, depreciation, and amortization) is a commonly used accounting-based measure of firm success.

is difficult to predict, and new situations arise regularly, which make it necessary for the players to adjust their behavior.

In such a situation, strategy is more likely to be a "pattern in a stream of decisions" [11]. Management may have originally implemented a strategy that was well conceived for its time, but when conditions in the company and its environment change, the firm's strategy may no longer be suitable to achieve the company's goals. In this fairly common situation, a realignment of the strategy, or possibly the underlying business objective, is necessary. The influence of the environment on the strategic process becomes particularly clear if a company pursues a strategy that is more a product of external events than internally generated goals.

Unanticipated moves from competitors or actions from important partners are examples of environmental changes that influence the viability of a firm's strategy. Taking advantage of an unforeseen opportunity, such as a merger or an acquisition, while considering its related risks, can also result in the implementation of a previously unplanned strategy.

An example for a strategic reaction to an unforeseen change in a company's environment is the evolution of Bayer. In 2001, Bayer faced a massive issue with one of its statins (a cerivastatin used to lower cholesterol and prevent cardiovascular disease) called Lipobay in Europe, and Baycol in the United States, respectively, which Bayer launched to compete with Pfizer's highly successful atorvastatin Lipitor. Reports of fatal rhabdomyolysis (a condition that may result in renal failure) forced Bayer to withdraw the product from the market worldwide. As a result of what came to be known as the "Baycol Crisis," Bayer was sued, but on the other hand this crisis led to an enforced restructuring of the entire company. Under the umbrella of the Bayer Holding company, four legally independent business divisions were established: Crop Science, Health Care, Polymers, and Chemicals [12]. This restructuring was meant to enable the independent divisions to react even quicker and more stringently to market changes, both in terms of opportunities and threats. In the following years, Bayer engaged in a number of strategic moves. By separating out the majority of the chemicals activities – basically by combining the Chemicals division as well as parts of the Polymers businesses – and divesting these businesses into a new company known as Lanxess, the remaining Polymer business became much more focused. These strategic alignments to Bayer's portfolio resulted in a reduction of the four initial business divisions to three, namely Health Care, Crop Science, and Material Science.

In addition, the position of Bayer's Health Care division in the over-the-counter (OTC) market was significantly strengthened by acquiring Roche's as well as Merck's OTC activities. The acquisition of Schering in 2006 marked another important strategic move to further strengthen Bayer's global growth and presence in the pharmaceutical industry, which is known for high R&D expenses. After the successful integration and a further portfolio concentration within the three businesses, Bayer is now targeting another move in terms of focusing its portfolio on active ingredients in the areas of Health Care and Crop Science. The company has announced establishment of its Material Science business as a legally as well as business-wise standalone company. This would complete a 10–15 years evolution towards a more Health Care and Crop Science focused company [13, 14].

The case of Bayer reveals great insights into the dynamics and directions of company strategies. But it also shows that the answer to a debate begun by strategy scholars such as Alfred Chandler in 1962, namely, asking whether structure follows strategy or vice versa, is difficult to give, as a strategic realignment might require an alignment of company structures but at the same time a change in structure could enable a strategic realignment [10].

Similar to how this paradigm of strategy has changed over time, the basic understanding of strategic management has also undergone an evolution. New theoretical and empirical insights have certainly contributed to its evolution, but continuous and even disruptive changes in the business environment have placed requirements on strategic management that simply have not been met by the traditional concepts in a satisfactory way. An overview on respective concepts is, nevertheless, given in the next section before more practical guidelines for strategy development processes are presented.

2.3 Historic and Current Trends in Strategic Management

The first concepts of strategy go back 2500 years and were developed to win wars. Strategies helped politicians in their decision-making and were often focused on avoiding war [15, 16]. Over time, many of these strategy concepts spilled over into the corporate world and business leaders became familiar with these topics. New ideas and actual issues have evolved, providing new facets to be considered and making strategy more complex.

During the late nineteenth century industrial revolution, many businesses concentrated their strategy on creating exclusivity and monopolies, where a company became the sole supplier of a particular product or service, such as in the case of railroad companies. Given the lack of competition, monopolistic firms could set higher prices and reap the greatest possible profit. The Standard Oil Company set one of the best-known examples for this type of strategy [17]. At the same time many leading global chemical companies were founded. Typically, the starting point for these foundations is marked by the development of a chemical-based product or a molecule triggering a consumer benefit in a certain application. For instance, the success story of P&G started in 1837 with the industrial production of soap for end consumers, BASF started its business as producer of tar-based colorants quickly followed by indigo-based colorants, and in 1897 Dow Chemical began producing chloride and bromide followed by bleaching agents.

At the beginning of the twentieth century, the concept of **scientific management** was developed: a company aims to increase productivity systematically by making workers and the flow of work as efficient as possible. This idea was later extended to all aspects of productivity and efficiency within a company [18]. During that time, growth and development of the first players in the chemical field accelerated significantly. Based on a continuous standardization of business processes, such as product development and production, they were able to enter new business areas rapidly and to further complement and grow their chemical product portfolios. While in the beginning this happened based on established chemical production processes, over time establishment of completely new processes based on newly discovered chemical backbones and technical production processes could be observed, leading to a further diversification of businesses.

In the following decades, strategy became more standalone and analytical and grew into a management discipline. By the mid-twentieth century, strategy had moved from a task just for senior executives to an area managed by **strategic planners**. As a result, many companies began to distinguish between those that develop and those that implement strategy – a distinction that would later be questioned by scholars such as Henry Mintzberg.

Nevertheless, important new strategy concepts were also developed during that period. The idea of the experience curve arose, which illustrates a direct relationship between cumulated production and production costs: the more units a company produces and the more experience a company has in the production process, the lower the production costs [19]. This phenomenon can be exploited for competitive advantage, with price setting and planning of production volumes and costs. For example, Hirschmann shows what respective learning curves in different branches and industrial sectors – including Process Industries – would look like [20]. Leading chemical players used this phase of the industrial evolution to further diversify their product portfolios, while at the same time strengthening the integration of processes as well as realizing scale efficiency. For instance, in this period Dow Chemical became known for its highly efficient production plants and BASF established highly integrated production sites – an advantage that later became known as *Verbund*.

In 1980, Harvard professor Michael Porter introduced the concept of generic strategies. In Porter's view, a company can gain a competitive advantage by achieving overall cost leadership, differentiation or focus [21].

The cost leader's competitive advantage in comparison with its competitors is that they can ideally offer their product or service at the lowest price within the relevant market. This competitive advantage is gained by production processes being as efficient as possible and by realizing cost savings throughout the entire production process. Such cost savings usually result from economies of scale and scope, which align with the already explained concept of the learning curve. Therefore, pursuing a cost leadership strategy regularly results in extending production capacities and seeking the highest market share among all companies that also try to sell their products or services at a comparably low price within the relevant market. Companies pursuing a differentiating strategy primarily rely on characteristics of their product or service other than price when they bring them into a certain market. They want to attract their customers by a specific bundle of product or service attributes, which are valued by the market as being unique. Such a differentiating characteristic towards a competitor's product or service can, for example, be achieved by quality attributes, technological advantages or design features. These two strategies – becoming the cost leader or adopting a differentiating strategy – can be applied in global markets or in more distinct niches, leading to four different strategic options to be applied in principle. For global strategies, the characteristics of the products and services have to be discernible for the whole market. For niche strategies, it is sufficient if these characteristics are valued by the market segment, which is chosen by the provider. A regional segment, which cannot be accessed by cheaper providers, can, for example, represent a respective niche, allowing for a cost leadership strategy. Satisfying the needs of other specific customer segments by product or service characteristics that differ from the other providers' products or services thus marks a differentiating strategy within a niche. Within the chemical industry, developing specific additives for small-scale customers can be given as an example of the latter.

Furthermore, Porter remarks that companies that do not clearly decide on and implement one of these four strategic options will suffer from a reduced profitability compared with others who have clearly focused on and executed one of the options. This situation usually occurs if cheaper offers, whose characteristics are equally valued by the customers, exist. This remark regarding efficiency implications when being "stuck in the middle" has been discussed with considerable controversy. However, the discussions have not refrained from applying Porter's theory as a common practice and as a basis for many strategy evaluation processes in the chemical industry as well as others.

In chemicals, cost leadership and differentiation can broadly be associated with commodity and specialty chemical businesses. The concept of commodities (e.g., acetic acid, butadiene, and ethanol) and specialties (e.g., adhesives, food additives, and lubricants) is often applied when deciding on a company's strategy. The strategy of cost leadership applies for chemical commodities in particular, since commodity producers are often confronted with the need to create economies of scale and thereby realize cost improvements and efficiencies (see Figure 2.2).

Because commodities (also known as bulk chemicals), offer – in contrast to specialty chemicals – little to no opportunity for product differentiation (e.g., customers demanding benzene can switch from one supplier to another relatively easily), commodity chemicals companies compete heavily based on price (i.e., the price/volume ratio is lower for commodities than for specialty chemicals). Survival in commodity markets, therefore, is strongly depended on cost leadership and operational excellence. In the case of specialty chemicals on the other hand, success depends strongly on a firm's ability to offer differentiated products or services. Offering differentiated products typically enables companies to charge a premium price to customers. Although one might get the impression that the specialty chemicals business is in general more attractive, being a commodity chemical company does not necessarily mean inferior

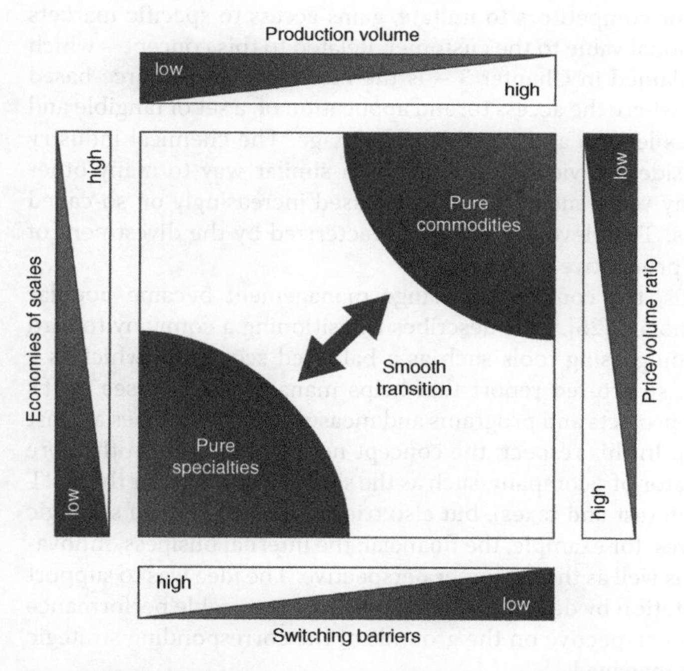

Figure 2.2 Distinction between commodities and specialties [22]. *Source:* Adapted from: Leker J and Herzog P. 2004. Marketing in der chemischen Industrie, in *Handbuch Industriegueternmarketing* (eds K Backhaus and M Voeth). Gabler: Wiesbaden, pp. 1172–1193

performance. In fact, companies such as the Brazilian petrochemical company Braskem or the China-based company Sinopec are very successful.

Based on Porter's concept, companies that cannot assert themselves in overall cost leadership or differentiation strategy need to focus on achieving cost leadership or differentiation in a particular segment. This third generic strategy is therefore called focus strategy. Consequently, the starting point for applying a focus strategy in a niche market is to define a manageable niche market in which a cost leadership position can be established, or the identification of a specific customer market segment in which requirements are not properly addressed by other globally active players. While a chemical company can achieve cost leadership in a niche (and hence achieve a competitive advantage), it will not have overall cost leadership in the entire industry.

In contrast to Porter's outside-in approach to strategy making (i.e., the external environment determines the strategy), other theories moved away from solely looking at external factors to an inside-out approach to strategy making. One of the most prominent examples for this school of thought is Prahalad and Hamel's concept of core competencies [23]. The company that can identify a combination of resources and skills that distinguishes it in the marketplace and makes it difficult for competitors to imitate, gains access to specific markets and provides additional value to the customer. Related to this concept – which will be further explained in Chapter 3 – is the more general resource-based view of a firm [24], where the access to, and application of, a set of tangible and intangible resources lead to a competitive advantage. The chemical industry has chased this inside-out view to strategy, in a similar way to many other industries, for many years and, as a result, focused increasingly on so-called core business areas. This development is characterized by the divestment of non-core and non-productive businesses.

In the mid-1990s, the concept of change management became popular among strategy scholars [25]. This describes transitioning a company toward its desired positioning using tools such as a balanced scorecard, which is a semi-standardized, structured report that helps management oversee staff's implementation of projects and programs and measure their outcomes against strategic goals [26]. In this respect, the concept not only focuses on the core performance indicator of a company, such as the shareholder value or the EBIT (earnings before interest and taxes), but also tries to display a set of strategic relevant perspectives, for example, the financial, the internal business, innovation, and learning as well as the customer perspective. The idea was to support strategy implementation by developing and applying measurable performance indicators for each perspective on the grounds of the corresponding strategic business area to be managed.

Around the turn of the millennium, new technological and economic trends such as the Internet, big data, social communities, and the global financial crisis in 2008, led to new strategic directions. Technologies emerging in the new economy allowed companies to create modular value chains, in which networked enterprises are built of Lego-block-like components [27]. This new structure completely reengineered companies, in addition to their value chains, transforming them from integrated to vertically aligned businesses [28, 29]. Companies were, thus, able to completely penetrate the individual value chains of all their business units and to facilitate processes of coordination within the single areas of value creation as well as among their interfaces. Thus, aligning own value chains with external partner's value chains and even directly integrating them became possible. While the importance of being vertically integrated along a value chain in times of more and more global and volatile markets is a major success factor, such an integration can nowadays be better achieved in networks or across players, segments, and even industry value chains.

Somewhat ironically, the increasingly complex economic and technological environment, combined with a proliferation of strategies, led to frameworks built on some simple cornerstones. The concept of mastering a competitive environment through authentic leadership and corporate culture, for instance, is based on strong leadership, clear direction, and an organization with powerful corporate values [30].

Some strategic concepts focus on very specific aspects of an organization. These are given, as follows, to provide an initial overview of the respective concepts as well as some references to which the interested reader can refer to get a deeper understanding:

- **Lean Six Sigma** [31] aims to minimize waste in all processes (e.g., defects, overproduction, overprocessing non-utilized resources, inefficient transportation, and excess waiting time) as well as to optimize inventory and logistics.
- **Strategic Business War Gaming** [32] is a method for simulating and analyzing business strategies in industries with high competitive intensity. It comprises the development and application of market models, which are based on game theory instruments, computer simulation, and scenario analysis. In a Business War Game, market interaction is simulated between teams (human-based simulation) and/or computer-based models enabling testing of competitive strategies and gaining experience on the drivers of competitors' behavior.
- The **customer centricity and loyalty concept** drives growth by focusing a company's activities on customer value and loyalty [33].
- The idea of **benchmarking** compares a firm to its competitors, identifying potential improvements and implementing rivals' best practices [34].
- Another approach to tackling complexity is to focus on a limited number of strategic aspects, such as targeting uncontested market spaces with a so-called **Blue Ocean Strategy**, thereby unlocking new demand and making competition irrelevant [35].

- One of the most recent strategic approaches involves running organizations based on agility, flexibility, and resilience. Rather than trying to formulate a lasting strategy in a rapidly changing environment, a company reshapes its culture for quick and flexible adaptation to change [36].

The concept "The Future of Strategy" also attempts to conquer complexity. This innovative strategic framework transforms an overall strategy from an incomplete game plan into comprehensive organizational energy. The concept is based on three principles (see Figure 2.3) [37].

First, companies move from formulating strategies based on a present-out to a future-in approach, that is, a process that combines scenario planning with design thinking techniques. **Scenario planning** involves shifting one's perspective from "what we know about our point of departure" to "what we know about the future." **Design thinking** takes place across a company's various functions (e.g., R&D, marketing) and capitalizes on user inspiration for the creation of innovative new business models, products, and services [38]. Understanding the products or services from the user's point of view not only leads to better innovations, but those that will likely remain distinctive and effective for longer time periods. Those innovations are also easier to launch, due to the involvement of different functions throughout the company. The combination of these two approaches delivers the creative and comprehensive strengths of design thinking and the new possibilities posed by scenario planning for the creation of a future-in strategy.

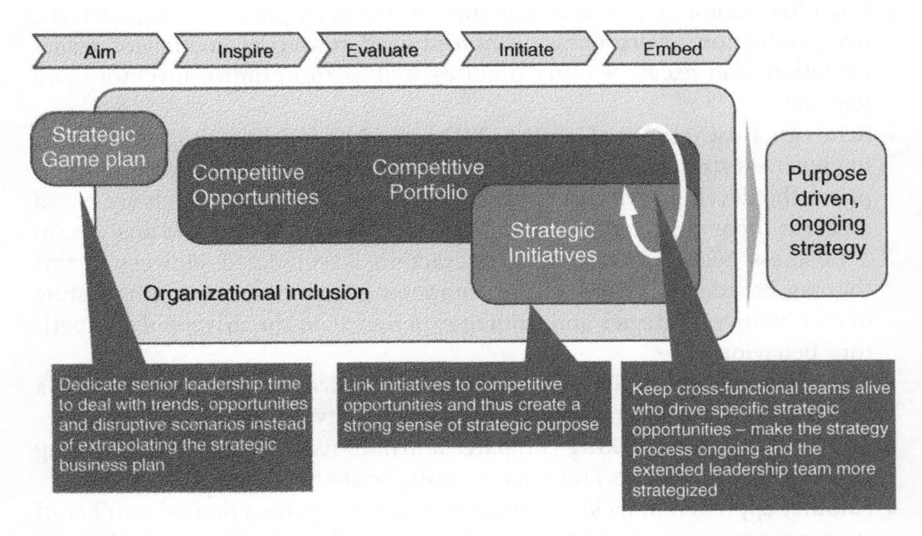

Figure 2.3 Principles of approach for future of strategy [37]. *Source:* Aurik J, Fabel M, and Jonk. 2015. *The Future of Strategy: A Transformative Approach to Strategy for a World that Won't Stand Still*. McGraw Hill Professional: New York

The second pillar of the concept involves moving from a cascading-down approach to organizational inclusion during the strategy-definition process. In fact, both design thinking and the future-in approach require more human capacity. Aided by the latest collaborative and social media, people throughout an organization can contribute to its success. They form a so-called shadow organization of employees willing to push the company's strategic frontiers with their input and involvement, combining the better of two worlds: a specially formed initiative determines the company's future, but it uses the best ideas and experiences from employees throughout the company [39].

What is distinctive about this approach is that it is easier to motivate employees to participate. In fact, employees are inspired to take on such cognitive tasks when they are given autonomy (i.e., having some say about the outcome), mastery (i.e., having a sense of personal growth), and purpose (i.e., having a sense of meaning). Giving employees the opportunity to influence strategy by working with colleagues on the issues arising has a profound effect. Inclusive setups offer another benefit: when strategies are created with employees' input, there is little to no transition to implementation because employees have had a personal stake in the process, are already familiar with it, and have implicitly bought into it from early on [40, 41].

The third pillar involves the development of an ongoing portfolio of competitive advantages instead of a single strategy focused on the creation of a particular competitive advantage that is intended to endure for several years. In fact, it is important for companies to realize that, in today's business climate, it is no longer possible to maintain an ongoing strategic advantage. Instead, competitive advantage has a beginning, a maximum, and an end. Knowing how to recognize these stages, capitalize upon them, and move on to the next competitive advantage are all part of being successful in this context. What's more, having sets of related or complementary competitive advantages will help companies manage their life cycles, categorizing them in such a way that when one advantage reaches its end, there are others identified and ready to build upon [42]. The strategy for managing competitive advantage then becomes a continuous process. Instead of a single undertaking, where management identifies a central competitive advantage for the company for the foreseeable future, it oversees periodic evaluations of the firm's advantages as the environment and competitors shift, staggering them at various stages of fruition. What unifies this process is the firm's overarching strategic plan.

Chemical companies can take a cue from smaller startups, which have learned that they can create more effective strategies in short order by testing them as concepts in the market before fully investing in them [43]. This early feedback can indicate what customers or the business climate will welcome or reject. In the spirit of this approach, companies also may want to consider replacing their single, all-encompassing business plan with a set of smaller, more dynamic plans [44].

The combination of these three pillars enables companies to tackle today's challenges and those coming in the future by continuously defining and defending their competitive advantage.

2.4 Strategy Development Process

Strategic planning is the process of defining a company's game plan (see Table 2.1) and allocating the necessary resources for its implementation.

It is interesting to note that even though a well-defined process greatly helps to develop strategy, many organizations still lack a consistent process or rigor to implement it. Given the rapidly changing competitive environment, a well-defined and executed process begins with regular reviews of a company's strategy. If no unexpected changes with major impact on a company's strategy occur, this is typically done once a year. This practice is useful for identifying changes that would affect operations, such as shifting customer needs, technology and regulatory developments, or competitors' moves. It allows strategy to be continuously adapted based on external events, rather than internal planning cycles.

To be sure that macro-economic trends, overall portfolio performance, industry dynamics, financial projections, and other necessary inputs are considered in strategy development, its definition usually incorporates different management processes, such as mergers, acquisition, divestitures, budgeting, operating reviews, approval and allocation of capital expenditures, and talent development and assignment.

Depending on the execution of strategic planning, the entire organization is affected: starting from the leadership, namely upper management levels of each business unit, as well as the core corporate function; following with the strategic planning process which is typically executed and orchestrated from a (corporate) function in close interaction with the strategic controlling (Figure 2.4).

Strategic planning can be divided into three main steps: (1) analyzing the environment in which the company operates, (2) identifying strategic options, and (3) assessing and selecting the best option. Some corresponding tools and methodologies that should be applied during the strategy development process will be outlined in more detail in Chapter 3.

To analyze the environment, the strategic planning team evaluates the organization's competitive context and positioning, which includes an analysis of stakeholders such as customers and competitors as well as the company itself, including its organizational structure, resources, liabilities, and capabilities. A megatrend analysis helps to understand mid- to long-term trends, starting from a macro point of view often carrying out 10–15 years' time horizons [45]. Generally, megatrends do not majorly change in the short term.

Table 2.1 Principle elements of a company's strategic "game" plan.

Overview	Financial targets	Business portfolio	Market and competition	Trends and drivers
• Strategy 20xx • Summary	• Key performance indicators • Ramp-up 20xx + 5 years	• Positioning • Target portfolio	• Developments • Positioning	• Opportunities • Growth strategies
Regional strategy	**Innovation roadmap**	**Functional strategy**	**Mergers and acquisition (M&A) roadmap**	**Strategic initiatives**
• Trends, sites • Regional strategies	• Themes • R&D focus, budget	• Potential • Functional strategies	• Deal book • M&A schedule	• Initiatives • Timetable

Source: own table

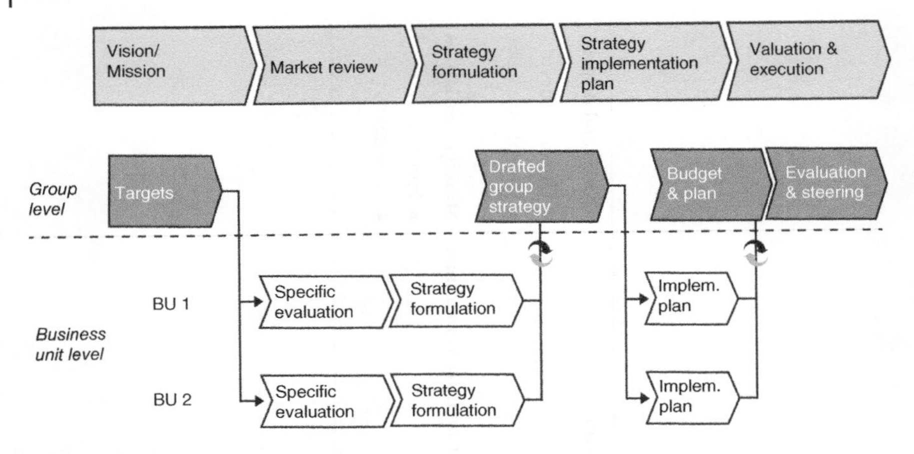

Figure 2.4 Simplified strategic planning process

Understanding these trends and its mid- to long-term implications on customer industry value chains helps to formulate the right questions and to define the right scenarios, which determine the boundaries of a company's strategy and strategic direction.

A SWOT analysis (strengths, weaknesses, opportunities, and threats) is useful for pulling the internal and external pictures together. A SWOT analysis starts with the assessment of a company's individual strengths and weaknesses. It is completed with a review of the core competences that highlight the company's unique skills and which set it apart from competitors. In a second step, identified strengths and weaknesses are opposed to the opportunities and risks, which are observed within the company's relevant environment. When performing a SWOT analysis, special attention has to be given in order to clearly differentiate internal and external aspects as well as to mark only strengths or weaknesses as such if they do not apply to competitors. Universal trends and expectations, which equally impact all competitors, have to be identified as opportunities or risks instead. Once all relevant aspects have been identified and opposed, the SWOT analysis can be completed by answering the following four questions:

1) Which strengths in particular can the company use to benefit from arising opportunities?
2) Which strengths in particular can the company use to react to arising risks?
3) Which opportunities cannot, or only marginally, be realized due to the company's weaknesses?
4) Which risks cannot, or only marginally, be reduced due to the company's weaknesses?

Different frameworks, such as PEST analysis (political, economic, social, and technological), can be applied to reveal where the company is positioned well or weakly within the larger environment [46]. Since these factors are usually beyond the firm's control, the company has to align its strategy with changes in the environment. There are many more tools and methodologies to be applied at this stage; concepts for the strategic analysis will be further illustrated and laid out in Chapter 3.

After conducting SWOT and PEST analysis, a firm has to take a close look at its customers – the firm's source of profits. Customer needs can be pinpointed through a stakeholder analysis. Market segmentation can further refine this understanding by dividing the market into segments, each with its own characteristics, such as different customer needs and price sensitivity. The customer analysis will be complemented by an environmental analysis with a close look at competitors, their products, and competencies. The company's unique selling proposition as its key differentiating point should be derived from this analysis. All of this lays the basis for identifying the firm's strategic options, which are sets of initiatives that can help the firm to achieve a competitive advantage while staying in line with its overall vision and mission. Based on this, the management continues with defining the market segments, product range, and marketing mix with which the company wants to compete. These extend to the business model that will be necessary to operate successfully, along with the required resources, assets, and investments the company wants to employ or has to build up.

In the final step of strategy development, the company delves into each strategic option in detail to select the best option to be implemented. Financial and non-financial criteria are used to select the best option. With regards to financial criteria, cost-benefit analysis, break-even analysis, net present value (NPV) and internal rates of return (IRR) calculations, or decision trees are commonly used. Decision matrix analysis is useful to bring financial and non-financial decision criteria together, because this considers subjective and objective features and weighs individual decision criteria. Ultimately, the chosen strategic option has to fit with the company's vision and mission, and needs to be implemented with available resources. The final outcome is a plan that describes the company's strategy and the steps necessary to achieve it. It should include an action plan and a set of measures or key performance indicators (KPIs) for monitoring its implementation. Respective KPIs should depict the relevant targets of the considered area and their level of achievement, and, therefore, make the implementation measurable. As we mentioned at the outset of this section, savvy companies will update their strategic plan periodically to account for changes in the environment and the actual performance of the organization.

2.5 Industry Dynamics, Signaling Systems, and the Effect of Trends

Global megatrends such as climate change or resource scarcity present significant challenges for the chemical industry, but at the same time represent opportunities with enormous potential [45]. In Chapter 1 of this book you heard about this important aspect to demonstrate in which areas particular challenges are recognized and to what extent they are seen as threats. Different views or estimations of successful and less successful companies presented in Chapter 1 offer additional insights. In the following, a short overview of industry dynamics and trends, demonstrating the strategic relevance of megatrends, will complement these insights.

An overview of megatrends and potential growth fields for chemical firms is presented in Figure 2.5, while strategically relevant estimations towards changes in the competitive environment of the chemical industry are highlighted in the following. In accordance with the SWOT analysis, megatrends represent the central opportunities and risks that companies in the chemical industry have to face over the forthcoming years. Timely identification of the

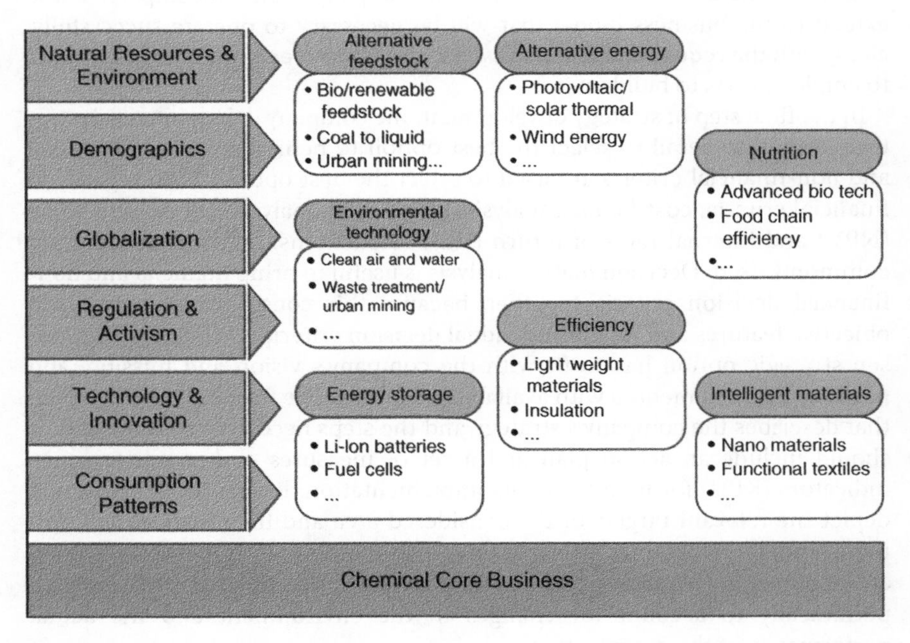

Figure 2.5 Future trends in chemicals, driven by global makro trends [47]. *Source:* von Hoyningen-Huene J, Rings T, Forrest R, and Schulz O. 2016. *Chemical Industry Vision 2030: A European Perspective*

respective impact of opportunities and risks is essential to enable the strategic management of a firm to act both proactively and reactively towards them.

Demographics, including aging populations and their mobility, as well as limited resources and growing demand for energy and water, have a dramatic impact on the global supply of chemical molecules and where they will be used in the future. Globalization is driving the increasing connectivity of the global economy, which has an amplified impact on chemicals. Global consumption as well as changing customer needs are becoming more evident but increase the complexity in product portfolios and service offerings of the chemical players, which represent an additional challenge for chemical firms. As a result, four trends have materialized that will affect the strategic management in the chemical industry in the forthcoming decade:

1) Growing supply of non-conventional feedstocks
2) Accelerating shift of customer industries to China, India, the Middle East, and Africa (the so-called CHIMEA region)
3) Competition for downstream market access
4) Increasing market volatility coupled with more difficult predictability.

Leading a chemical business over the next 10 years will be significantly different compared with today: the trends listed here will force players to redirect their focus from the typical 2–4 year strategic plans to shorter orientations and they need to transform their businesses to compete in an environment of game-changing, long-term developments.

Right before the financial crisis of 2008, chemical players struggled to secure cost effective access to strategically important raw materials and energy. Increasing worldwide demand for them, paired with high speculation about availability, led to a tremendous increase in raw material prices. Crude oil prices, for example, temporarily surpassed $150 a barrel. With the exception of the Middle East, this phenomenon affected chemical players in all regions. Shale gas as well as coal have become low-cost **feedstock alternatives** to conventional resources since then. With respect to shale gas, in effect, this has triggered a transient gold rush, in North America, as for coal gas in China, which is seen as the major backbone of future chemicals in Asia. Shale fuels are seen as a game changer for the global energy balance as well. North America will become a net exporter of natural gas shortly. The continent has already switched from being an importer to a net exporter of chemicals. Interestingly, the largest shale gas reserves are not in North America, but in China. In comparison, however, the Chinese reserves are not well quantified, oil and gas players in that country have not yet mastered the required exploration and production technologies to extract and process these reserves, and the industry needs significantly more infrastructure, especially pipelines and gas fractionation facilities [47]. Given the scarcity of water in China, there are significant barriers to the accelerated development of these resources.

Coal-based feedstock seems to be the more immediate option for China as it seeks to remain competitive in the global feedstock game, where it can leverage existing technologies such as coal-to-liquid, coal-to-gas, coal-to-synthesis gas, or coal-to-carbide.

Based on its significant shale-gas cost advantages, the United States will move along the cost curve for ethylene-based feedstock to reach a position second only to the Middle East. US players will likely realize US$500–600 per ton in cost advantage against naphtha-based supplies from Europe or Asia. They will also outperform coal-based feedstocks from China. With any shortage of raw materials that needs to be overcome by chemical players for the next decade, access to low-cost feedstock and the ability to leverage this cost advantage in a global market will become the main strategic advantage for Middle East suppliers, as well as those from North America.

Over time, the global chemical landscape has shifted from a supply-driven towards a demand-driven market, with players increasingly striving to secure a cost-competitive supply of feedstock on the one hand, while having to secure downstream outlets for their products on the other. The chemical industry follows its customers' industries, such as automotive, construction, and food, as they shift locations, most recently to the Far East. Today, automotive manufacturers, for instance, produce 50% of their vehicles in Asia. In 10–15 years, this rate will exceed 60%. Similarly, construction firms anticipate that more than 50% of the global demand for their services will come from the CHIMEA countries by 2025 [46]. Consequently, the battle for market and customer access and secure outlets for globally produced chemical molecules will intensify. The United States, and potentially China in the long run, will become major exporters of a broad range of chemicals, challenging current market leaders in the Middle East.

With slower economic growth, especially in China, and the simultaneous expansion of chemical capacities, overcapacities will be very likely. The impact on high-cost feedstock regions, such as Europe and North Asia, could be dramatic for olefin derivatives as well as for ethylene/propylene components. Chinese players, such as Sinopec or PetroChina, have sought to secure autarky for themselves by acquiring stakes in US shale-gas fields or taking over coal steam gas assets. The increasing need to secure downstream outlets is expected to drive the next wave of mergers and acquisitions from the foundation of new strategic partnerships.

In the past, Middle East-based chemical players focused on increasing value creation and building capabilities in their home region. Now they are targeting opportunities to move **downstream**, outside of the Middle East as a way to get closer to customers and markets. As a result, we will likely observe more collaboration along the chemical value chain and across countries. Most players still lack repeatable and well-deployed processes for collaboration and currently cannot capture the full value of supplier relationships, shared key accounts, and joint innovation. This situation is expected to change as most future value improvements may be captured at supplier and customer interfaces. Players that jointly design and develop new products, services, processes, and business models will reap the fruits of innovation. One of those fruits is access to better information, which opens doors to more efficient operations planning and inventory management and can optimize distribution. It also leads to better sourcing conditions along the supply chain.

In production, players may benefit from improved planning, capital spend, complexity management, joint specifications rework, and integrated process improvement. Collaboration and better information create sales and marketing opportunities as well. Companies that team up can tap into new joint value propositions and co-branding.

Globalization results in a highly interconnected worldwide economy. It has led to an extended period of prosperity, but at the expense of higher risk. When an economic problem appears in one region, the impact spreads rapidly throughout the world. As a result, volatility is increasing in amplitude and frequency. This makes feedstock prices particularly unstable. Short-term production fluctuations, ongoing macro-economic uncertainty, speculation, and a decoupling of different feedstock sources such as oil and gas, are all influencing feedstock price swings. **Predicting** these changes is a challenge, thanks to the complexities of an increasing number of feedstock sources. Each one has its own economics, value chain, and substitution trends. What is more, state-driven players do not necessarily follow short-term market patterns because they are more concerned with their nations' long-term economic success. As the last crisis demonstrated, chemical players that develop sophisticated volatility-management skills, such as early warning systems, can cope better with this increased volatility and reduced predictability than those that do not have these skills.

Looking back at the last crisis, by late summer 2008, key global indicators had been declining for a year – yet major chemical companies were caught by surprise when sales declined by up to 50%, orders were deferred or cancelled, and inventories increased. Still, the economic boom driven by the rise of the middle class in developing regions and countries such as Brazil, India, and China made a sudden downturn into recession seem unlikely. The risk of missing the associated opportunities appeared to be too big for chemical companies to shift away from their focus on growth. To an extent, this reluctance to embrace the potential crisis was proven right, when, at the beginning of 2009, incoming orders reached new heights. Again, many companies were caught by surprise, this time by the speed of the recovery. Depending on their exposure to more volatile customer value chains, and their regional footprint, chemical players were affected in various ways. Players that were more exposed to European markets benefited from government initiatives, such as the automotive scrapping premium in Germany. Supportive unions in Germany afforded other players high flexibility on temporarily reducing costs and labor. Others saw a quick recovery of exports to Asia. Those with more exposure to North America suffered from the depressed US housing market and limited exports. Chemical companies that successfully managed their businesses during the crisis have excelled for many reasons. They have structured their portfolios to continuously reduce exposure to volatility. They put early warning systems in place, coupled with centralized decision-making based on their own interpretation of business-specific indicators, and they have the processes in place to make decisions and take action quickly. What they also gained from the crisis was competitive advantage. They used the market downturn to make tough decisions about restructuring assets, reducing costs, or even divestments. Thus when the market picked up again, they were able to capture rebound upsides.

Looking forward, chemical companies that clearly define the business model they want to compete with are in a better position to ensure future success. Generally, these companies choose to be integrated players, asset-driven players, or specialty players. Certain factors apply for each of these models. For asset-driven players, which are typically found in upstream segments such as petrochemicals or basic chemicals, access to low-cost feedstock and manufacturing is crucial. In the best scenarios, assets are located close to the feedstock basis, and players enjoy some of the lowest manufacturing cost thanks to global scale or leading-edge technology. Given the shifting balance between global supply and demand for asset-driven players, access to downstream customer markets and the capability to manage global and regional supply and demand fluctuations are important factors as well. Furthermore, the ability to fund growth investments and explore global partnerships is key. Positioned further downstream, specialty players are successful if they have achieved customer confidence, technology leadership in niche markets, and cutting-edge capabilities in innovation and sustainability. Typically, to differentiate themselves from competitors, players in specialty chemicals markets have broad product and customer portfolios that require sophisticated complexity management and performance excellence in a variety of customer-industry value chains. For asset-driven players, the ability to fund growth investments is crucial, which puts a lot of pressure on more broad-based, specialty chemical players. These firms may be rigorous and focused in their portfolio management and still not be able to achieve critical scale for each of their businesses and customer industries. Finally, for integrated players, it is difficult to operate both asset-driven and specialty business models in one company, when each model requires different success factors and its own approach to management and governance. Those that succeed with this model link value chains end to end, from raw material and feedstock to downstream customer industries and markets. This type of vertical integration has proven to be very successful over previous decades.

To successfully and profitably reach 2025, chemical companies need to prepare themselves now for those trends that will affect the industry the most: the increasing supply of non-conventional feedstock, the accelerating shift of business to CHIMEA, the fight for downstream market access, and increasing market volatility. But, there is even more: successfully transforming an enterprise means coping with increasing diversity in many forms, including disciplinary, cultural, and gender-wise. Being more inclusive in these ways positions companies to become even more global in terms of corporate structures, languages, and ways of communicating. It also helps them capture value from a growing reservoir of international resources, knowledge, and intellect.

2.6 Summary

- **The management of a company should ask strategic relevant questions on a regular basis.** Owing to the fact that the landscape in the chemical and pharmaceutical industry is affected by radical change regarding, for example, changing customer needs, reconfiguration of value chains, emergence of new technologies, and globalization, asking the right strategic questions is important to sustain a successful business in the chemical and pharmaceutical industry.
- **The mission and vision of a company serve as a guideline for further strategy development from which more detailed corporate goals can be formulated.** As the environment changes, implemented strategies may need to be adjusted.
- **Chemical companies increasingly move beyond single strategies.** Starting with Michael Porter's outside-in approach to strategy making in general and with his three generic strategies of overall cost leadership, differentiation, and focus in particular, nowadays most companies follow a more dynamic approach of strategic management based on an inside-out approach. Companies should figure out their core competencies, focus on future scenarios, include the customer and all corporate functions in the innovation process, involve the whole company in the strategy formulation process, and keep a portfolio of several competitive advantages, instead of focusing on one strategy which pursues a particular competitive advantage.
- **Strategic planning starts with analyzing the company's environment.** Such an analysis (e.g., in the form of SWOT or PEST analysis, an analysis of customer needs, or competitor analysis) should reveal the unique selling proposition (USP) of the company, followed by figuring out a number of strategic options whereby, finally, the best is selected. This option, like every strategic decision, needs to be in line with the overall mission and vision of the company.

- **When making strategic decisions, managers within the chemical industry need to consider at least four important global megatrends.** These are the growing supply of non-conventional feedstocks, shift of customer industries to the CHIMEA region, competition for downstream market access, and increasing market volatility. Implementing warning systems will help managers to anticipate environmental changes (e.g., changing feedstock sources), thus supporting them to adapt to changing market conditions.

References

1 Macharzina K and Wolf J. 2008. Unternehmensführung: das internationale Managementwissen in *Konzepte, Methoden, Praxis*. Springer-Verlag: New York.
2 Hauschildt J. 1999. *Unternehmensverfassung als Instrument des Konfliktmanagements: jenseits von rechtlichen, institutionenökonomischen und soziologischen Überlegungen (No. 500)*. Manuscript from the Institutes of Business Administration at the University of Kiel.
3 Kenny G. 2014. *Your company's purpose is not its vision, mission, or values.* https://hbr.org/2014/09/your-companys-purpose-is-not-its-vision-mission-or-values (accessed 21 July 2016).
4 Procter & Gamble. 2014. *Purpose, Values & Principles.* http://us.pg.com/who-we-are/our-approach/purpose-values-principles (accessed 15 December 2015).
5 BASF. 2015. *We create chemistry – Our corporate strategy.* https://www.basf.com/documents/corp/en/about-us/strategy-and-organization/BASF_We_create_Chemistry.pdf (accessed 15 December 2015).
6 Dow Chemicals. 2015. *Mission & Vision.* http://www.dow.com/en-us/about-dow/our-company/mission-and-vision (accessed15 December 2015).
7 Clariant. 2013. *What does Clariant stand for today*? http://www.annual-report.clariant.com/2013/annual-review/the-clariant-story/what-does-clariant-stand-for-today.html (accessed 15 December 2015).
8 Henkel. 2011. *Vision and Values.* http://www.henkel.at/blob/20054/0761524dad1d9ef772850214ec84a8ba/data/vision-and-values.pdf (accessed 15 December 2011).
9 Markides CC. 2013. *Game-Changing Strategies: How to Create New Market Space in Established Industries by Breaking the Rules.* John Wiley & Sons Ltd: Chichester.
10 Chandler AD. 1990. *Strategy and Structure: Chapters in the History of the Industrial Enterprise.* Vol. 120. MIT Press: Cambridge, MA.
11 Mintzberg H. 1978. *The Nature of Managerial Work.* Prentice Hall: Upper Saddle River, NJ.
12 Bayer. 2015. *History 2001–2010.* http://www.bayer.com/en/2001-2010.aspx (accessed 10 March 2016).

13 Bayer. 2014. *Media News 18 September 2014*. http://www.press.bayer.com/baynews/baynews.nsf/id/Bayer-plans-to-focus-entirely-on-Life-Science-businesses?Open&parent=news-overview-category-search-en&ccm=020 (accessed 4 September 2015).
14 Handelsblatt online. 2014. *News article about Bayer*. http://www.handelsblatt.com/unternehmen/industrie/boersengang-fuer-sparte-material-science-bayer-verabschiedet-sich-vom-plastik/10718422.html (accessed 4 September 2015).
15 Tzu S. 2011. *The Art of War*. Shambhala Publications: Boulder.
16 Von Clausewitz C. 1873. *On War*. Vol. 1. N. Trübner & Company: London.
17 Tarbell IM. 2009. *The History of the Standard Oil Company*. Cosimo, Inc.: New York.
18 Taylor FW. 1998. *The Principles of Scientific Management*. Dover: Mineola, NY.
19 Henderson B. 1974. *The Experience Curve Reviewed*. John Wiley & Sons, Inc.: New York.
20 Hirschmann WB. 1964. Profit from the learning-curve. *Harvard Business Review*, **42**(1): 125–139.
21 Porter ME. 1985. *Competitive Advantage: Creating and Sustaining Superior Performance*. Free Press: New York.
22 Leker J and Herzog P. 2004. Marketing in der chemischen Industrie, in *Handbuch Industrieguetermarketing* (eds K Backhaus and M Voeth). Gabler: Wiesbaden, pp. 1172–1193.
23 Prahalad CK and Hamel G. 1990. The core competence of the corporation. *Harvard Business Review*, **68**(3): 79–91.
24 Barney J. 1991. Firm resources and sustained competitive advantage. *Journal of Management*, **17**(1): 99–120.
25 Kotter JP. 1995. Leading change: Why transformation efforts fail. *Harvard Business Review*, March–April: 59–67.
26 Kaplan RS and David P. 1992. Norton, The balanced scorecard-measures that drive performance. *Harvard Business Review*, **70**: 71–79.
27 Evans P, Wurster TS, and Bits BT. 1999. *How the New Economics of Information Transforms Strategy. Ideas at Work*. Harvard Business School Publishing: Cambridge, MA.
28 Hammer M and Champy J. 1993. *Reengineering the Corporation*. HarperCollins: New York.
29 Hagel J and Singer M. 1999. Unbundling the corporation. *Harvard Business Review*, March–April: 133–144.
30 Drucker PF, Christensen CM, Grant AM, Govindarajan V, and Davenport TH, 2011.10 Must reads on leadership. *Harvard Business Review*, January: 1–352.
31 George ML. 2002. *Lean Six Sigma*. McGraw-Hill Education: New York.
32 Oriesek DF and Schwarz JO. 2009. *Business wargaming. Unternehmenswert schaffen und schützen*. Gabler: Wiesbaden.

33 Reichheld FF and Teal T. 2001. *The Loyalty Effect: The Hidden Force Behind Growth, Profits, and Lasting Value*. Harvard Business Press: Cambridge, MA.
34 Kreuz W. 1995. *Mit Benchmarking zur Weltspitze aufsteigen*. Verlag Moderne Industrie.
35 Chan Kim W and Maurborgne R. 2005. *Blue Ocean Strategy – How to Create Uncontested Market Space and Make the Competition Irrelevant*. Harvard Business School Press: Boston, MA.
36 Doz YL and Kosonen M. 2008. *Fast Strategy: How Strategic Agility Will Help You Stay Ahead of the Game*. Pearson Education: London.
37 Aurik J, Fabel M, and Jonk G. 2015. *The Future of Strategy: A Transformative Approach to Strategy for a World that Won't Stand Still*. McGraw-Hill Professional: New York.
38 Brown T. 2009. *Change by Design: How Design Thinking Transforms Organizations and Inspires Innovation*. Harper Business: New York.
39 Kotter J. 2012. How the most innovative companies capitalise on today's rapid-fire strategic challenges – and still make their numbers. *Harvard Business Review*, **90**(11), 43–58.
40 Pink D. 2009. *The puzzle of motivation*. TED *Talk*. https://www.ted.com/talks/dan_pink_on_motivation?language=de (accessed 18 May 2015).
41 Ariely D. 2012. *What makes us feel good about our work. TEDxRiodelaPlata*. http://www.ted.com/talks/dan_ariely_what_makes_us_feel_good_about_our_work. html (accessed 18 May 2016).
42 McGrath RG. 2013. *The End of Competitive Advantage: How to Keep Your Strategy Moving as Fast as Your Business*. Harvard Business Review Press: Cambridge, MA.
43 Ries E. 2011. *The Lean Startup: How Today's Entrepreneurs Use Continuous Innovation to Create Radically Successful Businesses*. Crown Books.
44 Blank S. 2013. Why the lean start-up changes everything. *Harvard Business Review*, **91**(5): 63–72.
45 Laudicina P. 2004. *World Out of Balance: Navigating Global Risks to Seize Competitive Advantage*. McGraw-Hill Professional: New York.
46 Gupta A. 2013. Environmental and pest analysis: An approach to external business environment. *Merit Research Journal of Art, Social Science and Humanities*, **1**(2): 013–017.
47 von Hoyningen-Huene J, Rings T, Forrest R, and Schulz O. 2016. *Chemical Industry Vision 2030: A European Perspective*. https://www.atkearney.com/chemicals/ideas-insights/article/-/asset_publisher/LCcgOeS4t85g/content/chemical-industry-vision-2030-a-european-perspective/10192?_101_INSTANCE_LCcgOeS4t85g_redirect=%2Fchemicals%2Fideas-insights (accessed 18 April 2016).

3

Strategic Analysis: Understanding the Strategic Environment of the Firm

Jens Leker[1] *and Manuel Bauer*[2]

[1] *University of Münster, Department of Chemistry and Pharmacy*
[2] *LEDVANCE, Innovation Management*

> *Strategic management is not a box of tricks or a bundle of techniques. It is analytical thinking and commitment of resources to action. But quantification alone is not planning. Some of the most important issues in strategic management cannot be quantified at all.*
>
> Peter F. Drucker (1909–2005), Writer, professor, and management consultant

This chapter deals with analytical methods and concepts in strategic analysis, which form the basis of strategic planning and decision making. Readers are becoming familiar with strategic analysis and enabled to select the relevant technique appropriate to the specific problem in hand. Given the large range of methods and concepts for strategic analysis available, we will focus on what we perceive to be the most useful ones for the chemical industry, while the overall outline of the chapter follows recent developments in the field of strategic management theory.

In the course of this chapter, we firstly define what is really meant by "firm performance" and clarify the role of the firm's shareholders (owners) as well as other stakeholders relevant to this definition. Secondly, we will introduce various tools for strategic analysis that deal with analyzing the external environment of the company. In so doing, we are giving hints on how to position the company within its environment (relative to competitors, suppliers, and customers), by also illustrating potential ways to deal with changes in this environment (e.g., due to new disruptive technologies, substitution, or new entrants). Thirdly, we outline how managers can identify, build, and leverage internal resources to enhance the firm's performance, both in an intentionally

Business Chemistry: How to Build and Sustain Thriving Businesses in the Chemical Industry, First Edition. Edited by Jens Leker, Carsten Gelhard, and Stephan von Delft.

planned manner based on the firm's existing set of resources, as well as in an emergent manner that is reactive to changes in the firm's external environment. Therefore, in Section 3.3, we will introduce the logic of the so-called resource-based view (RBV) of the firm and how it can be applied in the context of a chemical company. In Section 3.4 we focus on how to recognize changes in the environment that may affect the company's competitive position, including the introduction of tools such as the S-curve (assessing strategically relevant technological changes) as well as concepts of how to recognize industry convergence. Furthermore, a concept to describe strategic re-direction of competitors is presented, allowing managers to depict first courses of action on the basis of resource-based analyses and competitor behavior in the relevant market. Section 3.5 focuses on the strategic management theory of "dynamic capabilities" and introduces current management tools on how to alter the firm's resource and capability base in order to cope with changes in the external environment.

3.1 Strategic Analysis to Improve a Firm's Performance

Defining the purpose of "strategic analysis" first requires a common understanding of "What is strategic management?" This, however, is a difficult task, as already pointed out in Chapter 2, and also by scholars such as Costas Markides from the London Business School, who explains that "despite the apparent simplicity of this question, it is one of the most controversial in the field of management. People seem to disagree about almost everything contained in this question: about what issues are relevant; about the process that a manager should go through to develop strategy; and about the actual physical output that should emerge at the end of a strategy process" [1: 1]. To create common ground for the following discussion, we will follow Nag, Hambrick, and Chen (2007) and define strategic management as follows: "Strategic management deals with the major intended and emergent initiatives taken by general managers on behalf of owners, involving utilization of resources, to enhance the performance of firms in their external environments" [2: 944]. Relating to our discussion on strategy in Chapter 2, we will take a broader view on strategy in this chapter, since owners (i.e., shareholders) solely represent one of various stakeholders that influence the firm's overall decision making, including the firm's employees, suppliers, customers, the government, or NGOs (non-governmental organizations).

The methods for strategic analysis (Figure 3.1) covered in this chapter can be organized in a simple matrix. We distinguish between two fundamental dimensions of how to perform strategic analysis: the horizontal dimension distinguishes between a static and a dynamic analytical perspective. From a static

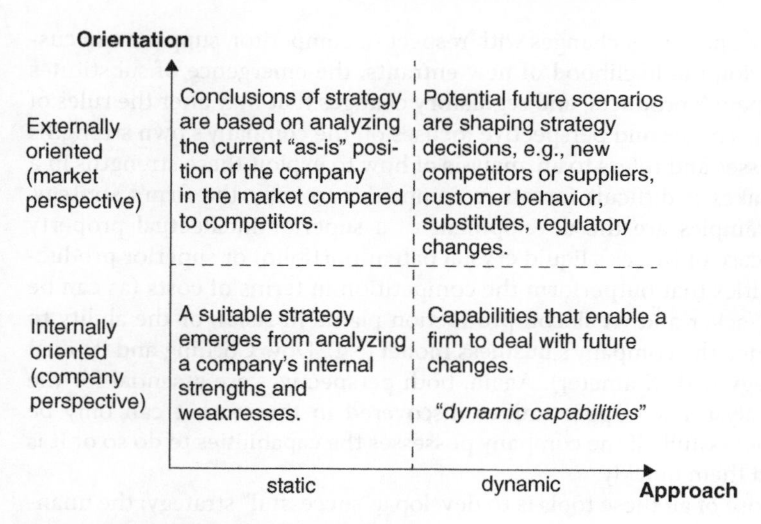

Figure 3.1 Methods for strategic analysis

point of view, the analyst looks at the current situation and draws conclusions for strategy formulation based on the currently observed behavior of market participants, for example, analyzing what assets and production capacities competitors possess, what are the current price and margin levels, what business models dominate the market, and what are the current strengths and weaknesses of the focal company. The conclusions on what strategic moves should be pursued are based on facts and actual observations. In contrast, from a dynamic perspective, the analyst is interested in the potential future scenarios that may change the current status quo in order to develop a strategy that may be used to cope with or even shape these future changes. Questions arise, such as: "How will competitors behave if my company adopts a cost leadership strategy?"; "Will my company's technological capabilities enable my company to satisfy future customer needs (e.g., towards more sustainable products)?"; and "Does my company have strategic capabilities to quickly adapt to unforeseen changes in the market place?" Since these questions lead to predictions about the future, the derived strategy has to be flexible and it emerges while the company is learning whether its initial assumptions were correct or not. It goes without saying that both analytical perspectives are relevant and important for strategy formulation, since the former clarifies the actual starting point while the latter develops projections and a judgment of their likelihood about how to win in the future.

The second dimension on the vertical axis of Figure 3.1 distinguishes between an externally oriented market perspective and an internally oriented company capabilities perspective. Following the first perspectives, managers analyze the firm's current positioning and prospective changes within the industry, including the analysis of changes with respect to competitor, supplier and customer behavior, the likelihood of new entrants, the emergence of substitutes for the company's products, and regulatory changes that may alter the rules of the business. The second perspective focuses on the company's own strengths and weaknesses and refers to an analysis of how to exploit these strengths in a way that makes it difficult for other companies to copy the firm's strategy. Classical examples are the development of a superior intellectual property basis (as in case of Merck's liquid crystal patent portfolio) or superior production capabilities that outperform the competition in terms of costs (as can be seen with Wacker and its silicon production plants in Asia), or the ability to quickly change the company's business model (e.g., Dow Corning and its dual brand strategy with Xiameter). Again, both perspectives are essential for the strategic analyst since opportunities discovered in the market can only be exploited successfully if the company possesses the capabilities to do so or it is able to build them quickly.

The purpose of all these tools is to develop a "successful" strategy; the unanswered question though is: What is a **successful** strategy? A strategy is conventionally considered to be successful if it allows the firm to increase its performance, which, again, is the observable outcome of the firm's value creation. In general, a firm adds value if the revenue generated from selling its products is larger than the costs of the raw materials required to produce those products. If the value added to the firm is larger than its obligations to all of its stakeholders it creates **economic value**. Examples of stakeholder obligations (besides serving raw material suppliers' claims) are employee salaries, maintenance costs of production equipment and service contracts, bank interest, government taxes as well as dividend payments to shareholders. The excess revenues that remain after serving all stakeholder obligations is the retained profit, which the firm's senior management can decide either to re-invest in the firm or to pay out to shareholders. Looking at retained profit, not just over a single business year, but rather over a longer period of time (e.g., the economic cycle of the specific industry), the retained profit becomes sustainable. Developing a competitive advantage, thus maximizing retained profit, is the primary objective of strategic management. This retained profit also represents the principle indicator of the firm's performance.

We want to specifically emphasize that we generally prefer a sustainable-oriented consideration of value generation, even if we are totally aware that the prediction of the expected sustained retained profit is subject to various difficulties (e.g., lack of valid information). Here, we aim to extend the proposition by Nag *et al.* [2], that strategic management is to maximize firm performance on behalf of the owners by adding stakeholders (e.g., customers, employees, suppliers) in general. This is in line with the appraisement by Grant (2015), who argues that "Management cannot create stock market value – only the stock market can do that. What management can do is to generate a stream of profits that the stock market capitalizes into market value" [3:415]. The stakeholder-oriented perspective applied in this chapter leaves open how to distribute retained profits among the various stakeholders of a company. A high positive economic profit is an indicator of superior firm performance since firms that earn high economic profit obviously generate more economic value than is needed to pay its stakeholders and to produce their products. Many financial indicators have been developed to measure economic profit but the concept is difficult to capture in one number. The reason is that companies may forgo generating financial profit, as it is stated on the income statement, out of economic profit for the sake of re-investment in, for example, market share expansion, capability building, or strengthening the brand image. In this context, the strategic analysis tools presented in this chapter become significantly important. However, we would like to mention that strategic analysis does not and cannot replace strategic decision making, but it provides information for decision making as well as a structure for the preceding thought process that leads to strategic decisions.

3.2 Industry Analysis

In this section we focus on the current situation of an industry and how it can be analyzed, before the subsequent sections then introduce you to the dynamic perspective and also consider the analysis of ongoing changes.

As the considerations and the introduction of the SWOT analysis in Chapter 2 have already demonstrated, each company has its own strengths and weaknesses and operates in an external space that provides chances and risks. The respective external space (i.e., the environment) comprises various elements. While some elements can be more or less directly influenced by the company (e.g., the choice of customer segments), other elements can be at best only indirectly influenced, such as the legal framework or a country's economic development. The European Union's (EU) REACH regulation, which came into effect in 2007, for instance, forced chemical companies located in the EU to adopt the respective regulations and, thus, changed the environment (e.g., prohibition of chemicals, stricter restrictions) in which companies operate. In addition, rising environmental awareness and demand for more social, ecological, and economic management in a geographical area are other developments that a company primarily has to respond to, although it cannot directly influence those developments through its own efforts. Both the REACH example and the politically intended German "Energiewende," which resulted in higher energy costs, show that the consequences can have a significant effect on a company's business. A thorough consideration and investigation of the environment, which cannot directly be influenced by a company, therefore obviously becomes a necessity.

In contrast to the global environment, which is, at least in certain areas (e.g., geographically), equal for all companies, a company is surrounded by a direct – more idiosyncratic – business environment. A company directly interacts with this environment or respective stakeholders, such as customers, competitors, suppliers, or employees. To analyze an industry or particularly the direct environment a company operates in, Porter's "Five Forces" represent a well-established tool to obtain a structured overview of an industry's activities and the prevailing forces that determine an industry's profit potential [4]. The Five Forces are depicted in Figure 3.2 and encompass: competitors, customers, suppliers, potential new competitors, and substitutes. When we apply Porter's Five Forces framework for chemical companies, it is essential to consider the characteristics of the chemical industry's predominant business-to-business (B2B) nature and that it is a process industry. For instance, in many cases the existing production processes, which, in turn, derive from the established process structures within the chemical industry, define existing opportunities for forward and backward integration, the associated definition of the customer and supplier structures, as well as the intensity of the competition.

The existing competition within the industry itself is central to its opportunities and overall attractiveness. Thus, we initially focus on various factors that determine the extent of rivalry among existing competitors. The total number

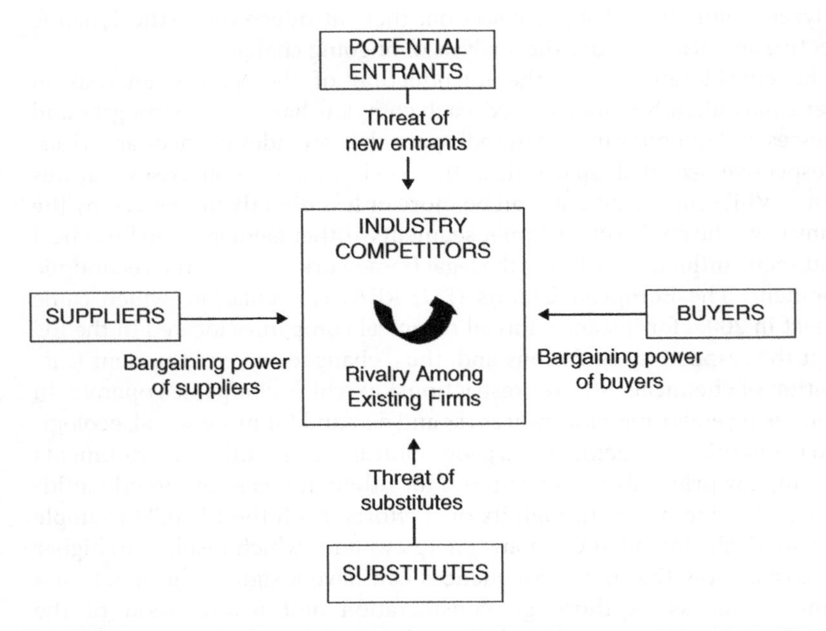

Figure 3.2 Porter's Five Forces framework. *Source:* Omerzu, https://commons.wikimedia.org/wiki/File:Branchenstrukturmodell-Five-forces.png. Used under CC BY 3.0

of competitors that offer more or less the same products has a substantial impact on the overall intensity of competition. Although nowadays classical monopolies are rare, concentrations of only a few companies (e.g., in the oil and gas industry) can lead to a kind of coordination and less intensive price battles. Whereas a strongly growing industry provides growth opportunities for each company by serving the additional needs, market decline analogously causes the increase in rivalry for market share. One example refers to the European market for photovoltaics, in which input producing companies such as Wacker with its polysilicon products suffered massive price competition driven by reduction in governments' subsidies for solar energy production, for example, in Germany and Greece. As demand in the global photovoltaic market rose at the same time, this is another example that demonstrates the importance of the global environment for an industry's activity [5]. High fixed costs of production and the stage of the industry's differentiation represent two other factors that drive competition by, for instance, cutting prices, which can be observed in particular in the production of commodity chemicals.

Setting up a production site for several thousand tons of a commodity product (e.g., ethylene or polypropylene) is a massive investment and designed for the respective bulk production. Commodities are well understood by market players (e.g., customers) and, therefore, do not allow for product differentiation. As this, in turn, fosters price competition, the profits to be earned in this industry segment might decrease. Furthermore, overcapacities, especially if they originate from large steps in capacity augments, as in chemical bulk production sites, increase the rivalry within the industry in the same way.

Besides existing competitors, new companies are attracted by profitable markets, thus they try to enter these markets and then represent a threat to existing players in the industry, since they add additional capacities and demands for market shares to the existing competition. New companies consequently represent a second threat that might comprise similar mechanisms (e.g., the risk of potential price battles) as mentioned earlier. The extent of a new entrant's threat is not unique to all industries. If market entry barriers are low, the threat is comparatively large, while industries with large market entry barriers are impacted to a lesser extent. The high investment costs of building a bulk production plant for a commodity chemical, for instance, denote one of several such barriers. Other barriers that can inhibit new entrants are non-accessibility to a certain technology (e.g., due to comprehensive intellectual property (IP) protection), the great significance of scale effects or of strong established brands, regulatory governmental activities (e.g., required quality standards), and the extent of customers' switching costs, which refer to transaction costs, learning costs, and artificial or contractual costs [6].

The third force, which is associated with the extent of customers' switching costs, is the customers' overall bargaining power. In general, if switching costs between different providers are low, the customers' bargaining power increases.

For instance, with regard to commodities such as hydrochloric acid or ethylene, the barriers and associated costs for switching the supplier are quite low. On the other hand, if customers demand specific materials, that is, for instance, particularly tailored to their idiosyncratic needs or materials only available in that form from one company (e.g., due to IP protection), the barriers and associated costs for switching suppliers are substantially higher. These costs, for example, might be associated with the need to adjust processes on the customers' side or a lower level of quality of the newly sourced product. Thus, the extent of a contribution to a product represents a crucial factor that influences the bargaining power.

In addition to the total number of competitors offering the same or easily substitutable products, the total number of existing customers also influence the customers' overall bargaining power. If there are a lot of customers, their individual bargaining power decreases as the provider might be able to easily turn to other customers. Furthermore, if it is likely that customers may be able to produce the desired product by themselves, that is, if the risk of backwards integration is high, the customers' bargaining power also rises. Vice versa, if the customers' job could be performed by the relevant industry, that is, if the risk of forward integration is high, then the customers' bargaining power decreases.

As in the case of the customers' bargaining power, similar mechanisms – with a changing perspective from a forward to a backward view along the supply chain – apply for the fourth force: the industry's suppliers. For instance, if only a small number of relevant suppliers exists, their bargaining power increases as the industry depends on the respective products. Other factors include switching costs to other suppliers or the extent to which a supplier contributes to the overall product quality. Further, a high level of suppliers' bargaining power becomes evident by referring to: (i) the more general example of focusing on the regional supply of rare earth materials due to their deposits and degradability, and (ii) the specific example of Merck's well-protected liquid crystal technology. In the first example, the supplier of rare earth materials possesses the access to unique and very scarce resources, which can only be exploited at distinct geographical locations. Thus, the buyer of rare earth materials is committed to a small number of suppliers, which then possess a high bargaining power. This is similar to the second example, where Merck artificially creates a bottleneck of knowledge to a specific technology by maintaining an extensive patent portfolio. Customers who are dependent on this technology are forced into obtaining the relevant technology by asking Merck for permission.

Finally, the threat of substitutes represents an additional force that impacts the profitability of an industry. This basically follows the same mechanisms as the threat of new entrants. The more aggressive and efficient or convincing their value for money is the higher is the resulting threat. Again, the costs of switching can obviously have a large impact. In general, substitutes fulfill the same or similar features of the product or service which is already established in the market [7]. Thus, for example, drop-in chemicals are closely related to substitutes. Within this drop-in approach, similar chemicals can be produced, but based, for example, on renewable rather than fossil resources.

With regard to all of these Five Forces, companies generally seek to realize profits by forming and exploiting any type of competitive advantage. For the formulation of an appropriate strategy to achieve this kind of competitive advantage, managers can apply Porter's concept of the internal value chain, which supports them in visualizing and analyzing all activities that might contribute to the firm's competitive situation. As the firm's value chain is embedded into the additional value chains of the companies' customers, suppliers, and competitors, a holistic approach can be considered by referring to the so-called value system (the agglomeration of various interrelated value chains). Figure 3.3 illustrates such a value system with regard to the battery system for electric cars. As the example shows, the presented battery system allows – depending on the degree of complexity of the underlying technology, the required production steps, as well as the field of application of the end product – a high degree of differentiation.

Firms operating in this value system have to decide which areas of the system they intend to cover and, in doing so, they have to consider in particular their internal strengths (and weaknesses) in relation to the overall competitive situation. Once firms have decided about their positioning within the value system, they can continue to design their internal value chain. Porter's concept of the internal value chain is depicted in Figure 3.4.

Primary activities are directly related to the production process or to a respective service provided to a customer, while support activities cover enhancing activities, such as providing essential inputs, including necessary human resources and technologies. More precisely, the support activities can be found within every primary activity and ensure that the value activity can be performed. Owing to the fact that primary activities comprise a competitive advantage for the company over its competitors, the examination of each value activity is of particular significance. The importance of each activity and therefore its influence on the margins that can be realized differs greatly between industries, but it can also differ within an industry due to different competitive strategies.

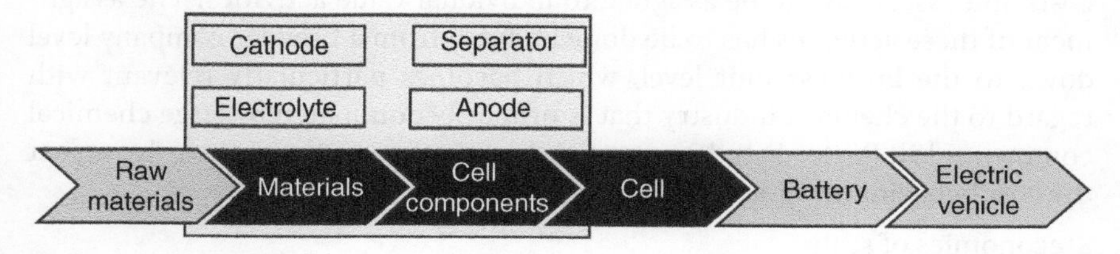

Figure 3.3 Battery value chain

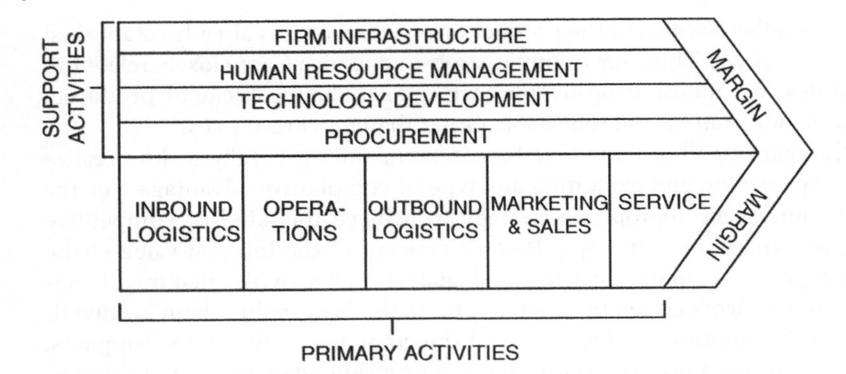

Figure 3.4 A firm's internal value chain. *Source:* Frog Thomas, https://commons.wikimedia.org/wiki/File:Das_Modell_einer_Wertkette.svg. Used under CC BY SA 4.0

Within the chemical industry, chemical distributors might primarily compete by using effective in- and out-bound logistics as well as marketing and sales activities. Here, the operations themselves might be of less importance since those firms rarely produce a physical product. Within chemical production, operations might be a lot more differentiating, for instance, as the historical example of ammonia production illustrates. The Haber–Bosch process enabled BASF to make use of a revolutionary and new source of ammonia, through which the industrial production of ammonia was established.

The previous examples further demonstrate the importance of "technology development" as a particular source of advantage of support activity. Before BASF was able to industrialize the synthesis of ammonia, the ammonia required for use as a fertilizer was obtained from natural ligated ammonia. The new process had a significant influence on all primary activities and enabled BASF to meet the increasing demand for ammonia. While technological developments as well as human resource management and procurement – at least in its subunits – often refer to a primary activity, the infrastructure of the firm has to be considered in a more global context. Firm infrastructure covers activities such as accounting, planning or legal activities, as well as general management.

The resulting framework can eventually be used for value chain analysis or, more precisely, for a cost or differentiation analysis. To this end, operating costs and assets have to be assigned to individual value activities. The assignment of these activities has to be done in detail from a broader company level down to the business unit level, which becomes particularly relevant with regard to the chemical industry that is primarily dominated by large chemical companies [3]. Porter lists 10 potential drivers of costs that primarily impact the cost behavior of value activities:

- economies of scale
- learning
- the pattern of capacity utilization

- linkages
- interrelationships
- integration
- timing
- discretionary policies
- location
- institutional factors [8].

In the following, some of these drivers will be outlined in greater detail. Linkages refers to influences between different value chain activities or even between different value chains. For instance, co-products such as phenol and acetone from the cumene process can show potential cost advantages: if a company can make good use of both co-products in its resulting volumes, it should receive an advantage over those who have a greater need for only one of the products.

With regard to the chemical industry, location represents another important cost driver, since profits of chemical companies strongly depend on energy and raw material costs (see also the discussion of shale gas in Chapter 2). For example, in Germany the government is forcing the development of renewable energies, which leads to higher cost for energy production compared with energy produced by means of fossil fuel. In addition, institutional factors, that is, governmental incentives or restrictions, are often closely related to location choices and can constitute cost differences. For instance, factors such as government and tax regulations may build barriers for competitors to get into the market, due to the fact that it is simply forbidden for competitors to enter the market.

With Porter's internal value chain analysis, it is possible to reveal not only cost advantages compared with competitors, but also the main differentiating characteristics of a company's product portfolio.

The effective pursuit of a cost leadership strategy finally demands that the accumulated costs of all relevant value chain activities are lower than those of the relevant competitors. Thus, it is important for companies to not only analyze their own value chain, but also the competitors' value chains in order to identify cost advantages and the sources of cost differences. Based on this information, firms can eventually assess whether existing cost advantages are sustainable. Porter suggests two primary ways to gain or improve cost advantages: (i) controlling the cost drivers and (ii) reconfiguring the value chain to improve its cost position relative to competitors. By doing so, existing differentiation (advantages) should not be destroyed, unless on a well-considered basis.

Analysis of a firm's value chain activities can be further support for identifying a differentiation strategy by specifically disclosing potential sources of uniqueness – basically, each activity that constitutes the firm's value chain might be a source of uniqueness and thus lead to differentiation. Following Porter, policy choices are emphasized as the single most prevalent uniqueness driver.

The technology employed, the quality of inputs, or the skill and experience level of personnel employed in a certain activity can constitute differentiating policy choices. Further, many of the factors that are considered as cost drivers also coincide with potential drivers of uniqueness. For example, timing represents a relevant differentiator since being first to market can make a product unique (provided it cannot be directly imitated by competitors). In addition, a follower strategy can also result in differentiation advantage since the technology used for a product in the first place might not be the best technology in the long run. Moreover, the generation of a differentiation advantage can very often also occur with increasing costs (e.g., differentiating by quality leads most of the time to higher cost for the product or service that is offered by the company). Thus, the analysis of the value chain should consider both uniqueness and cost drivers, with the aim of choosing the most valuable differentiation value chain activities relative to the associated costs.

Besides analyzing their own, internal value chain, an analysis of all activities throughout the internal and external value chain (i.e., the whole industry's value chain) supports firms with a more holistic picture of their total profit margins [9]. Here, analysts might make use of a four-step process following the so-called profit-pool mapping from the management consultancy Bain & Company, namely: (1) defining the pool, (2) determining its size, (3) determining the distribution of profits, and (4) reconciling the estimates (see Figure 3.5). In general, profit-pool mapping reveals the location and size of profit concentrations within an industry and sheds light on how those concentrations might shift [9]. It shows where the highest value is created in the value chain and may therefore facilitate thinking on how to participate in these value pools. Vertical or horizontal integration steps are often the consequence. But further analysis of why particular value pools are as big as they are may lead to the development of new customer engagement models, or even new business models. IBM's transformation from a hardware to a software and now to an analytics company may serve as an appropriate example.

The first step refers to defining the value pool by identifying value chain activities that are relevant for today's and tomorrow's business. Instead of limiting the analysis to the traditional value chain perspective, for example, industry definitions of suppliers and customers (i.e., along the value chain), analysts should rather apply a broader view that covers the company's internal value-creation activities (e.g., R&D, production or assembly, design, and

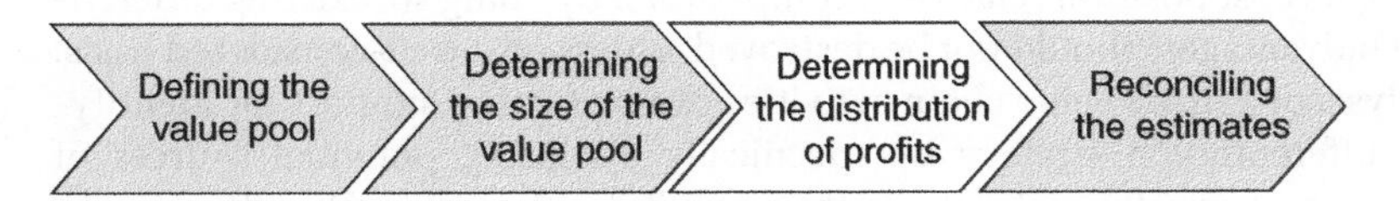

Figure 3.5 Steps in creating a profit pool

marketing), the competitors' value-creation activities as well as the value-creation steps upwards (i.e., towards suppliers) and downwards (i.e., towards customers) along the value chain. Furthermore, the perspective should not be limited to today's situation but should include visible trends that may change the map of potential future value-creating activities (e.g., based on new technologies or new entrants). Value pools can be defined not just along the value chain of an existing industry, but also across industries, for example, across components of a final product, such as value contributions to a PC from microprocessors, hard drives, other components, software, and peripherals. This requires companies to think beyond their traditional activities. A good example is IBM, which transformed from a hardware to a software company after having identified that the profit pools will shift from hardware to software. The final test for ensuring that all relevant value-generating activities are captured requires adopting a customer perspective: from the perspective of the customer, which activities along the consumption chain create value for me? The consumption chain follows five discretional steps: Awareness, Purchase/Access, Usage, Service, and Disposal. Even though a company may not consider some of these steps relevant, its customers might care deeply about them. Take for instance the delivery of highly moisture-sensitive polymer material. While a chemical company might regard the logistics (a part of "Access") as a peripheral activity that falls into the responsibility of a contracted logistics company rather than to its core business, the company's customers may see high value in this if they receive a large proportion of damaged bags with spoiled material from any supplier.

Given this multitude of perspectives, it is also vital to distinguish relevant from irrelevant activities when defining the value pool. The key question to distinguish the two is: What activities are valued the most by downstream users? For instance, while farmers value the product R&D activities of agrochemical companies, because they assure effective pesticides against ever-increasing resistances, commodity polymer customers will likely disregard their suppliers' product R&D activities, since performance requirements hardly change. Another example can be taken from the decorative paints and coatings industry: if a producer of colorants considers downstream integration into point-of-sales (POS) paint formulation by putting so-called "tinter machines" and tinters into paint shops, the producer may consider offering financial services (e.g., leasing, renting, or financing options) to enable such machine purchases. Depending on the type of target customers (small paint shops in emerging economies versus large DIY retail chains in mature economies), the value contribution associated with financial services will differ for the two customer types (high for the former, low for the latter) [9].

The second step considers the determination of the profit pool's size, which refers to cumulated profit of all players in the relevant segment. Profitability in this respect can be calculated in various ways and typically depends on the availability of data. Theoretically, the most relevant profitability measure is return on invested capital (ROIC), but also EBITDA margin or net income margins can provide sufficiently accurate information. All three figures are relative numbers stated as percentage of return on invested capital (ROIC) or sales (e.g., EBITDA, net income). The profit-pool size is then calculated by multiplying the respective segment's cumulated revenues from all players by the average profitability number of all players in the segment. While this sounds easy in theory, it is more difficult in practice, since the data are hardly ever provided by the segment one chooses in the first step. At this point of the analysis it is important to keep the objective of this step in mind: the determination of the profit-pool's size should just give a baseline estimate, against which to check the reliability of the more detailed, activity-by-activity calculations in the later steps. To do so, a useful estimate can be derived by identifying a "segment-pure" player, that is a player that is only active in the business segment one wants to investigate or whose financial figures (revenues and profitability) are reported in a segment-representative way. The more such players one can find the better. Investigating their profitability (e.g., from financial statements in the case of public companies) and identifying their position in the industry (are they leaders, mid-tier players or laggards?) is essential for this step. The segment's overall revenue pool can often be estimated (e.g., from analyst reports). Deducting the cumulated revenues of the identified players from the segment's overall revenue pool and multiplying it with the remaining revenues by an adjusted profitability number relative to one's segment-pure players (i.e., benchmarks), for example, by reducing the profitability slightly if your benchmarks are segment leaders or increasing it if they are rather third- or fourth-tier players, leads to the size of the profit pool. To crosscheck the validity of the analysis, the same approach should be applied from several perspectives, for example, from a player's, a product's (using product reports) or a regional perspective. If all analyses point towards a similar size of the value pool, the baseline will likely be sufficiently correct. For example, if the segment of analysis is PET (polyethylene terephthalate) packaging, the analysis begins by identifying packaging firms with the largest share of PET in their portfolios; or it starts by adding up all profits from the product's perspective, such as pure resin, fiber enhanced and recycled material, and then stripping out all those products that are not relevant for packaging purposes. The comparison of the two differently derived numbers for the profit-pool size should not differ by much more than 20%.

The next step is to analyze the distribution of profits across all relevant value chain activities. Analysts should start from their own company's perspective. Let's assume the analyst of a chemical company that produces specialty polymers wants to analyze the profit pools of the 3D printing "industry." First, the analyst needs to ask: "What is the profit contribution from my company's relevant activities?" Several polymers may be appropriate for 3D printing applications but not all. Therefore, the analyst needs to derive sales and profitability for the relevant products. This is not always straightforward, since fixed cost sharing with non-relevant polymers needs to be considered or appropriate overhead cost allocation should be taken into account. In other words, the analyst needs to disaggregate the chemical company's profit and loss (P&L) statement into the relevant activities. The same holds true for the pertinent activities of other stakeholders in the value chain. For instance, the chemical company may want to know the profit pools downstream from the polymers, namely printing cartridge production, designing relevant parts, digitizing the parts, printer production, and printing and finishing. While some of these activities may be performed by players whose activities and therefore their profit pools are uniform for all kinds of materials (e.g., design and digitization), other players perform activities whose profitability is considerably dependent on the printed material (e.g., profitability of printing cartridge production, printer production and finishing differ a great deal between metal, ceramics, organics such as food, and polymer printing). Consequently, when defining the relevant profitability along these activities for printing polymers, the analyst either has to aggregate the profits of "pure players" in each activity or to disaggregate the activities of mixed players (i.e., those who process polymers and other materials), and sum up only those profits derived from polymer-specific activities. Getting access to these data can be difficult and making "educated guesses" can often not be avoided. But some typical data sources should always be considered: annual reports, 10-K filings, and stock-analyst reports (for public companies), as well as company profiles by research organizations such as Frost & Sullivan or ICIS, and reports from industry associations and trade magazines often provide useful information. As in the previous step, if pure players in each relevant activity can be identified, their profitability could possibly be a good benchmark. Furthermore, there is no need to be exhaustive. In many industries the largest 20% of the players often account for roughly 80% of revenues (as illustrated in Figure 3.6). An analysis of these top 20% provides a sufficiently detailed picture about the relevant activities' profitability, and extrapolation based on reasonable assumptions will provide sufficiently accurate profit-pool data. At the end of this step, the profit-pool map should be complete. The analyst will know the revenues and profit margins of each value chain activity so that it is possible to compare the chemical company's own economics against these activity averages.

Finally, the results from the two former steps should be compared, representing the fourth step, namely reconciling the estimates. Here, the analyst needs to add up the profit estimates for each activity, and compare the cumulated segment profitability with the overall estimate of segment's profit pool, which was estimated in step two. If the whole profit pool does not align with the sum of all the profits of the value chain activities, revisions of the assumptions made during step two, but also primarily step three, are required.

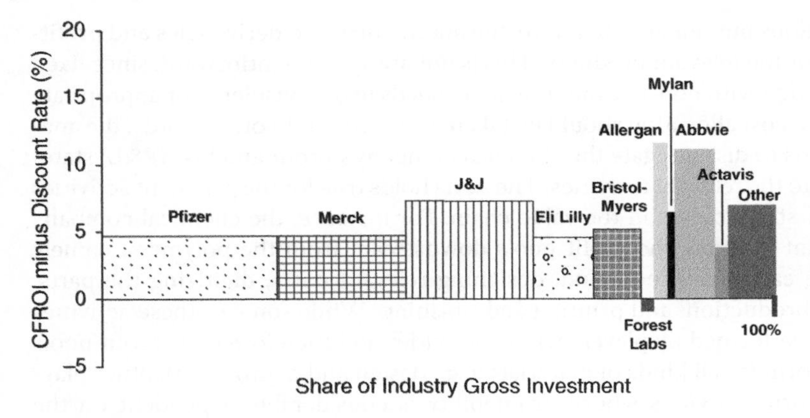

Figure 3.6 Profit pool per competitor

Additional data sources or expert interviews may be helpful to pressure test the assumptions [9]. Again, when the numbers match within a 10–20% range, this should usually be sufficient to derive strategic implications from this analysis.

3.3 The Resource-based View in the Context of Strategic Analysis

Whereas in the previous section, the perspective of the strategic analyst was focused on finding a strategic "sweet spot" in the industrial environment – while considering characteristics of the firm's value system – we now turn the perspective much more towards the internal resources and capabilities of the firm and analyze how they can be utilized to generate economic profit. This perspective is distinctly different from the previous one.

Despite its undisputed usefulness and success, two main weaknesses of the competitive positioning approach to strategy, for example as reflected in Porter's Five Forces framework, need to be stressed. Firstly, the perspective is rather static since many of the underlying assumptions are based on a given industry structure. Taking the chemical industry as an example, a strategy based on economies of scale is an effective way to keep new entrants out of the industry, since entry barriers are quite high due to the required high capital intensity necessary to erect a chemicals plant. However, frameworks such as Porter's Five Forces remain relatively quiet about what strategy companies should adopt when the basis of the industry structure changes. Keeping the chemical industry as an example, what should incumbents do if competition shifts to a decentralized, small-scale production (e.g., due to a shift in process technology such as a shift to micro-reactors and decentralized raw material supply from local bio-refineries)? Frameworks such as Porter's Five Forces lack a dynamic perspective that is relevant when the rules of competition change.

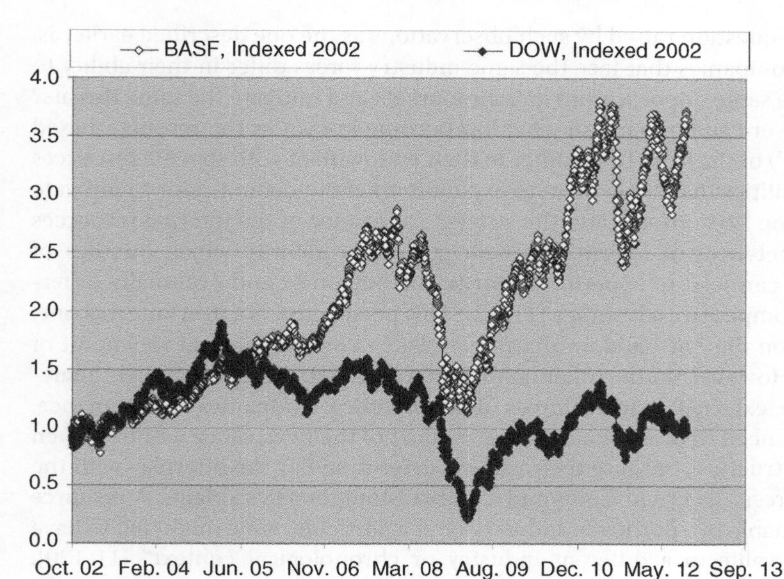

Figure 3.7 Share prices of BASF and Dow Chemical

The second critique focuses on the framework's explanatory power for intra-industry differences in performance. Little explanation can be found in the competitive positioning literature about performance differences of similar companies with similar scale, competing in the same product markets with a comparable positioning within the same value chain. There are several such cases in real-world industries, including the chemical industry: Dow Chemical and BASF, Wacker and Dow Corning, Merck and Bayer, to name just a few; the companies in each of these pairs are similar in size and structure and compete largely in the same product markets. The Five Forces framework, for instance, offers no convincing arguments that explain why such companies can show significant differences in their performance even though they face the same underlying market forces. Using the indexed share price of BASF versus Dow Chemical as a proxy for their performance,[1] BASF's performance evolved much better than that of Dow's over the last decade; more exactly, nearly four times better (see Figure 3.7). Industry structure, value chain analysis, and competitive positioning reasoning fail to explain these performance differences.

---

1 It is well known that the share price is only one among several perspectives on a firm's performance and that it depends on factors beyond the companies' control (for a more detailed discussion see Grant (2016) [10]); however, since both companies compete in very similar markets, their share price difference is expected to be driven, in the largest part, by investors' expectations on the companies' different abilities to exploit the same market opportunities and mitigate the same threats.

The key question raised by such observations as the one described earlier is: Why do companies that face the same industry forces differ in their ability to exploit the same opportunities in their markets and mitigate the same threats?

An answer can be found in what has become known as the resource-based view (RBV) of the firm: firms differ in their endowment with specific resources and capabilities that allow them to exploit market opportunities or to mitigate threats. The RBV investigates the strategic relevance of the internal resources and competences of a firm. Its main goal is to identify which bundles of resources can lead to superior performance outcomes and eventually differences in competitive advantage [11, 12]. This perspective is a firm-internal one, which is, on the one hand, in sharp contrast to Porter's external viewpoint of the firm. However, on the other hand, it complements Porter's strategic analysis of the external market forces that influence companies' performance. Resources need to be evaluated in the context of their relevance within a given industry structure, because their value is determined by the interplay with the market forces. As David Collis and Cynthia Montgomery explain, "A resource that is valuable in a particular industry or at a particular time might fail to have the same value in a different industry or chronological context" [11: 120]. Therefore, to build a successful strategy based on a thorough assessment of a firm's internal resources and capabilities, a firm first needs a deep understanding of the market forces it is facing. In this section, we explain in detail the key characteristics of the RBV approach towards strategy, including how to identify competitively relevant resources and capabilities.

As indicated previously, the key difference between the RBV and earlier strategic management concepts is that the RBV of the firm follows a firm-internal perspective on resources and capabilities as the main source of competitive advantage. However, not all resources and capabilities that a firm possesses are equally relevant for deriving value from them. Certain criteria need to be fulfilled and the capabilities need to be embedded in an appropriate organizational structure to allow the firm to exploit them and generate economic profit from them.

3.3.1 Underlining Assumptions for the Resource-based View

To understand the logic of the RBV, we provide a set of definitions and outline the reasoning behind the RBV by answering the following questions:

i) What are resources and capabilities?
ii) What is a sustainable competitive advantage?
iii) What criteria make resources strategically valuable?

(i) In his fundamental work on the RBV, Jay Barney defines resources as "All assets, capabilities, organizational processes, firm attributes, information, knowledge, etc. controlled by a firm that enable the firm to conceive of and implement strategies that improve its efficiency and effectiveness" [13: 101]. This very broad definition mainly emphasizes the immobility aspect of resources, which means strategically relevant resources reveal their value only if they are inextricably tied to the specifics of an organization [11, 14]. For example, a single scientist, who knows the exact synthesis recipe to form a unique material, is of limited strategic value to another firm as long as the production of this material is inextricably related to a specific production process that the first firm owns proprietarily. This scientist is a tremendously valuable resource for the first firm but of little value to another firm.

Furthermore, Kathleen Eisenhardt and Jeffrey Martin explicate that resources include tangible and intangible assets as well as capabilities: "[Resources] are those specific physical (e.g., specialized equipment, geographic location), human (e.g., expertise in chemistry), and organizational (e.g., superior sales force) assets that can be used to implement value-creating strategies [...]. They include the local abilities or 'competences' that are fundamental to the competitive advantage of a firm such as skills in molecular biology for biotech firms or in advertising for consumer products firms" [15: 1105–1106]. Resources and capabilities are deeply rooted in the organization. They are built upon the foundation of physical assets as well as employees' professional activities or habits, so-called routines, and the organizational processes that guide or steer these activities. When the physical resources (e.g., employees, labs, plants, etc.) are utilized in the routines (R&D, scale-up, production) in accordance with an organizational process (e.g., a stage-gate process) then we speak about a capability (e.g., new product development). Every company has resources, routines, processes, and capabilities. While most of the capabilities that a company possesses are necessary to act and compete in a given industry, a few are sufficient to really create a competitive advantage. Moreover, it is usually not only one distinctive resource or capability that allows a company to outperform its rivals. It is a bundle of complementary resources and capabilities that form so-called core competences. These core competences are finally the distinctive set of resources and capabilities that – when taken together and aligned with the relevant market forces – allow a firm to outperform the competition. Figure 3.8 outlines a hierarchical map of resources/routines (level 0), capabilities (level 1), and core competences (level 2) [16]. The key challenge is to identify the right set of resources and capabilities in an organization that allow the formation of distinctive core competences (details are given later). To make this a little more tangible, let's take the example of a specialty chemical company whose core competence is the fast and low-cost development of improved next-generation products. Let's disaggregate this core competence into its fundamental capabilities. The fast development of a next-generation product portfolio requires (without claiming completeness): (a) great customer insight and foresight of customer needs to know early on the performance requirements for the next

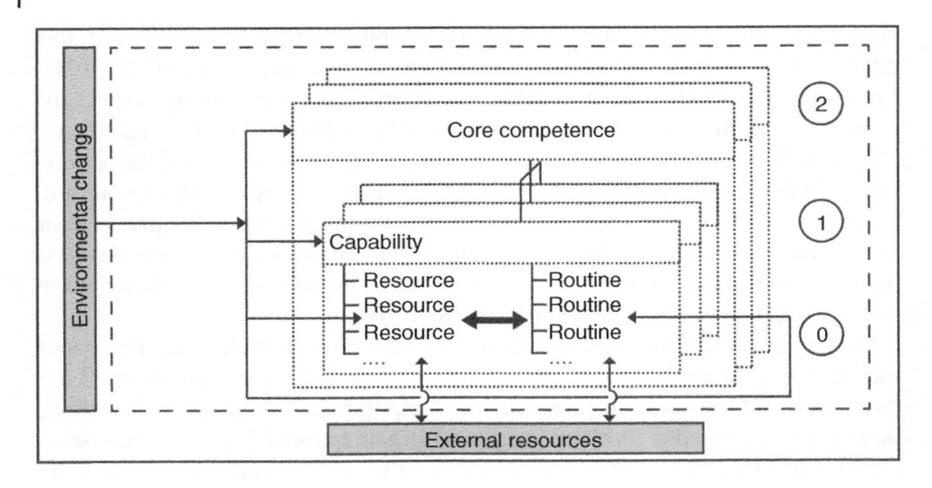

Figure 3.8 The relationship between resources, capabilities, and core competences

generation products; (b) fast formulation development capability; (c) integrated scale-up capability; and (d) rapid application testing capability. Among the vital resources that form these capabilities are key account managers with close customer relationships, excellent lab chemists equipped with state-of-the-art development facilities (labs, mass-screening analytics, pilot plants, etc.), process engineers experienced in the focal chemistry with flexible production equipment to accommodate varying formulations, and testing protocols approved by the customer. All this has to be integrated through a process that aligns the routines and the availability of resources for just-in-time handovers and processing.

(ii) Value (i.e., economic profit), in the logic of the RBV, is created when firms achieve sustainable competitive advantage (SCA), which is when a firm "can improve its efficiency and effectiveness in ways that competing firms cannot" [17:44]. This can be done by implementing a strategy "not simultaneously being implemented by any current or potential competitors and when these firms are unable to duplicate the benefits of this strategy" [13:102]. Sustainability in this definition does not mean unrivaled, everlasting competitive advantage, since every competitive advantage will eventually abate. It refers to a period of relative stability of the current economic and market conditions the company faces. Disruptive shifts in the means of competition, however, can change the viability of a certain resource-based strategy and therewith the value of the according resource bundle.

(iii) After this brief summary of what qualifies as resources and capabilities in the RBV sense and what is the RBV's strategic objective, an important question arises: How can management identify those valuable resources and capabilities within their firm that allow them to build a superior strategy that leads to sustainable competitive advantage? "No two companies are alike because no two companies have had the same set of experiences, acquired the same assets and skills, or built the same organizational cultures. These assets and capabilities determine how efficiently and effectively a company performs its functional activities" [11: 120]. Identifying the right set of resources and capabilities that are uniquely valuable to the focal firm out of the plethora of resources that every company carries along and building a strategy that focuses on their optimal deployment to achieve SCA are the key objectives of the RBV. In the following, we will provide tangible guidance on how to perform each of these two steps.

3.3.2 VRIN/O Characteristics

To identify value-creating resources and capabilities, a set of criteria has been developed which has become known as so-called VRIN/O criteria: resources need to be **v**aluable, **r**are, **i**nimitable, **n**on-substitutable, and coupled to an **o**rganization that supports their leverage. Let's go through each of these to understand their deeper meaning [11, 13, 18].

- **Value:** A resource is valuable if it contributes to the production of products that customers need at costs that are below the price that customers are willing to pay. This is obvious or at least easy to understand for most managers and therefore it needs little further explanation. However, managers need to keep in mind the acquisition costs of their valuable resources, which are sometimes difficult to quantify. An often-asked question by senior managers, for instance, is: Are the investments into our R&D capabilities justified compared with the returns from our newly developed products? This question led to a multitude of academic articles and practical proposals on how to answer it and still does, so far without providing a conclusive answer.[2] Furthermore, and as indicated earlier, the value of a resource depends on its relevance in competing under certain market forces. For instance, the capability of fast new product development and market introduction is much more relevant and therewith valuable if downstream customer markets have short product lifetimes than in a case where customers' products remain almost unchanged for many years in the market. Take the example of a synthetic colorants company: downstream customer industries include the automotive coatings industry as well as the printing industry. While customer requirements regarding colorants for automotive coatings have remained

2 Among the many documents, here is a short selection of recommended readings on the topic: Schwartz, Miller, Plummer, and Fusfeld (2011) [19], Adams, Bessant, and Phelps (2006) [20], McKinsey & Company (2013) [21].

almost unchanged for 10 years or more, the printing industry is changing rapidly and demands new colorant raw material specifications with a frequency of 2–3 years. New product development capabilities with short time-to-market are almost invaluable for suppliers of the printing industry.

- **Rareness:** The second criterion relevant for building a resource-based strategy is that the underlying resources need to be rare. The scarcity attribute is important since it relates to the definition of competitive advantage. If a resource is not rare, meaning it is available to any competitor, the possession of this resource would not distinguish a firm from its competitors in its ability to implement a value-generating strategy. A typical example for a scarce tangible resource in the chemical industry is cheap raw material feedstock. Take, for instance, the large Saudi Arabian petrochemical company SABIC: its rise into the top 10 league of the largest chemical companies in the world is primarily based on its proprietary access to cheap natural gas, which comes as a by-product from the country's oil production and that is offered to the company at a below market price by the country's government.[3] Such cheap feedstock is a rare resource, which only very few companies have access to. In addition, intangible resources may also be scarce and can serve to build successful resource-based strategies. Among intangible resource are IP rights. Patents, by definition, are scarce since they are granted to only one patent holder and their purpose is to prevent others using the same technological inventions. The German chemicals company Merck leveraged this scarcity effect to become the world's leading liquid crystal producer. In the 1990s, Merck acquired the liquid crystal patent portfolio from the Swiss company Roche, which complemented and completed their own liquid crystal patent portfolio. Through this and some further patent acquisitions, Merck held the largest and most comprehensive IP portfolio of liquid crystal technologies (more than 2000 liquid crystal patents in the 1990s). This unique access to IP rights was the basis for Merck to develop a liquid crystal strategy that helped them sustain their global leadership position over three decades.
- **Inimitability:** IP rights also fulfill the third important criterion – they are difficult to imitate. More accurately, they protect against 1:1 copying, preventing competitors from exactly imitating the substances or processes under IP protection. However, IP rights cannot fully protect companies from the threat of imitation. Competitors may find ways to circumvent single patents or simply build new but similar chemical molecules that are not covered by the firm's IP rights. Only large, extensive IP portfolios can serve as reliable protection against imitation. The requirement for a resource to be inimitable directly connects to the sustainability logic of the RBV. A resource needs to

3 Primarily in order to build jobs within the country of Saudi Arabia.

be "imperfectly imitable" [13] in order to generate competitive advantage sustainably, as otherwise competitors can quickly copy the value-generating strategy that is built from these resources. In order to build a strategy that prevents short-term imitation and that prolongs the profit streams generated by it, the underlying resources need to fulfill at least one of four characteristics [11, 13]:

a) *Being physically unique,* which means the resource exists in only one place and it is entirely owned by the firm; for instance, the Israel-based chemicals player ICL has a unique access to halogens (particularly bromide) through its proprietary chemicals exploitation rights of salt from the Dead Sea. This ultra-highly concentrated brine is probably the most cost effective source of halogen mining in the world.
b) *Being path dependent* in its creation, which means lots of time is required to gain the same experiences or to develop the same capabilities as the to-be-imitated firm. The reason is that "These resources are unique and, therefore, scarce because of all that has happened along the path taken in their accumulation. As a result, competitors cannot go out and buy these resources instantaneously. Instead, they must be built over time in ways that are difficult to accelerate" [11: 123]. Consequently, if a firm wants to imitate a strategy that is built on path-dependent resources or capabilities it would require either immense effort or it may even be impossible to extract the full value from such resources (e.g., building brand-value and customer loyalty) in the short-term if the copycat did not possess them already. The chemical company BASF can serve as an example. In Europe, BASF is the most well-known chemical company with the highest reputation among chemistry students; this is predominantly due to its sheer size, being the world's largest chemical company and its ubiquitous presence in almost all fields of chemistry. Consequently, BASF is strategically leveraging this path-dependent resource when it comes to recruitment of Europe's best talented chemists. Building a strategy that is based on superior chemistry know-how and R&D capabilities as a strategically relevant resource is thus possible for BASF, but hardly imitable in the short- to mid-term for a mid-sized chemical company that has little popularity among the best talented chemists.
c) *Being causally ambiguous,* which means the causal relationship of how a resource contributes to the value-generating strategy is too complex to be understood by externals. This makes it difficult for competitors to identify which resources to imitate or how to leverage them in order to copy the focal firm's strategy. Take for example W.L. Gore: the firm is known for its successful innovation capabilities and its ability to extract value from new products. While the individual products can be imitated, it is very difficult to identify what exactly makes W.L. Gore such a successful innovator. Causal ambiguity often resides in organizational

capabilities; at Gore, for instance, it is likely that its innovation-oriented culture, its project-based organization, and its family business style of leadership structure are fused into a successful system that fosters innovation. The complex interactions among these individual ingredients that probably complement each other with respect to successful new product development can hardly be analyzed and duplicated by other companies.

d) And finally *being economically deterring*, which refers to situations where a first-mover advantage erects entry barriers that are too high for followers. The cement business may serve as an example here. Cement is largely a commodity for local construction markets, meaning it is usually consumed close to where it is produced. Cement production is very energy intense and production costs as well as the quality of the cement are strongly dependent on the available raw materials, that is, primarily limestone rock, shale, and clay. Therefore, where to build a cement plant is determined by the nearby availability of the raw materials (since transportation costs are imperative) and sufficient local market demand. Finally, the profitability of cement production is strongly dependent on economies of scale: you need a certain minimum viable production capacity to dilute fixed costs sufficiently to become cost competitive. Taking together these prerequisites for building a cement plant, it becomes clear that once a cement plant is built in a favorable location (including the ownership of a key raw materials quarry) it will be very difficult or even impossible for a competitor to build another plant close by, since access to raw materials may be constrained and the local demand is probably satisfied by the one plant. Consequently, being a first mover erects entry barriers so high that they are economically deterring for any competitor thinking about imitation.

- **Non-substitutability:** The last criterion refers to the relevance of a resource for gaining SCA, which means it must not be replaceable by another resource. This criterion is important due to the same reasoning as in the case of inimitability. If competitors can deploy a substitute of a resource so that the firm builds its value-creating strategy and can therefore gain a similar advantage, again competition will quickly take away any initial advantage. Taken together, both criteria – inimitability and non-substitutability – in essence boil down to the same requirement: durability. "The longer lasting a resource is, the more valuable it will be" [11: 125]. However, the speed at which the value of certain strategic resources depreciates is strongly dependent on how fast-paced the industrial environment is. While some chemical market segments are stable for years, others are so dynamic that the resources chosen for building a strategy on them today are obsolete one or two years later. Even within one and the same product segment, resources such as market and customer access can be of significantly varying durability, depending on the pace of change in the downstream customer markets. Take as an example the chemical product segment of synthetic colorants. Chemical companies producing pigments and dyes are often supplying both the automotive coatings and the printing industry. While a supplier position for the automotive coatings industry is rather a safe place due to extremely high entry barriers into this market, such as very long testing and approval times (for instance, the industry standard "Florida test" takes 5–6 years for approval), market access to the traditional printing industry is of rapidly diminishing value, due to the fast and significant changes this industry is currently undergoing. Since the industry is shifting gradually from traditional offset printers towards digital printing, the technology and thus the type of companies competing in this market are shifting. This requires colorant suppliers to come up with new product formulations and opens up inroads into the printing market for new, previously not established colorant suppliers.

Finally, once strategic analysts have pressure tested their firm's capabilities against these criteria, they have to ask one more question: Who owns the resource(s) on which the strategically relevant capabilities are built? Owning refers to the ability to control the utilization of the resource and therewith its appropriation. Companies have to make sure that strategically critical resources cannot be simply snatched away by competition. If the strategic analyst realizes that a strategically relevant resource can be hired away (employees) or simply bought (e.g., mining rights on land you don't own) by the competition, the company may reconsider its strategic choice. Knowledge, for instance, resides within the brains of scientists: if they leave, the resource leaves with them. A prominent example in the synthetic colorant industry was the development, rise, and fall of the diketopyrrolopyrrole (DPP) pigment. When Ciba-Geigy developed and commercialized it in the 1980s, it quickly became famous in the industry as the "Ferrari Red" because of its brilliant coloristic and superior stability properties. According to industry tales, one of the two chemists who had been in charge of developing and bringing the new pigment to commercial scale left Ciba in the 1990s and acted as a consultant to the newly rising chemical companies in China. This was at a time when European IP rights were hardly enforced in China; according to the tale, a Chinese competitor to Ciba started up with the help of the former Ciba employee, building a DPP plant in China almost concurrently to Ciba, but finished it several months before Ciba was able to do so. This instance marked both a dramatic net-loss investment for Ciba and a rapid commoditization of the DPP chemistry business, and it is an excellent example to demonstrate the importance of testing and assuring the applicability of strategically relevant resources.

Some years after his groundbreaking work, Jay Barney extended his concept by admitting "That in addition to simply possessing [VRIN] resources, a firm also needed to be organized in such a manner that it could exploit the full potential of those resources if it was to attain a competitive advantage" [22, 23: 124]. In other words, the organizational structure of a firm needs to allow exploitation of the synergies between its various capabilities. It is not sufficient for the strategic analyst to identify two or three capabilities as being strategically relevant; the routines and processes also need to be organized in a way that the identified capabilities are deployed in a coherent and complementary manner.

The superior product development capabilities of the company W.L. Gore, well known for its Gore-Tex® products, may serve as a good example. Gore has been awarded a multitude of prestigious innovation awards and is frequently used as a shining example of an innovative company. But what is it that allows Gore to deploy its market research, R&D, scale-up, and product launch capabilities more effectively than other companies? The answer probably lies within Gore's unique organizational structure. Gore is organized in an extraordinarily flat, "lattice" structure with no official job titles or line reporting structures [24]. What sounds esoteric for the usual representation of a functionally organized corporation is probably Gore's key to fast decision making, cross-functional collaboration, risk taking, corporate entrepreneurial spirit, and ultimately innovation success. "Because there are no bosses, there are no hierarchies that push decision making through the organization. Because there are no hierarchies, there are no pre-determined channels of communication, thus prompting associates [= employees] to communicate with each other. And because associates don't have titles, they are not locked into particular tasks, which encourages them to take on new and challenging assignments" [24]. It is just such an organizational setting in which Gore's undoubtedly superior product development capabilities are embedded that complements them to become so utterly effective.

Through the organizational setting, capabilities are bundled and deployed in a coherent manner aimed at achieving a common organizational target (e.g., the generation of economic profit). In this way, capabilities can complement each other. Bauer and Leker (2013), for instance, found that chemical companies can enhance their new product sales significantly if they organizationally combined their new product development capabilities (radical as well as incremental) with process innovation capabilities [25]. In general terms, companies need to build the organizational processes that enable the alignment and coherent execution of a set of strategically relevant capabilities in order to exploit them most effectively. Such organizational processes can be check lists, flow charts, or managerial handbooks that act as a reference point for all employees involved in executing single tasks that belong to the capabilities. These reference points offer guidance on how to integrate and time-wise align the various tasks so that they complement each other to achieve results in a fast and synergistic way.

Having brought together all the factors that are considered in the literature to be relevant for appraising the strategic usefulness of resources and capabilities, it is now time to provide a brief outlook on how to bring the resource-based view into action in the strategy process. A simple four-step process is suggested here.[4]

1) **Identifying the firm's capabilities**
   In a first step, the analyst defines the right level of analysis; usually this is at the corporate or the BU level for strategy making. This defines the granularity and scope of the analyst's assessment. Next, the analyst performs a functional or value chain analysis to highlight the main capabilities and then lists them. A functional analysis starts with listing all business functions, such as marketing, sales, production, and R&D, as well as the distinctive capabilities that these functions are supposed to have. A value chain analysis captures the path of value creation throughout the corporation, starting from identification of market and customer needs, through product development, procurement of raw materials, scale-up, production, warehousing, marketing, sales, and distribution up to inventory management to after-sales service, and support activities such as HR and IT, finally adding the capabilities required to create the value along this path. The advantage of this perspective is that it also explicitly takes into account those capabilities required to manage the functional interfaces.
2) **Relevance**
   Once the capabilities have been listed, the analyst scores the identified capabilities on a dashboard that is based on the above-mentioned VRIN/VRIO criteria: value, rareness, inimitability (including an understanding for the reason), non-substitutability, integration within the organization, and a cross-check for the appropriateness of the resource or capability for the company. A qualitative 1–5 scale is usually sufficient with the scores 1, 3, and 5 being verbally anchored (e.g., for inimitability: 1 = each industry player has or can buy or build this resource/capability; 3 = this resource/capability is only accessible to a few industry players; 5 = this resource/capability is unique to us and can only be imitated with a massive effort or after a long time). The assessment should be done at the company's senior management level with the support of some external industry experts to ensure an unbiased assessment.
3) **Competitive strength**
   The analyst then benchmarks the top 8–12 capabilities against the competition by using unbiased evaluators (e.g., external consultants, customers). The result of the benchmarking is a differentiated picture of the key strengths and weaknesses of the company in comparison with its competitors along with the most relevant capabilities and resources in your respective business.

4 As according to Grant (1991) [26].

4) **Deriving strategy from key strengths and weaknesses**
Finally, the analyst takes the key strengths and develops a strategy that: (a) systematically fosters and extends them to become really exceptional and (b) aligns the organization around them (e.g., builds them into the brand image, prices them into the products, segments the market by customers valuing particular strengths, focusing the sales force on those customers). With respect to the key weaknesses, there are in general three ways for how to deal with these. Firstly, the company could work to improve them. While this is theoretically the most attractive option, it has two downsides in reality. (1) Building or improving capabilities takes rather a long time and is typically quite resource intensive. (2) Capability building absorbs not only personnel resources but also a considerable amount of management attention, which by itself is a scarce resource. Spending this resource on building new or improving existing capabilities means devoting less managerial attention to existing strengths. Since a strength is usually a strength because management deliberately nurtured and enhanced it, it can quickly deteriorate if management attention is taken away. Consequently, trying to improve weak capabilities will be a lengthy journey along a fine line, requiring the balance between maintaining existing strengths while spending sufficient managerial attention and resources on improving the weaknesses.

A second and potentially easier way is to analyze whether the existing weaknesses in the capability portfolio can be outsourced. The answer to this question depends on the degree of current integration of this capability within the organization. The more integrated the capability is in other organizational processes the more difficult it will be to outsource it while still maintaining effectiveness. Logistics, warehousing, maintenance, and sometimes even inventory management are classical examples in the chemical industry that have been outsourced to third parties to reduce costs on the company balance sheet while at the same time increasing flexibility, response time to customer requests, reducing equipment downtime, and so on.

Finally, the third option is to analyze if there is a business model that does not require those capabilities that are currently weaknesses in the company portfolio. For instance, many chemical companies struggle with being stuck in the middle between commodities and a specialties business model.[5] They have the structures of a specialty products provider with high technical service, R&D, and marketing costs, which are required to maintain the customers for the high-priced specialties portfolio; while at the same time a large part of their product portfolio is commoditized and under competitive price pressure, which requires a large-scale, low-service business model.

5 See also Chapter 7 on business model design.

The high selling, general, and administrative expenses (SG&A) cost structure is often a weakness for the commodities business of many chemicals players. The silicon company Dow Corning, for instance, realized this weakness and implemented a radically new business model for their commodity products. They divided their product portfolio into commodities and specialties and started selling the commodities products exclusively through their online sales portal "Xiameter," which reduced their SG&A costs to a minimum and made these products price-competitive to their Asian competitors [27].

To summarize, the RBV of the firm enables the strategic analyst to bridge the gap between the external environment and the firm-specific capabilities by means of using internal strength and weaknesses to get a positional advantage within the value chain. By considering the external environment in combination with the RBV a stronger and longer-lasting strategy evolves, instead of focusing on one strategic segment alone.

3.4 Dynamism of Markets

In this section, we focus on how to recognize dynamic changes in the environment that may affect the company's competitive position. Certainly at this point the trends displayed in the introduction, as well as those checked in Section 2.5 regarding their strategic relevance, are not going to be considered in detail again. However, it is worth mentioning that various drivers and trends affected market dynamics within the chemical industry in the past decade. In this aspect the chemical industry is not particularly different from other B2B industries. These market dynamics will also be observed in the future. Moreover, the general notion is that there will be stronger dynamics, or even disruptive changes, in markets and market mechanisms. Within this context it is of particular interest for research-intensive industries (e.g., the chemical industry) to illustrate the relationship between technological progress and the resulting change in value chains and markets, as well as the behavior of competitors. Therefore, in the next section selected concepts, where the main strength is the early anticipation of corresponding dynamic changes in technology and competitor behavior, will be considered: (1) S-curve concept, (2) convergence analysis, and (3) strategic reorientation.

Firstly, we want to examine the impact of emerging new "alternative-technologies." In this regard the S-curve concept has gained remarkable significance [28]. This concept is a simplified two-dimensional representation of any technological development. A measure of effort for the further development of the considered technology is applied on the x-axis, usually measured in money, whereas a measure of technology performance is applied on the y-axis. Depending on the considered technology, multiple

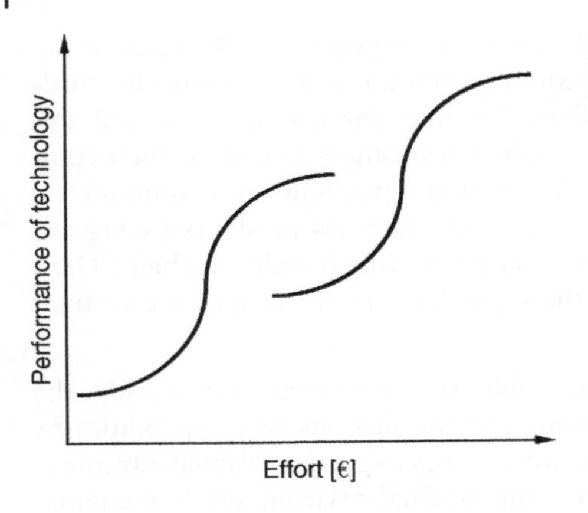

Figure 3.9 S-curve concept. *Source:* https://commons.wikimedia.org/wiki/File:Abb5_S-Kurven-Konzept.png. Used under CC BY 3.0

performance indicators are possible. Owing to the fact that many technology developments show an S-shaped curve in this method of presentation, it was obvious to name the concept similarly. The reason for such an S-shaped occurrence lies in the fact that, especially in the beginning, it requires considerable effort to develop a new technology with respect to its performance. Only when the basic principles of the technology and the essential control variables for an increase in performance caused by appropriate research effort have been known and understood, is it possible to achieve significant performance improvements with relatively little effort. The extent to which these improvements take place depends on the performance potential of the technology, until it comes to a weakening of this trend, as the fundamental and obvious levers for a further increase in output are widely exploited. From that moment it is again necessary, in relative terms, to put in more effort to achieve an enhancement in performance. Therefore, the curve has passed its turning point and, finally, the considered technology has reached its performance limits. Nevertheless, marginal performance enhancements are only attainable, if at all, with significantly higher efforts. This concept, as it is depicted in Figure 3.9, is often used to compare two technologies: one well-established technology and one emerging alternative technology.

In this context, the question of whether and when it is advantageous for a company to cross over from an established to an emerging technology is of strategic relevance. Or, less radically formulated, when it is strategically intelligent for a company to operate at relatively higher costs for the development of emerging technologies. Although the decision seems easy to take into consideration for an ideal curve and with knowledge of the future curve progression, in practice it is associated with high uncertainties. At the time of making the decision to either jump on the next curve or invest in the incumbent technology, the future curve progression cannot be predicted, thus, the corresponding forecasts are of a theoretical nature. Besides, the future behavior of competitors, which have a tremendous influence on the development of technologies, is also unknown.

As a consequence, for a company the question arises of whether to focus on the development of new technologies early on in comparison to its competitors and establish a strategy as a technology leader, or to pursue a restrained strategy as a technology follower. Regardless of this decision, which has to be made at the corporate level, retrospectively it can often be shown that most of the competitors pursue a wait-and-see strategy. The competitor does not use its power of research for the development of new technologies, but for further development of established technology, even if the potential is limited. In this context, the "sailing-ship-effect" is the right description [29].

The aforementioned decision based on the S-shaped curve is of strategic relevance for companies within the chemical industry because it determines the future position of a company in the market. If the company fails to have a timely shift to a superior technology, it will lose its competitiveness in this field. Vice versa, if a company only focuses on emerging technologies, which cannot fulfill the performance requirements in a timely manner, or at all, and neglects well-established technologies, it will also fail to hold its competitive edge.

If the new technology is not an alternative that replaces or competes with the established one, but is a combination of technologies from formerly separate industries, this therefore creates totally new product/service offerings, ultimately representing new challenges for the company. In this context one speaks of technology convergence [30]. The new technology caused by the convergence of two industries usually threatens product/service offerings in both industries, which can be seen as an indicator for technology convergence. While this form of technology convergence has been solely discussed for the example of smart phones, within the telecommunications sector or the IT and electrical engineering industries, nowadays a variety of particularly relevant technology convergence fields for the chemical industry can be examined. "Functional foods," for example, constitute the convergence of the food industry, the pharmaceutical industry, and the chemical industry [30, 31]. "Electric cars" are the result of the convergence of the automobile industry, chemical industry and IT [32, 33]. "Biopolymers," as with the last example, represent the convergence of the agriculture and chemical industry [34].

The uniqueness about these convergence situations is that they result in a new competitive situation for all players, which requires core competencies of companies to be reconsidered, business relations as well as their own position with respect to the new value chain caused by the convergence of industries. Usually a company is unable to map this emerging and complex value chain on its own, whereas at a very early stage the question of strategic suitable business partners arises. In this context, an early identification of emerging technology convergence is of particular importance. The sooner a company anticipates the development of a new convergence field the greater becomes the design space with respect to the newly created value chain and the freedom in the choice of available cooperation partners.

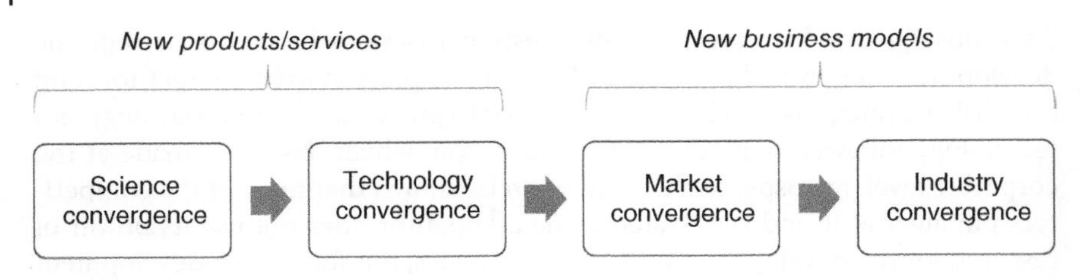

Figure 3.10 Convergence model

To forecast technology convergence one usually takes a step back and tries to detect science convergence based on publication analyses. In the next step, patent applications are particularly useful to detect the first signs of technology convergence [35]. This approach has proven its applicability in research-intensive industries, like the chemical industry, with relatively long innovation processes. For industries with significantly shorter innovation processes, such as service industries, the systematic analysis of information from the World Wide Web is more appropriate instead [36]. These types of information enable companies to analyze market convergence, which is an antecedent in a theoretical framework for industry convergence and the logical consequence of technology convergence (see Figure 3.10).

However, at this late stage convergence analysis serves companies more in the sense of a dynamic strategic competitor analysis than in terms of a forward-looking determination of a technology strategy. This dynamic strategic competitor analysis, next to a technology-induced strategic change analysis, should be used to evaluate the strategic moves of competitors in a structured way. In addition to this dynamic competitor analysis, it is useful to classify competitors, thus enabling a company to get a better overview of its own strategic position within an industry and recognizing changes in a timely manner. One approach for classification is the systematic analysis of competitors related to their strategic re-orientation [37]. Here, four different types can be distinguished:

- **Strategic Re-orientation Type 1: "Expansionist"**
  The competitor is buying and integrating new fields of business mostly through acquisition. As an outcome the overall turnover significantly increases and, if successful, previously unavailable new areas of expertise are directly accessible to the company. The German chemical company Merck recently pursued such a strategy. In 2010 it acquired Millipore, a company in the life science business with a focus on biotechnology research and production. In the following years, several other acquisitions, such as Biotest and Biochrom, demonstrated Merck's ambitions to extend its portfolio into the life science business.

- **Strategic Re-orientation Type 2: "Innovator"**
  Competitors following an "Innovator" type aim to establish new technologies and new fields of business by creating new production facilities. As a result, new technologies and/or new innovative products or services will be introduced to the market. The competitor, if successful, will be able to develop new areas of competences independently on its own.
- **Strategic Re-orientation Type 3: "Re-allocator"**
  Competitors following a "Re-allocator" type aim to transfer their competencies and know-how to countries with specific strategic location advantages. For the chemical industry, the immediate access to large markets or the immediate access to essential raw materials would be an example. As a result, new production facilities in new countries are created. If successful, the competitor generates new strategic advantages in global competition. The German chemical firm Süd-Chemie (today a part of the specialty chemicals company Clariant) was pursuing such a strategy. In the past Süd-Chemie had acquired several companies from all over the world, for example, Chemetron (USA), Airsec SA (France), and two smaller catalyst manufacturers and a casting-concrete manufacturer in Asia.
- **Strategic Re-orientation Type 4: "Concentrator"**
  Competitors following a "Concentrator" strategy focus on already established competencies by concentrating on the main, basic fields of business. In this context, it usually comes down to significant divestitures. As an outcome the turnover decreases significantly. This focus, if successful, leads to a relatively higher profitability than before. Evonik Industries, for example, is currently trying to get back to its core competencies by focusing on its special chemistry business. In 2015, the company sold all of its shares in Vivawest, a real estate business.

Principally, all four described types of strategic re-orientation can be observed at the same time within a competitive environment, each performed by a different company. For a dynamic strategic analysis, it is essential to recognize similarities across the competitive behavior. These behaviors might not only have a tremendous influence on competition in the market, but also could change the position of their own company in the market landscape in terms of an unintended new emerging strategy [38].

3.5 Dynamic Capabilities

This section deals with the integration of the three previously mentioned approaches to strategic analysis. In essence, dynamic capabilities are those special capabilities (see Section 3.3) within a firm that allow a faster or more effective response than the competition to changes in the external industrial environment (see Section 3.4), and, thereby, build a competitive advantage from adapting to or even driving the change in the competitive forces of the market (see Section 3.2).

To understand the logic of dynamic capabilities, we will first describe what we mean by a "significant" change in the market forces of the chemical industry and why organizational agility is key to survival under such circumstances. Secondly, we will outline the generic description of dynamic capabilities, supported by some examples where chemical companies have demonstrated that they possess such capabilities. Finally, we will set forth a few recent strategy ideas that all build to some extent on the idea of dynamic capabilities.

The question: "Why is it so critical for companies to be able to adapt swiftly to changes in their industrial environment?" can be answered as follows. When the rules of the competitive game change, companies need to very quickly learn how to play under the new rules or they will be kicked out of the game. The *Chemical & Engineering News* annually published list of the largest chemicals companies provides a glimpse into the dynamic changes in the chemical industry: half of the global top 10 chemicals companies in the year 2003 were not present on this list just 10 years later. Companies that played in the top league of chemicals companies were either acquired (like ICI) or were forced to file bankruptcy (like LyondellBasell, at least temporarily).

As with any other industry, the chemical industry is constantly influenced by external forces such as oil and energy price volatility, regulatory changes, new technologies, changes in customer demand, raw material availability, or increasing low-cost competition from emerging players. For instance, oil price volatility is certainly one of the most relevant for the chemical industry. Coming from a steady-state price of around $40 per barrel before 2004, the price of Brent (European oil standard) and WTI (US oil standard) began fluctuating between a high of almost $140 in 2008 before the financial crises let it plummet to around $50 per barrel in the same year, and it went up to over $110 between 2010 and 2014, while dropping down to below $40 again more recently (see Figure 3.11). Since crude oil is still the single most important raw material for the production of chemicals, it is obvious that those players who are better capable of dealing with these extreme raw material price fluctuations have a competitive advantage over those who don't. Capabilities such as backward integration (see, for instance, BASF with its subsidiary Wintershall), better forecasting capabilities, and raw material price hedging or diversification are some examples of specific capabilities that help absorb shocks in developments in the oil price.

The second most important input factor for the chemicals industry is energy; here, the competitive forces have also shifted significantly in the recent past. In the United States, shale gas has become available at competitively low prices compared with sources in Russia and the Middle East, providing a cost advantage for US-based chemicals players compared with Europe-based producers. China is leveraging the most advanced western technology to exploit its vast

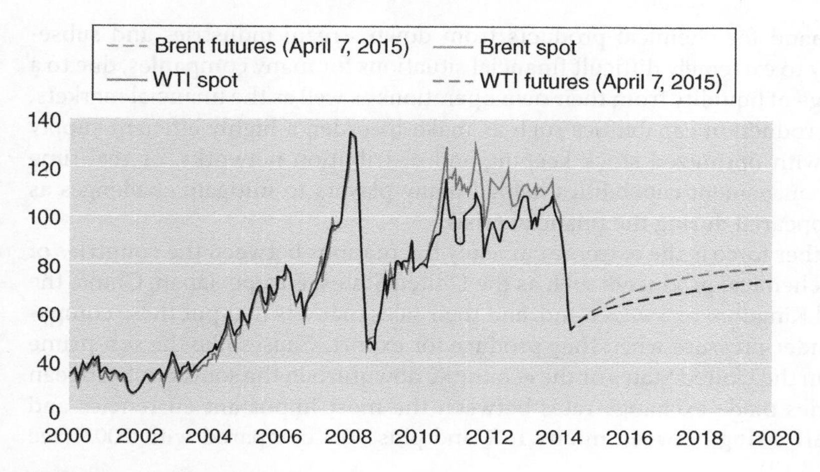

Figure 3.11 Brent and WTI

coal reservoirs most efficiently. To make the situation even more severe for European chemicals players, energy prices in Europe are increasing due to governmental regulations favoring, from the current point in time, expensive renewable energy sources, which are widening the energy cost gap between Europe and other countries. Resources and capabilities, such as the very energy efficient production facilities (e.g., BASF's Verbund structure), or the possession of and ability to run their own energy production plants (like Wacker) in an optimally integrated way to support the core business can be ways of protecting the business from regional energy cost imbalances.

Changes in customer demand as well as regulatory changes can have severe influences on chemical companies' business performance. For instance, increased public concern about the negative health impact from phthalate plasticizers in poly(vinyl chloride) (PVC) during the 1990s caused industrial users of PVC, such as toy or medical device producers, to switch to other polymers or use phthalate-free PVC. While the reduced demand for PVC and phthalates put some chemicals players under significant pressure, others with strong R&D capabilities were able to quickly develop new alternatives to phthalate or PVC and proliferate under the new market conditions. When the European Union banned the use of certain phthalates in 2009, a development race started among Europe's plasticizer producers to come up with suitable alternatives. Those players that sensed the need for alternatives early on, as well as those with superior and fast new product development capabilities, were placed in a competitively advantageous position since they could capture market share from their competitors.

However, other, more indirect, forces can also shake the industry's market balance. The financial crises of 2008/2009, for instance, led to dramatic declines in demand for chemical products from down-stream industries and subsequently to extremely difficult financial situations for many companies, due to a shortage of liquidity from their own operations as well as the financial markets. Lean production capabilities such as make-to-order, a highly efficient supply chain with optimized stock keeping and distribution networks, or real-time cash management capabilities helped many players to mitigate challenges as they appeared during the financial crisis.

Another force is the extreme currency fluctuations between the countries of major chemical producers such as the United States, Europe, Japan, China, the United Kingdom or Switzerland, and their sales markets that put these companies under pressure when they produce for export. Causes like the sub-prime crises in the United States or the economic downturn in the southern European countries made exchange rates between the most important currencies and the USD swing between minus 15% and plus 70% compared with 2003 (see Figure 3.12).

Obviously, no company can maintain a huge pool of qualitatively outstanding capabilities that would allow them to cope with any of the aforementioned difficult-to-predict changes in the market forces at any time. This is due to the fact that capabilities can only be maintained at a high quality level if they are used frequently [39]. Since companies need to work with economic efficiency with their limited stock of resources, they cannot retain capabilities that are not critical under current circumstances and whose necessity for the future is uncertain. For instance, backward integration to hedge against unforeseeable price upswings of certain raw materials is hardly beneficial in times when the

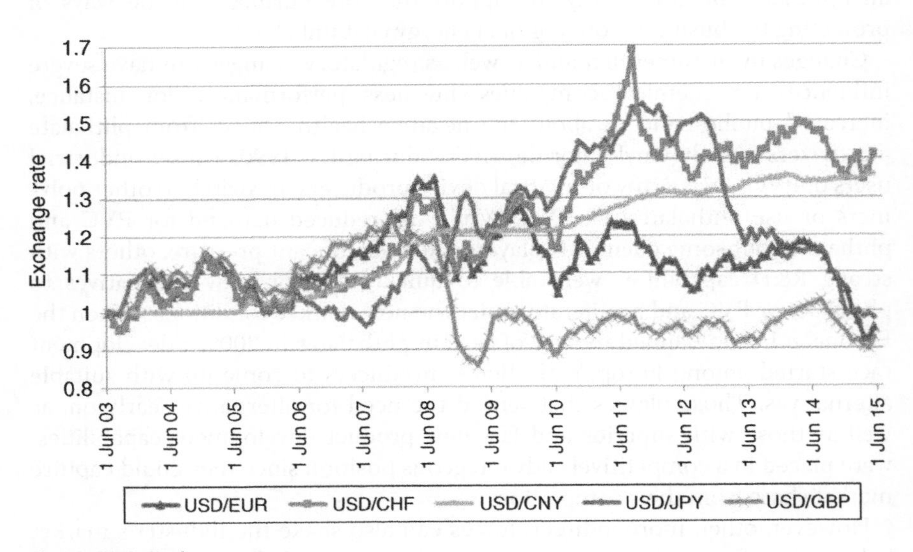

Figure 3.12 Exchange rates

raw materials are cheap, since suppliers are usually more efficient in producing and selling them to a large number of customers compared with the focal firm that usually has only one outlet for the raw material. On the other hand, building capabilities when needed is difficult since the development of capabilities usually takes a long time [40]; but when the competitive forces of the industry are shaken up as quickly as, for instance, in the case of the financial crises in 2008, there is no time left for building up appropriate capabilities. In such situations, companies need to react swiftly, focusing their resources (e.g., management attention, work force activities, etc.) on the truly critical task of survival. However, when the changes in the industry are also progressing gradually, steadily, and visibly, companies often have difficulties in developing new capabilities in a timely manner; many companies react only when their business model starts to deteriorate at an accelerated pace and the business performance indicators are falling, resulting in shareholders raising pressure on the executive management. Polaroid is a case in point: the world's leading producer of instant photography films and cameras actually decided to continue competing in the arena of photography during the age of rising digitization in order to stay close to what the company considered to be its core business, instead of searching for new business models that could have leveraged its core capabilities in silver halide chemistry (amongst others). Caught in the development race for more and more mega pixels in digital cameras at lower and lower prices, Polaroid could not win against electronics giants like Sony, Canon, etc. [41].

But what capabilities are there that allow some companies to react more quickly and more effectively to changes in the market forces than others? What type of capabilities are needed to enable a large chemical corporation to cope with or even create an advantage out of severe changes in the competitive market forces?

Many answers have been given to these questions, but they all have one fundamental concept in common: companies that are successful under changing market forces showed what has become known as "dynamic capabilities." Before we highlight some of the most influential management concepts, we will first describe the generic concept of dynamic capabilities, which is the foundation of most of the subsequent strategic approaches that target helping companies to deal with the increased dynamism of today's markets.

Dynamic capabilities refer to a firm's "ability to achieve new forms of competitive advantage" by utilizing changes in the competitive market forces [42]. This conceptualization particularly emphasizes "two key aspects that were not the main focus of attention in previous strategy perspectives. The term 'dynamic' refers to the capacity to renew competences so as to achieve congruence with the changing business environment. [...] The term 'capabilities' emphasizes the key role of strategic management in appropriately adapting, integrating, and reconfiguring internal and external organizational skills, resources, and functional competences to match the requirements of a changing environment" [42: 515].

In short, all management concepts that are based on the dynamic capabilities framework try specifically to identify and improve those top-management, strategic capabilities that allow a firm to improve or renew its operational capability (see Figure 3.7) and resource base in order to remain competitive when the market forces are changing. Through this definition it can be seen that dynamic capabilities and operational capabilities are closely related to each other. While operational capabilities represent a company's ability to engage in daily business, dynamic capabilities ensure a firm's aspiration to alter these processes and activities, due to a changing business environment [43].

But what is so difficult with "renewing the operational capability and resources base" of a firm? There are several challenges for the strategic analyst related to this task. Basically, the first and maybe the most critical factor for successful dynamic capabilities is the ability to foresee the need and the direction for change in the operational capability base at a point in time when the change in the market environment still allows for sufficient time to do so. This is the first of three capacities a firm needs to have that make up what is described as dynamic capability: "Dynamic capabilities can be disaggregated into the capacity (1) to sense and shape opportunities and threats, (2) to seize opportunities, and (3) to maintain competitiveness through enhancing, combining, protecting, and, when necessary, reconfiguring the business enterprise's intangible and tangible assets" [44: 1319]. In the following, we describe in detail what exactly each of these three capacities refers to and underpin them with practical examples from the history of the German specialty chemicals firm, Wacker Chemie (see also Box 3.1).

3.5.1 Capacity (1): Sensing and Shaping Opportunities and Threats

Sensing and shaping opportunities and threats implies that the top management team needs to be able to foresee changes in the competitive forces of the market at a point in time when there is still enough time left to react before either competitors sense the opportunity too, or competitive threats destroy the firm's basis of operations. As a case in point, let's consider the German specialty chemicals firm Wacker Chemie. When Wacker started in 1903, it began as a producer of acetylene as a lighting source for rural areas that had no connection to the electrical grid. However, the fast progression of electrification of private households during the early years of the twentieth century in Germany quickly reduced the demand for acetylene-based lighting, which soon became a major threat for the acetylene industry of that time. While many acetylene companies collapsed, the founder, Alexander Wacker, leveraged the firm's strong R&D capabilities to investigate alternative uses for acetylene. The chemists at Wacker soon came up with a series of economically viable processes that used acetylene as a raw material for other chemicals. Amongst others, Wacker chemists developed the world's industry standard process for acetic acid production, one of the world's major chemical base products. Besides making use of these processes in their own production, Wacker licensed these new processes to other chemical companies, which not only created the basis for survival but also the prosperity of the company.

The founder, Alexander Wacker, showed two decisive capabilities that allowed him and his company to react dynamically to changes in the market forces. Firstly, he sensed early enough that electricity was the new technology that would soon replace acetylene technology as the source for lighting, which gave the firm enough time to react. Secondly, out of the many capabilities that every company undoubtedly had, he identified R&D as the core competency of his firm under the new market conditions. With their R&D know-how in acetylene chemistry that allowed the company to investigate alternative uses of acetylene, Wacker shaped the threat to the industry into an opportunity for the company, by making use of the cheap availability of acetylene from overcapacities and the declining number of competitors due to the industry shake-out.

From a conceptual perspective, dynamic capabilities require the top management's capacity to: (a) realize a threat to the company's core business early enough so that reacting is still possible and (b) identify which of the firm's existing capabilities can be leveraged for a new purpose in order to turn a threat for the industry into an opportunity for the company. In practice, mature corporations often install dedicated organizational units and processes for sensing and shaping opportunities and threats. For instance, the corporate foresight units at Siemens, or BASF's "New Business" unit have the task of gathering information about technological as well as user demand trends and of deriving potential changes in the market forces. New business development functions, as they are present in most chemical companies in one way or another, are usually deployed to invest into a portfolio of potential future opportunities in order to gain a deeper understanding of the potential future markets or technologies and to be prepared in case a certain opportunity becomes a major industry development.[6] Open Innovation platforms are often used to tap into a widespread web of external knowledge with the hope of getting early hints on major technological or customer preference shifts. The list of such explicit capabilities that large corporations deploy to improve their opportunity-and-threat-sensing ability can be expanded at will; the key step towards an outstanding dynamic capability at this stage, however, is to be able to collect, connect, integrate, and make sense out of this plethora of information in order to create a solid foundation for decision making. Information processing procedures, such as presentation templates, score cards, or reporting guidelines can help with this job. But at the end, top managers' cognitive and entrepreneurial capacities will still remain crucial when it comes to decision making.

6 See also the Chapter 6 in this book on new business development.

3.5.2 Capacity (2): Seizing the Opportunity

While many CEOs may realize the threat to their company's core business that comes with fundamental changes in the market forces, far fewer CEOs are able to turn such a threat into a sizable opportunity. Wacker decided to leverage the firm's R&D capabilities in acetylene chemistry (instead of continuing to compete in the lighting business) to develop new chemical processes. But besides making the right choice of capabilities, Wacker also nurtured his firm's core capability of acetylene chemistry R&D by complementing it with new resources and capabilities. Firstly, amidst the decline of the German acetylene industry, Alexander Wacker bought three acetylene plants, which were cheaply available at that time. In this way, the young company quickly increased in size and they could apply the newly developed processes on a larger scale to exploit economies of scale. Secondly, he built an alliance network with the leading German chemical producers of that time, Bayer and Hoechst, by licensing out Wacker's newly developed production processes to them. Thus he built strong ties to decision makers in these companies, which allowed him to win them over for co-investments into some key assets, including his company's own electrical power plant. Since the chemical processes invented were very energy intensive, Wacker convinced Hoechst to co-invest (as a minority shareholder) into a hydropower plant. While this move secured Wacker's production cost base, at the same time it became a key resource for the expansion into various other, energy-intensive chemical products.

The key capabilities that Alexander Wacker showed at this stage were courageous investment decision making and risk taking (buying three acetylene plants in a time of market decline), risk balancing (finding co-investors), complementing the asset base (backward integration in energy production), and business model design (process development for own usage as well as licensing out). These are key ingredients of dynamic capabilities, but they are rarely found in modern enterprises. As David Teece notes: "The capacity to make high-quality, unbiased but interrelated investment decisions in the context of network externalities, innovation, and change is as rare as decision-making errors and biases are ubiquitous. [...] One should not be surprised, therefore, if an enterprise senses a business opportunity but fails to invest" [44: 1326–1327].

Large, established corporations often struggle to execute these capabilities because decision makers favor investments into improvements of well-established products and processes over the creation of new, un-proven products; they favor the continuation of existing business models that have proven successful in the past over the experimentation with new alternative business models with an uncertain outcome; they prefer investing small amounts into a large portfolio of many opportunities, which often results in under-critical funding of each opportunity instead of betting on a small number of opportunities – sometimes just one opportunity – and funding them with whatever it takes to achieve, successfully, and in time, the envisioned returns. In the end risk aversion instead of risk taking, short-termism instead of long-term planning, experience-based decision making instead of visionary decision making are often the management principles that hinder the implementation of effective dynamic capabilities. We will list some more recent concepts from the strategic management discipline at the end of this section, the objective of which is to help managers overcome these hurdles.

3.5.3 Capacity (3): Reconfiguring

Once an opportunity is seized, top management needs to make sure that the larger organization's management system, processes, and capabilities are (re-) configured in a way that they optimally integrate and support the new business. This is a continuous challenge: when the old meets the new, organizational change will be requested from the old, established organization (not the new) to provide the new one with sufficient room, flexibility, and also efficiency to flourish.

As constant change in the market place (e.g., emergence of new technologies and competitors, new regulatory boundaries, and changes in the availability of raw materials) drives the need for continuous adaption of the existing organization, senior management has to evaluate and re-evaluate over and over again the existing configuration of practiced capabilities, organizational structures, management systems, productive assets, product offerings, and target markets in order to ensure that they complement each other under the given and future market conditions. The underlying reason for such ongoing self-evaluation is twofold: on the one hand – as already noted – it is required to accommodate and integrate the organization's new business opportunities. On the other hand, it is required to mitigate the threat that once successfully practiced core capabilities can become core rigidities [45]. Mental models like "the way we operate was successful in the past – so it will be in the future" can easily trap companies into becoming stagnant in the status quo, without them realizing the fast pace of change around them. Success can create inertia: and this can rule out the larger organization appreciating, optimally supporting, and integrating previously seized new business opportunities. But once competitive pressure rises and the environmental change becomes apparent to managers through falling performance KPIs, a typical human reaction is to do more of the same – more of what is proven, more of what has already been experienced and has served well in the past, instead of pursuing fundamental **organizational** change. However, fundamental organizational change requires knowing the new direction, which in turn requires first knowing the root cause of why the established system is no longer working as well as it did in the past. Such analyses, in addition to setting the new direction and implementing fundamental organizational change, are very time and resource intensive and thus costly. When companies realize the need for organizational adaption and change too late, that is, when performance KPIs have fallen to a critically low level, then the ongoing business may fall short in generating sufficient cash to finance fundamental change. Severe restructuring, take-over by a competitor or default, and bankruptcy are all too often typical consequences. Prominent examples in the chemical industry in recent years are the specialty chemical company Clariant, which went through massive restructuring after an unsuccessful acquisition, or the pigments producer Ciba, which was taken over by competitor BASF after years of performance decline, or the base chemical giant LyondellBasell, which was forced to file for chapter 11 bankruptcy protection by US law in 2009, after having been taken by surprise by the probabilities of the financial crisis.

Taken together, the target of the dynamic capabilities concept is to avoid the need for short-term fundamental organizational change and restructuring. It actually starts with avoiding corporate inertia, that is, that core capabilities become core rigidities. Through constant adaption of the company's capabilities, organizational system, and objectives to the requirements those new business opportunities under the changing market environment put forth, companies can remain agile and flexible. Thus they become less vulnerable to disruptive environmental change since they can either react swiftly or they can even shape the change in their favor. In other words, the dynamic capabilities concept tries to describe how companies can change and renew themselves continuously in line with the ongoing environmental change while avoiding massive restructuring programs.

Box 3.1 Wacker as an example for applying the dynamic capabilities approach

Wacker continuously sought new business opportunities, seized them, integrated them into its larger organization, and consecutively nurtured them by enhancing, combining, protecting, and – whenever needed – re-configuring the underpinning capability base that was required to run the newly added as well as the established businesses. By doing so, Wacker re-invented itself several times during the course of its 100 years of existence. After its transition from an acetylene producer towards a diversified base chemical producer (with products like PVC, acetone, halogenated organic solvents, etc.) and production process licenser (as described earlier), the company's next transition started shortly after World War II. In 1941 while the world's first commercially relevant process for silicone production was invented on a lab scale by German scientist Richard Müller and in parallel by Eugene G. Rochow at General Electrics in the United States, Wacker concentrated its R&D efforts on this new opportunity and started up its first silicone production plant in 1949. This was Wacker's launch into its silicon age. The silicone business expanded quickly, utilizing resources from the existing organization, such as the cheap access to energy from its own power plant.[7] It also combined its existing marketing capability of having access to

7 Silane production, as a necessary precursor for silicones, is very energy intensive.

rubber and PVC customers with its new technological capability of silicone chemistry. To exploit the synergies, Wacker developed silicone-based release agents to initially sell to its existing rubber and PVC customers and later on to an expanded customer base in various other industries. But Wacker also backed up its new silicone business with capability-enhancing resources. It bought, for instance, a previously leased salt mine to secure access to the required raw material, chlorine. Chlorine was produced through electrolysis of brine, requiring vast amounts of energy. Through their backward integration into its own energy production, Wacker could establish a competitive advantage through the utilization of its access to cheap raw materials (brine and energy). This enabled it to become the world's leading silicone producer[8] with extraordinarily high profit margins by chemical industry standards.

With its growing know-how about silicon as a new and versatile raw material, Wacker early on realized that silicon's natural properties as a semi-conductor material might have significant potential for the electronics industry – and this insight paved the way for its next business transition. It consequently invested in solid state physics R&D, a discipline distinctively different from its former roots as a chemical synthesis company. While some existing capabilities, such as organic polymer synthesis and production, were of little help in this new business, Wacker's historically strong process innovation capabilities, access to cheap energy, and the recently acquired experience in handling small, volatile silicon compounds (silanes) helped it to quickly seize this new business opportunity. High-purity polycrystalline silicon, so-called polysilicon, turned out to be the material of choice for producing the newly developed integrated circuits parts that drove the miniaturization in the electronics industry. Even in the early1950s, Wacker had decided to invest in production facilities for polysilicon, resulting in the installation of the first polysilicon plant in 1953. The rest is history: silicon became the dominant material in the electronics industry during the 1960s and demand for high-purity silicon was spurred on in particular by the use of silicon wafers for integrated circuits for the upcoming computer industry. Wacker consequently exploited this opportunity by systematically expanding its high-purity silicon capabilities and assets, and continuously led the race in producing larger and larger silicon wafers.

Today, the company has reached another transition point. With the announcement of it exiting the silicon wafer business[9] and with the entry into the biotechnology business, Wacker is on the verge of re-inventing itself again.[10]

---

8 In a head-to-head race with US company Dow Corning.

9 This is after going through difficult times with its polysilicon business due to rapid demand and price cycles in the solar industry and strong commoditization of the business driven by rising Asian competition.

10 However, it is worth noting that Wacker's last true transformation is several decades ago; given that capabilities deteriorate over time and can get lost if they are not practiced [46], the future will have to prove whether Wacker still possesses its dynamic capabilities.

The history of Wacker shows that the concept of dynamic capabilities is more than abstract theory. Dynamic capabilities are real and they can be the source of competitive advantage and long-term prosperity if practiced regularly. However, continuous organizational change and renewal can have its downsides. It can counteract the establishment of fast and fluid organizational routines that require constant patterns and repetition to become effective and efficient. Routines are crucial for achieving efficiency in organizational processes [47]. Doing both, reconfiguring and changing the organizational set-up while enhancing and protecting it at the same time, is difficult to do [48]. This is why many large organizations struggle to find the right balance between the two. Managerial concepts have been developed to help managers overcome what seems to be a contradiction. Most concepts so far target the organizational structure [49–51], management, and incentive systems [52, 53] as well as top management's leadership capabilities [54, 55]. These approaches are outside the scope of a chapter on strategic analysis, but they are critical for the successful adoption of dynamic capabilities [56].

Finally, we will briefly present a selection of three concrete strategic management approaches that build on the concept of dynamic capabilities and that we consider worth highlighting here.

A must-read among the strategic management concepts are Harvard Business School professor Clayton Christensen's two seminal books *The Innovator's Dilemma* [57] and *The Innovator's Solution* [58]. Christensen's key message is that companies first need to understand their customers' changing needs at a level of detail that goes way beyond what companies, and especially chemical companies, usually do today. This insight is key to sensing changes in customer demand early enough to effectively seize a new business opportunity out of it. Furthermore, he outlines how companies can assess if the upcoming changes in customer needs and technical solutions are competency enhancing or competency destroying for the focal company and how companies should react in both cases.

Another one of the powerful and widespread strategic management concepts is the *Blue Ocean Strategy* approach by INSEAD professors Chan Kim and Renée Mauborgne [59]. They describe how established organizations can create new winning business models on the foundations of their existing business. Their strategy canvas approach stresses that companies willing to escape the arena of bloody competition need to first identify the basic capabilities on which their industry is competing. In a next step, they outline how companies can identify those capabilities that need to be kept, enhanced, reduced, and abandoned in order to serve a broad customer base more effectively and efficiently, while at the same time they distinguish themselves from the competition.

Most recently, Columbia Business School professor Rita G. McGrath's two books *Discovery Driven Growth* [60] and *The End of Competitive Advantage* [61] most explicitly single out how the concept of dynamic capabilities can be transferred into the strategic management tool. Her first recommendation is to transform today's strategic investment decision-making process into a portfolio management process of investing into real options that can be classified as core business enhancing, expanding and disrupting – and each of these classes should receive considerable investments. The second recommendation is to turn strategic planning into strategic learning. To do so, you build your business plan backwards from a clearly defined target and derive key assumptions that are critical for reaching the target; once they are identified, you deliberately spend resources on verifying or falsifying the assumptions to learn how you need to adapt your business model to reach your target. Finally, make your established business agile enough to accommodate the needs of your new business opportunities by considering disengagement from parts of the existing business as a healthy act rather than failure, conceiving organizational change as daily business rather than a forced, one-time exercise, and making decisions fast and frequent on a roughly right basis rather than waiting for a precisely developed fact base.

The concepts that are put forth in these strategic management textbooks are already shaping the agendas of many executives and strategists. Environmental change is likely to accelerate and the more it does the more important those concepts that are based on dynamic capabilities will become.

3.6 Summary

- **Successful strategies are the result of a systematic approach to analyzing a company's resources and their utilization by also considering all conceivable stakeholders in the environment.** The success of such an analysis and its interpretation for strategic decisions, done by the top management team, can be measured by the extent to which a company obtains retained profits, whereby the maximization of such profits should be a company's first and foremost goal.
- **Industry analysis can be done from an outside-in or inside-out perspective.** Porter's "Five Forces" represent a tool for considering a company's position in the external environment. Therefore, all of the important forces that put pressure on a company's profit potential need to be considered, namely the supplier, buyer, new entrants, substitutes, and industry internal competitors. Whereas an internal value chain analysis reveals not only a company's activities that contribute to enhancing the value of the end-product or service, it also enables a firm to calculate costs caused by each value activity and exposes the main differentiation characteristics. Thus, a strategy can be derived.
- **A profit-pool analysis exceeds the internal value chain analysis and considers the allocation of profits along the industry's value creating activities.** By (1) defining the profit pool, (2) determining its size, (3) determining the distribution of profits, and (4) reconciling the estimations,

changes in an industry's value chain can be recognized in a timely manner and highly attractive profit pools can be focused.
- **While the aforementioned approaches draw strategic decisions rather on their position in a given industry structure, internal (valuable, rare, imperfectly imitable, non-substitutable) firm resources, and their optimal processing may also constitute a sustainable competitive advantage.** In practice, managers can use the resource-based view by first identifying a firm's capabilities, sorting them by relevance, comparing them with competitors and deriving strategic choices based on determined strengths and weaknesses to foster and strengthen these capabilities.
- **The S-curve concept, a convergence analysis, and a strategic reorientation enables a firm to anticipate environmental changes early, thus making it possible to pursue the right strategic objectives.** To respond to these changing external factors adequately, dynamic capabilities are essential. The ability to sense opportunities and threats, seize opportunities, and foster or even reconfigure tangible or intangible firm assets in order to readjust operational capabilities leads to a competitive edge for every company. Taking the outside-in and inside-out approach of all aforementioned techniques together should result in a profound analysis, upon which strategies can be built.

References

1 Markides C. 2001. Strategy as balance: From "either-or" to "and." *Business Strategy Review*, **12**(3): 1–10.
2 Nag R, Hambrick DC, and Chen MJ. 2007. What is strategic management, really? Inductive derivation of a consensus definition of the field. *Strategic Management Journal*, **28**(9): 935–955.
3 Grant RM. 2015. *Contemporary Strategy Analysis 9e Text Only*. John Wiley & Sons Ltd: Chichester.
4 Porter ME. 1979. How competitive forces shape strategy. *Harvard Business Review*, **57**(2): 137–145.
5 AG WC. 2015. *Geschäftsbericht*. http://www.wacker.com/cms/de/wacker_group/wacker_facts/annual-report/annual-report.jsp?cid=11:geschaeftsbericht (accessed 1 March 2016).
6 Klemperer P. 1987. Markets with consumer switching costs. *The Quarterly Journal of Economics*: 375–394.
7 Porter ME. 2008. The five competitive forces that shape strategy. *Harvard Business Review*, **86**(1): 78–93.
8 Porter ME. 1980. *Competitive Strategy: Techniques for Analyzing Industries and Competitors*. Free Press: New York.
9 Gadiesh O and Gilbert JL. 1997. Profit pools: A fresh look at strategy. *Harvard Business Review*, **76**(3): 139–147.

10 Grant RM. 2016. *Contemporary Strategy Analysis: Text and Cases Edition.* John Wiley & Sons Ltd: Chichester.
11 Collis DJ and Montgomery CA. 1995. Competing on resources: Strategy in the 1990s. *Harvard Business Review,* **73**(4): 118–129.
12 Prahalad C. and Hamel G. 1990. The core competence of the corporation. *Harvard Business Review,* **68**(3): 79–91.
13 Barney J. 1991. Firm resources and sustained competitive advantage. *Journal of Management,* **17**(1): 99–120.
14 Wernerfelt B. 1984. A resource-based view of the firm. *Strategic Management Journal,* **5**(2): 171–180.
15 Eisenhardt KM and Martin JA. 2000. Dynamic capabilities: What are they? *Strategic Management Journal,* **21**(10–11): 1105–1121.
16 Bauer M. 2014. Managing a Portfolio of Innovation Capabilities – A Systems Approach in and for the Chemical Industry. Doctoral dissertation.
17 Barney JB. 2001. Is the resource-based "view" a useful perspective for strategic management research? Yes. *Academy of Management Review,* **26**(1): 41–56.
18 Kraaijenbrink J, Spender J-C, and Groen AJ. 2010. The resource-based view: A review and assessment of its critiques. *Journal of Management,* **36**(1): 349–372.
19 Schwartz L, Miller R, Plummer D, and Fusfeld AR. 2011. Measuring the effectiveness of R&D. *Research-Technology Management,* **54**(5): 29–36.
20 Adams R, Bessant J, and Phelps R. 2006. Innovation management measurement: A review. *International Journal of Management Reviews,* **8**(1): 21–47.
21 McKinsey & Company. 2013 *Chemical Innovation: An Investment for the Ages.* May issue.
22 Barney JB and Wright PM. 1998. On becoming a strategic partner: The role of human resources in gaining competitive advantage. *Human Resource Management (1986–1998),* **37**(1): 31.
23 Newbert SL. 2007. Empirical research on the resource-based view of the firm: An assessment and suggestions for future research. *Strategic Management Journal,* **28**(2): 121–146.
24 Anfuso D. 1999. Core values shape WL Gore's innovative culture. *Workforce,* **78**(3): 48–53.
25 Bauer M and Leker J. 2013. Exploration and exploitation in product and process innovation in the chemical industry. *R&D Management,* **43**(3): 196–212.
26 Grant RM. 1991. The resource-based theory of competitive advantage: Implications for strategy formulation. *California Management Review,* **33**(3): 114–135.
27 Johnson MW, Christensen CM, and Kagermann H. 2008. Reinventing your business model. *Harvard Business Review,* **86**(12): 57–68.

28 Foster RN. 1986. *The S-curve: A New Forecasting Tool.* Macmillan: London.
29 Ward W. 1967. The sailing ship effect. *Physics Bulletin,* **18**(6): 169.
30 Hacklin F. 2007. *Management of Convergence in Innovation: Strategies and Capabilities for Value Creation Beyond Blurring Industry Boundaries.* Springer Science & Business Media: New York.
31 Bröring S, Martin Cloutier L, and Leker J. 2006. The front end of innovation in an era of industry convergence: Evidence from nutraceuticals and functional foods. *R&D Management,* **36**(5): 487–498.
32 Song. 2015. *Früherkennung von konvergierenden Technologien.* Springer Verlag: Berlin. Doctoral dissertation.
33 von Delft S. 2013. Inter-industry innovations in terms of electric mobility: Should firms take a look outside their industry? Letter from the Editor. *Journal of Business Chemistry,* **10**(2): 67.
34 Preschitschek N, Curran C-S, and Leker J. 2011. *The importance of access to resources in a setting of industry convergence: The case of agriculture and chemistry* in *Technology Management in the Energy Smart World (PICMET),* 2011 Proceedings of PICMET'11. IEEE.
35 Curran C-S and Leker J. 2011. Patent indicators for monitoring convergence – Examples from NFF and ICT. *Technological Forecasting and Social Change,* **78**(2): 256–273.
36 Simon H and Leker J. 2016. Using startup communication for opportunity recognition – An approach to identify future product trends. *International Journal of Innovation Management,* **20**(08).
37 Leker J. 2001. Reorientation in a competitive environment: An analysis of strategic change. *Schmalenbach Business Review: ZFBF,* **53**(1): 41.
38 Mintzberg H. 1978. Patterns in strategy formation. *Management Science,* **24**(9): 934–948.
39 Zollo M and Winter SG. 2002. Deliberate learning and the evolution of dynamic capabilities. *Organization Science,* **13**(3): 339–351.
40 Winter SG. 2003. Understanding dynamic capabilities. *Strategic Management Journal,* **24**(10): 991–995.
41 Tripsas M and Gavetti G. 2000. Capabilities, cognition, and inertia: Evidence from digital imaging. *Strategic Management Journal,* **21**(10–11): 1147–1161.
42 Teece DJ, Pisano G, and Shuen A. 1997. Dynamic capabilities and strategic management. *Strategic Management Journal,* **18**(7): 509–533.
43 Helfat CE and Winter SG. 2011. Untangling dynamic and operational capabilities: Strategy for the (N) ever-changing world. *Strategic Management Journal,* **32**(11): 1243–1250.
44 Teece DJ. 2007. Explicating dynamic capabilities: The nature and microfoundations of (sustainable) enterprise performance. *Strategic Management Journal,* **28**(13): 1319–1350.
45 Leonard-Barton D. 1995. Wellsprings of knowledge: Building and sustaining the sources of innovation. *University of Illinois at Urbana-Champaign's*

Academy for Entrepreneurial Leadership Historical Research Reference in Entrepreneurship. Available at SSRN: https://ssrn.com/abstract=1496178 (accessed 23 June 2017).

46 Helfat CE and Peteraf MA. 2003. The dynamic resource-based view: Capability lifecycles. *Strategic Management Journal*, **24**(10): 997–1010.

47 Szulanski G and Winter S. 2001. Replication as strategy. *Organization Science*, **12**(6): 730–743.

48 March JG. 1991. Exploration and exploitation in organizational learning. *Organization Science*, **2**(1): 71–87.

49 Brown NAL. 2011. *Advanced messaging system and method*. U.S. Patent Application 13/171, 545.

50 Maine E. 2008. Radical innovation through internal corporate venturing: Degussa's commercialization of nanomaterials. *R&D Management*, **38**(4): 359–371.

51 O'Reilly CA and Tushman ML. 2004. The ambidextrous organization. *Harvard Business Review*, **82**(4): 74–83.

52 Gibson CB and Birkinshaw J. 2004. The antecedents, consequences, and mediating role of organizational ambidexterity. *Academy of Management Journal*, **47**(2): 209–226.

53 Klein KJ and Sorra JS. 1996. The challenge of innovation implementation. *Academy of Management Review*, **21**(4): 1055–1080.

54 Kotter JP. 1997. Leading change: Why transformation efforts fail. *IEEE Engineering Management Review*, **25**(1): 34–40.

55 Smith WK and Tushman ML. 2005. Managing strategic contradictions: A top management model for managing innovation streams. *Organization Science*, **16**(5): 522–536.

56 O'Reilly CA and Tushman ML. 2008. Ambidexterity as a dynamic capability: Resolving the innovator's dilemma. *Research in Organizational Behavior*, **28**: 185–206.

57 Christensen CM. 2013. *The Innovator's Dilemma: When New Technologies Cause Great Firms to Fail*. Harvard Business Review Press: Boston, MA.

58 Christensen C and Raynor M. 2013. *The Innovator's Solution: Creating and Sustaining Successful Growth*. Harvard Business Review Press: Boston, MA.

59 Chan Kim W and Mauborgne R. 2005. *Blue Ocean Strategy: How to Create Uncontested Market Space and Make the Competition Irrelevant*. Harvard Business School Press: Boston, MA.

60 McGrath RG and MacMillan IC. 2009. *Discovery-driven Growth: A Breakthrough Process to Reduce Risk and Seize Opportunity*. Harvard Business Review Press: Boston, MA.

61 McGrath RG. 2013. *The End of Competitive Advantage: How to Keep Your Strategy Moving as Fast as Your Business*. Harvard Business Review Press: Boston, MA.

4

Management of Business Cooperation

Theresia Theurl and Eric Meyer

University of Münster, Münster School of Business and Economics

> *Those who work alone are adding, those who cooperate intelligently are multiplying.*
>
> Joachim Milberg, former CEO of BMW

Today, companies are part of a business ecosystem with mutual dependencies rather than independent stand-alone fighters. The world has moved from large, ponderous conglomerates to smaller, focussed and swiftly adapting corporate structures. Cooperation may be part of these new structures that have become highly relevant in the chemical industry [1]. Nevertheless, many companies are lacking a comprehensive approach to managing the boundaries of the firm and their partnerships and have neither sufficient management capabilities to manage cooperation nor the ability to quantify the value of these partnerships to their companies.

In this chapter you will learn what a business cooperation is, how it can help to achieve corporate objectives and how the management of business cooperation works. You will also learn about the basic characteristics of cooperation and most common types of cooperation. In addition, you will learn why managing a cooperation is different from routine management and consequently requires specific management tools that are adapted to these peculiarities. Finally, this chapter introduces a five-step management process that addresses the peculiar characteristics of cooperation and provides instruments for coping with these specific characteristics.

Business Chemistry: How to Build and Sustain Thriving Businesses in the Chemical Industry, First Edition. Edited by Jens Leker, Carsten Gelhard, and Stephan von Delft.

4.1 Cooperation and Corporate Strategy

4.1.1 What Does Cooperation Mean?

In order to explain the idea of cooperation, we have to look at the smallest units of economic analysis: the transactions between individuals and/or companies. To carry out a transaction, individuals and companies can use markets. Pricing mechanisms efficiently direct resources to the uses where they are most value creating. But there are other ways of carrying out a transaction. Many transactions can be found within the boundaries of a firm, which is quite intriguing, if it is true that markets organise transactions efficiently. It was Ronald Coase who elucidated this mystery of the firm by asking the very simple question: Why do companies exist, if the pricing mechanisms of markets work so efficiently? Coase explained that using the markets for a transaction is associated with specific transaction costs. Similarly, carrying out a transaction within the boundaries of a firm also instigates transaction costs that differ from the costs of market transactions. He concluded that the appropriate mode for carrying out a transaction is determined by these different transaction costs. In consequence, an optimal organisational governance mode exists to carry out a transaction [2]. Decades later Oliver Williamson extended Coase's work by introducing hybrids to the transaction costs calculus. Williamson analysed the advantages and disadvantages of using markets or hierarchies, respectively. He was able to explain that there are hybrid forms for carrying out a transaction between markets and hierarchies, that is, organisational modes that combine the advantages of markets and hierarchies. These hybrid forms of governance can be considered as a business cooperation of two companies [3–5].

In order to understand the conditions that lead to the emergence of cooperation we have to analyse these advantages of markets and hierarchies (i.e. a solution within the boundaries of the firm). Using the market for a transaction has two evident advantages. Firstly, by combining similar demands producers are able to realise much larger volumes in production and consequently will achieve much lower production costs due to economies of scale. This applies in particular to very general products or products that contain a significant share of the general parts. Secondly, owing to the pricing mechanisms, market participants have high-powered incentives to improve their efficiency (process innovation) and to invent new products (product innovation) to escape temporarily from market pressure. These incentives will also lead to a company's flexibility to swiftly adapt to environmental changes [4].

However, markets are not without disadvantages if specific investments are necessary to carry out a transaction. Consider, for example, a transaction where investments are necessary to carry out this transaction: for example, a chemical company has to invest in production facilities that can only be used to produce a specific type of catalyst that the company aims to sell in this transaction. Therefore, the company has to invest specifically in these facilities. This specificity refers to the transaction carried out with a specific transaction partner. The transaction partner will recognise its powerful position because after the (specific) investment in the facilities it is the only market participant for whom the newly built facilities can produce. In a complex and uncertain environment, the transaction partner will attempt to use its powerful position to renegotiate the terms of the transaction to increase profits from this transaction, for example, by renegotiating prices. The producer, which has to invest specifically, will anticipate this possible hold-up and will try to reduce the risk by negotiating contracts that protect it from being exploited due to the specific investments. With increasing uncertainty and complexity of the transaction, the products and the environment, these contracts will become more and more complicated and expensive. The associated costs may turn the transaction into an unprofitable one and consequently the transaction cannot be carried out, to the detriment of both parties. In this situation, a hierarchical solution may be advantageous because no contracting is needed and there will be no contracting costs if the transaction is carried out within the boundaries of the firm [4, 6–8].

On the other hand, companies can exhibit some negative characteristics. Using the hierarchical governance mechanisms within the boundaries of the firm is usually subject to bureaucracy costs, longer decision procedures and organisational slackness. Although some companies try to overcome these problems by implementing "market mechanisms" within their organisation, these mechanisms typically do not perform as well as real markets and expose severe incentive problems (Figure 4.1).

Advantages of using markets	Advantages of using hierarchies
▶ Lower production costs → Economies of scale ▶ High-powered incentives ▶ Easier coordination → Use contract law for standard transactions	▶ Reduced costs for contracting ▶ Increased (micro-) adaptability ▶ Increase of specific investments → Better protection

Disadvantages of using markets	Disadvantages of using hierarchies
▶ Under-investment in specific assets ▶ Costs for protecting specific investments	▶ X-inefficiencies ▶ Less maintenance due to inferior incentives and monitoring ▶ Insufficient assigment of costs ▶ Insufficient incentives

Figure 4.1 Advantages and disadvantages of using markets and hierarchies

To sum up, both governance modes (markets and hierarchies) have advantages and disadvantages that result in different transaction costs. On the one hand markets provide efficiency and high-powered incentives, which lead to immediate reactions to market signals; on the other hand hierarchies are typically associated with superior control mechanisms. Therefore, the two governance mechanisms are characterised by:

- different costs for carrying out a transaction
- different systems for executing control
- different mechanisms to protect the company against hold-up and exploitation
- and different risk structures.

Cooperation is an organisational option to combine these different characteristics, creating an efficient solution for carrying out a transaction. Typically, certain trade-offs apply when using these hybrid types of organisation. If the management increases its control over the transaction in order protect itself against exploitation, it will lose some cost advantages (e.g., it reduces the economies of scale).

Now we will derive some basic characteristics of business cooperation. A business cooperation exhibits:

- The participation of at least two partner companies.
- A certain exchange intensity between the cooperating partners. Consequently, one-off transactions will never qualify for a cooperation. The intensity refers to the frequency (quantity) and to the quality of the exchange. A more frequent transaction requires more contact and can be carried out better in a cooperation environment. The quality of the exchange refers to the involvement of the partners. It is determined by the complexity of the transaction, the amount of information exchange and the monitoring of the transaction.
- Some kind of formal (contractual) or informal stipulations for cooperation. This is an immediate consequence of the control requirements to protect the company against exploitation by the cooperation partner.
- Some involvement for parts of the processes of the partner companies. In some cases, a company's processes have to be adapted to the cooperation requirements, in other cases it suffices to define cooperation needs and conditions in order to make the cooperation processes fit with the internal processes.

The phenomenon of business cooperation is described in many ways and one can find numerous synonyms in the literature dealing with cooperation: alliances, collaboration, partnerships, joint ventures, networks or supply chain relationships to name a few. Some of these synonyms describe specific types of cooperation (e.g. alliances or joint ventures), others are more general descriptions for the cooperation phenomenon. But all of these descriptions and synonyms share – to a different extent – the common characteristics elaborated here.

4.1.2 Why Is the Management of Cooperation Different?

From the preceding discussion it should have become evident that managing cooperation is part of a company's management and it can be described as managing the boundaries of the firm, by using hierarchical solutions of integration (e.g. mergers and acquisition), market relations and cooperation as instruments for organising these boundaries. Cooperation management enriches the usual decision of "make or buy", that is, of carrying out a transaction within the firm using hierarchy to govern the transaction or of carrying out a transaction in the market, with a multitude of new arrangements to carry out transactions, combining elements from these two governance mechanisms. This implies that cooperation management needs additional capabilities to identify the circumstances that require cooperation as an organisational solution and the ability to reasonably combine the governance elements from markets and hierarchies mentioned previously to develop new suitable cooperative solutions. Understanding these governance elements and sufficient knowledge of when and how they can be implemented are new management requirements. Management no longer just focuses on the management of the company's (internal) processes, it is extended to organise the boundaries of the firm intelligently and understand the processes and needs beyond these boundaries, since they influence the company's own business processes and value chain through these cooperative solutions. Moreover, monitoring the interfaces to the partner and deriving appropriate actions from the signals received at these interfaces is also an aspect of cooperation management.

But the management of cooperation is not only extending the way we organise and develop corporate value creation, it also changes the way we manage the processes of value creation. The fact that at least two participating companies join forces and cooperate in order to achieve common goals implies that the management no longer has complete control over processes that are carried out jointly. Management, thus, moves from "command and control" to new techniques that emphasise incentives for the partner. This move was brilliantly described by Thomas Malone who argues that companies have to move from management based on "command and control" to "coordinate and cultivate" types of management. Within the boundaries of a firm, management ideas can be implemented top-down, that is, the management decides on certain activities that will be implemented by the employees. The management gives commands and controls the execution of these commands. Although this perspective is – for explanatory reasons – fairly extreme and admittedly many companies encourage the discussion of decisions, it should become clear that in the end the management bears the responsibility for its decision and will use its commanding power to execute the decisions. Moreover, the complete discretion implies that the management is able to collect all the information considered necessary to monitor and control the execution of its decision [9].

This changes completely when cooperation is part of the organisational solution. Since it is almost impossible to give commands to a cooperation partner to implement desired management actions, the management lacks a substantial part of its usual power to enforce the execution of its decision. In addition, the management does not dispose of unrestrained access to all parts of the value chain where it joins forces with a cooperation partner, because it needs the consent of the cooperation partner. This leads to new management challenges that have to be addressed through appropriate cooperation management:

- How to collect information that is necessary for management decisions but that is produced within a partner company?
- How to develop solutions jointly with the cooperation partner that do not violate the interests of the company and contribute to the company's profits?
- How to implement such solutions jointly with other companies?

Nevertheless, implementation of management actions is also necessary in cooperative arrangements. On the one hand these actions pursue the individual interests of the cooperating company, but on the other the actions have to take into consideration the position of the partner company. This can be accomplished by taking into account the second part of Malone's idea: management in cooperative arrangements move towards cultivating a positive and collaborative atmosphere that facilitates joint decision making and actions (e.g. providing adequate information to enable the cooperation partner to find solutions that suit the company's own interests or creating negotiation solutions for acceptable management actions). Thus, the idea of "cultivating" means creating a positive environment for the cooperation, which induces the decisions and actions by the cooperation that are, at minimum, not in contrast with a company's interests, and that at best coincide with its interest. This replaces the command component in managing through hierarchies.

In a similar way the actions of the cooperating partners have to be coordinated instead of being controlled. Coordination requires different management instruments that use the self-interest of the cooperating partner to promote the achievement of the cooperation's common goals by creating environments that guide the decisions and actions of the parties in an appropriate way. Such instruments could be "cooperative" transfer prices (i.e. prices for exchanging goods and services between the cooperation partners) setting adequate incentives for the partners or guidelines for the behaviour of the partner companies. A description of these instruments will be part of Section 4.5, which describes the cooperation management process.

4.2 How Cooperation Can Help to Achieve Corporate Objectives

As mentioned in the preceding section, cooperation is an instrument for organising transactions efficiently. Cooperation objectives are derived from corporate objectives. Thus, their purpose is to help to achieve these corporate objectives, such as growth. Cooperation is not an end in itself but will contribute towards achieving corporate objectives. There are at least five objectives where cooperation might prove to be a helpful instrument [10].

4.2.1 Cost Advantages

Realising cost advantages through cooperation with other companies is driven by economies of scale (i.e. cost advantages obtained due to size, output or scale of operation). These economies of scale occur when the costs per unit of output decrease with increasing scale, while fixed costs, for example, large investments that are necessary to produce or develop a product, are spread out over more units of output. In the pharmaceutical industry, for example, the costs for the discovery and development of blockbuster drugs are enormous, forcing pharma companies to spread their research and development (R&D) expenditure across a greater volume of sales. As a result, total average costs decrease and production will be more profitable, and more units can be produced from these investments. These cost advantages make it worthwhile to analyse a company's value chain, searching for parts that exhibit these economies of scale. Such parts of the value chain, where large investments are necessary, are candidates for cooperation. Through cooperation, the companies may decrease the investment costs (e.g. joint R&D) or they may increase the sales of products, that is, they increase the production volume while keeping the investments constant. Technically the cooperation can be implemented by establishing a new company that is jointly owned by the partners or by selling the production capacities of one partner (or both of the partners) to a new legal entity that is owned by the partners. Alternatively, the cooperation partners might use the production capacities of one of the partners, which could be used for the production for both cooperation partners. This is an especially appropriate option where one of the partners disposes of several production sites, of which one could be dedicated to the joint production.

Following these examples, business cooperation aimed at generating cost advantages are mainly found in production and in large-scale R&D. Production also includes the provision of administrative services where cost reductions can be accomplished by cooperation. Networks that are used to provide services to customers are another area where cooperation may be suitable for cost reductions.

There are two main management challenges to cost-advantage-based cooperation. Firstly, the sub-processes that are carried out jointly with a partner company have to be integrated into the processes of the cooperating companies, that is, standardisation of the products and/or the processes have to be negotiated between the partners. The challenge clearly increases with the complexity of the products and their production processes. Secondly, the contributions from the partners have to be negotiated. This can be difficult if the partners' contributions are asymmetric. Then, complex valuation procedures have to quantify the mutual contributions in order to achieve a fair sharing of the burden and distribution of the profits.

4.2.2 Access to Resources, Know-how and Technologies

Cooperation can be an option in order to achieve much quicker access to relevant (new) technologies or specific knowledge compared with generating these technologies or the know-how internally within the company. In these cases, a company identifies significant deficits in its own know-how or its own technology base that is necessary to create and commercialise products. There are three options to obtain these technologies: buying the technologies that are available on the market; in-house development of the technologies or acquiring and integrating a company that can provide the technology (hierarchical solution); developing a cooperation to gain access to the technology. The hierarchical solution of developing the technology in-house will often be time consuming and costly, especially if the company is inexperienced in the respective technology field. Acquiring a technology-providing company can be very costly, particularly if the company is not a one-technology firm and also sells other technologies and services. Thus, entering into a cooperation could open a route into the required technologies; see for example Box 4.1.

There are basically two types of business cooperations for accessing technologies or know-how: companies could agree to exchange know-how or technologies in a cooperation; or the technology is provided by one partner and is used by the other partner who pays a monetary compensation for its usage. Both types can be combined, for example if the mutual exchange of technologies or know-how is asymmetric.

For the first type of cooperation, the main challenge is determining the value of the contributed technologies. It is frequently observed that the cooperating companies agree on some kind of exchange of their technology or strength of know-how without carrying out protracted valuations of these strengths for the partner company.

The second type in particular seems to be hard to discern from a market transaction. The crucial distinguishing feature is the extent and intensity of exchange between the two companies. Providing complex technologies to the production line of a partner company frequently demands complicated adaptations of the technology, or the know-how can be used within the partner

Box 4.1 BASF cooperates in R&D for new battery technologies [11]

BASF is a multi-product chemical company with sales of about €74.3 billion in 2014. The firm's product portfolio covers basic chemicals as well as specialized chemicals for application in numerous industries and technologies. In addition, BASF is active in oil and natural gas exploitation and production, and in providing products for use in agriculture.

In a strategic process, BASF determined e-mobility as being one major future trend that requires sophisticated new chemical compounds and production technologies for the development and production of energy storage solutions, such as batteries. The company estimates that innovations that allow production of batteries with high-energy density and low weight will be a future generator of profits. Unfortunately, as with any new technological development, it is unclear which technology will prevail in the end. Thus, diversifying the risk by having access to numerous new energy storage technologies is a suitable strategy. A parallel development of different technological solutions would be one option to spread this risk, which would be associated with multiplying the development costs – an investment that would be unlikely to be covered by the profits from the "winning" technology. Instead, through cooperation a company can get access to multiple new technologies while reducing the costs for the development. BASF perceives new battery technologies as a growing market and intends to provide the appropriate chemical technologies. Therefore, it works in cooperation with other companies to develop these technologies.

One of BASF's cooperation activities was to partner with EnerG2, a Seattle-based manufacturer of materials for energy storage solutions. According to BASF, "EnerG2 has developed a unique approach that engineers the molecular structure of a polymer precursor in order to customize the nanostructure, and, therefore, the performance of the resulting carbon" [11]. These new carbon materials can be used in electrodes of lead-acid or lithium-ion batteries to significantly improve the batteries' storage capacities. Through this partnership, BASF gains access to EnerG2's patented carbon technology platform, while BASF provides funding, additional expertise from its R&D and marketing know-how for the distribution of this technology. Together they intend to scale-up production and enhance the market penetration of the new materials.

This partnership is quite typical of a cooperation of a large partner (here: BASF) seeking innovative technologies and a smaller innovator (here: EnerG2) seeking additional funding, technical assistance and access to new distribution channels. Both partners benefit from the cooperation: BASF can save initial R&D costs and utilise its existing distribution network, while EnerG2 gains additional opportunities for selling the carbon materials to new customers, enabling the firm to increase the value of its innovation.

company to further develop its products or production lines. Thus, the processes of providing and using the technology become interwoven, constituting a cooperative arrangement. The main challenge is, arguably, to integrate the interwoven and connected processes of the partnering companies.

4.2.3 Access to Markets

A similar cooperation objective is the access to markets, where market refers either to distinct local markets or to certain customer groups; see for example Box 4.2. The partner company has extensive knowledge of customer groups or is deeply rooted in a particular country and has extensive experience in operating in this market. A company can use cooperation to exploit these competences for its own sales, that is, entry into new markets in other countries or new customers. In some countries laws restrict the operation of foreign companies in that country and demand that foreign companies must have local partners for their operations abroad, for example, by forming a joint venture with a local partner. Here legal restrictions force companies to cooperate, if they intend to enter the market in such a country. Examples of this type of cooperation are mainly found at the sales and marketing level.

Box 4.2 Lanxess and Mito Polimeri distribution partnership in Italy [13]

Lanxess is a German specialty chemicals company that emerged from a spinoff of Bayer's plastic rubber and specialty chemicals segments in 2004. Mito Polimeri is a privately owned company with a turnover of about €30 million, which specialises in the distribution of chemicals in Italy. More than 30 years ago Lanxess' predecessor Bayer established a distribution partnership in Italy with Mito Polimeri. In October 2013 Lanxess extended this cooperation to other specialty chemicals, for example, Durethan.

In order to access new (geographical) markets, Lanxess could have chosen an internal solution, that is, a special unit or subsidiary that markets the products or collaborates with a local distributor in the Italian automotive and electronics industry. While integrating sales activities has the advantage of superior control of the processes and the quality of information channelled to the sales unit, it lacks the local knowledge and trusting relationships with local customers. A distribution partnership therefore integrates the local ties of a local distributor with specialist know-how of the products that are to be sold. Thus, from Lanxess' perspective the partner profile for the distribution partnership requires technical, industry and market expertise and sufficient service and logistic capacities, which they found in Mito Polimeri. The main challenge in organising such a cooperation is identifying the relevant interfaces between these two companies, that is, the information that has to be exchanged between the two partners and the logistics for the distribution of Lanxess products.

The mutual benefits of such a cooperation arise from the rather different demands and assets. One company is seeking access to the market and the partner company possesses the desired experience in operating in these markets and can provide the access to these customers. The main challenge is to provide the appropriate fit between the partner companies. Integrating a partner into the marketing and sales processes of a company implies that the partner has to meet pre-defined standards for selling these goods or services. More complex products require a much deeper integration of the sales partner than just providing a new point of sale [12].

4.2.4 Time Advantages

Time advantages are a consequence of the two previous objectives for cooperation. If a company notices changes in a market and concludes that it has to adapt swiftly to these changes, or if it is operating in markets that are generally fast moving and consequently always require quick adjustments, establishing a cooperation to access new technologies or new markets can be a reasonable strategy in comparison with developing these technologies internally or investing in a sales force to access the new markets. Being unable to react in an appropriate timescale to these changes immediately implies losing market share, which leads to lower profits.

4.2.5 Distribution of Risks

Starting large projects (e.g. developing a new compound or a new drug) may result in a company having to take enormous risks. Cooperating with other companies in such a project means sharing the risks associated with the project, especially the risks of a project failure. Distributing the risk over several cooperating partners translates into a reduction of the risk costs, and therefore can be interpreted as a sub-category of the cost advantages.

4.3 Morphologies of Cooperation

There are different ways of classifying cooperations. Some classifications directly refer to the objectives of a cooperation, others try to identify certain characteristics. The most relevant morphological descriptions of cooperation are derived from the companies' value chain and from the institutionalisation of cooperation.

4.3.1 Horizontal, Vertical and Lateral Cooperation

Companies have different options for creating their cooperation along the value chain: horizontal, vertical and lateral cooperation. This differentiation focuses on the activities of the companies and has implications for the main management tasks for these types of cooperation.

- **Horizontal cooperation:** This type of cooperation occurs between companies operating in the same industry at the same stage of the value chain, for example, two chemical companies producing silicones that cooperate to produce a new polysiloxane. Owing to the fact that the companies combine their efforts for one specific part of the value chain (e.g. in production or marketing), horizontal cooperation frequently focuses on economies of scale and the associated cost advantages. In other cases, one company holds superior technologies or know-how in one part of the value chain that a partner company is searching for. Then cost advantages might be an objective for the owner of this superior technology, but accessing this technology clearly is the objective of the partner company. Thus, their cooperation would be driven by different objectives. Such asymmetric horizontal cooperation is much harder to manage, since assigning the partners' contribution becomes much more complicated. On the one hand, how much the production of the technology owning company benefits from the achievable economies of scale must be identified, and how these additional benefits can be assigned to the cooperating companies; on the other hand the partner company's access to the superior production technology has to be assessed and the contribution to their profits has to be evaluated. A successful horizontal cooperation has to balance these two effects with the second effect usually outweighing the first, that is, the partner company benefits more from the cooperation than the technology owning firm. Two different solutions can be observed for coping with this asymmetry problem. Firstly, the asymmetric benefits can be compensated by payments by the net benefiter from the cooperation. Secondly, in a more complex operation the two companies agree on cooperating horizontally in two different parts of the value chain with opposite net benefits, so that they mutually compensate each other.
- **Vertical cooperation:** In a vertical cooperation two companies that operate in the same value chain enter into a cooperation involving two different adjacent steps of the value chain. The reason for cooperating vertically is either to gain access to technologies, know-how and resources (backward cooperation) or access to markets (forward cooperation). Vertical relationships between companies are particularly exposed to the risks of specific investments. Cooperation is part of solving the problems caused by these risks. A pure market relationship could be easily exploited by the company that observes specific investments, which have to be made by the partner company in order to carry out the transaction. Cooperation provides mechanisms such as contracts, specific forms of ownership or new interwoven and overlapping forms of production that reduce the risk of hold-up and thus allow vertical relationships in a hybrid organisation.
- **Lateral cooperation:** This type of cooperation refers to cooperating companies that operate in different value chains (e.g. different industries) and that are working together on similar steps of the value chain or on completely

different steps. Again access is the predominant objective for this type of cooperation. It usually extends a company's business beyond its current boundaries by providing access to completely new technologies, specialist know-how or to completely new customer groups.

4.3.2 Types of Cooperation

To implement a cooperation, companies may choose from numerous different forms of institutionalisation. Some of these institutionalisations are characterised by their value chain positioning, others focus on the internal structure of the cooperation. The forms of institutionalisation presented as follows allow overlaps, for example, a cooperative is always a joint venture, some strategic alliances can also be interpreted as networks and virtual networks may show similarities with project cooperation.

4.3.3 Strategic Alliance

A strategic alliance is a horizontal cooperation of actual or potential competitors. Owing to its horizontal origin, the cooperation involves the same stage of the participating companies' value chain. The horizontal nature means that the goal is usually to achieve economies of scale and the corresponding cost advantages. Sometimes companies agree to cooperate horizontally on two or more steps of the value chain. In such a case it is not only cost advantages but also access to a company's superior know-how or technology that could be a motive for cooperation. Strategic alliances can be observed in research and development, procurement activities, production and in sales; see for example Box 4.3.

Since a strategic alliance happens between two actual or potential competitors in the same industry, stabilisation of the cooperation is a main focus of the management. Cooperation with a competitor implies excluding some parts of the value chain from competition and thus forgoing an opportunity of getting an edge over the competitor. As a consequence, strategic alliances could reduce the leeway for differentiation in the market, so companies will carefully select parts of the value chain and the extent of the cooperation in order to protect their competitive advantage. Several criteria may help to identify suitable areas for cooperation. Firstly, sufficient similarities between the cooperating companies should exist to achieve the aimed for economies of scale, that is, jointly manufactured products should allow for standardisation or the companies have to implement these similarities in their processes and/or the interfaces between their own company and the jointly organised part of production. Secondly, since cooperation with a competitor could blur the differentiation between the companies from the customer perspective, a strategic alliance is easier to implement the further away the involved value chain steps are from the final customer. Thirdly, with only a few rare exceptions, a strategic alliance

Box 4.3 GlaxoSmithKline and Pfizer create the joint venture ViiV Healthcare [14, 15]

In November 2009 the British and the US pharmaceutical companies GlaxoSmithKline and Pfizer created a joint venture in which they combined their HIV medicines. This new joint venture of these competitors operating in the same market was expected to research, develop and commercialise new HIV medicines. The starting points for GlaxoSmithKline and Pfizer were quite different, which led to different interests in the creation of the joint venture. GlaxoSmithKline had numerous HIV drugs with expired patent protection or patent protections running out in the coming years. GlaxoSmithKline is experienced in commercialising and distributing these medicines, that is, they have a large distribution network that could easily be used for similar drugs from other producers. On the other hand, Pfizer had just a few HIV medicines, but with a much longer patent protection. Consequently, its distribution network was inferior compared with GlaxoSmithKline's. Thus, the mutual advantage of cooperating with a competitor is evident. GlaxoSmithKline would receive new HIV drugs, which it could channel through its existing distribution network, and Pfizer could leverage its returns on HIV medicines by using a much larger distribution network.

The two companies agreed to create a joint venture, to which they contributed all their HIV medicines (eight from GlaxoSmithKline, three from Pfizer). In addition, they forwarded six HIV medicines into the development pipeline. The equity structure of the joint venture approximated the drugs' sales volumes, namely, GlaxoSmithKline owned 85% of the joint venture and Pfizer had a 15% stake. In order to cope with the uncertainty of future sales of the pipelined products, the partners devised a flexible equity scheme by introducing benchmarks that the partners had to achieve. If both partners achieved their benchmarks, GlaxoSmithKline's share would shrink to 75.5% while Pfizer's share would increase to 24.5%. If just GlaxoSmithKline's pipelined medicines were successful, its share would increase to 91% and if just Pfizer's new drugs found their way to the market its share would rise to 30.5%. From Table 4.1 it can be seen that Pfizer's Selzentry increased its sales in the joint venture by more than 50% from 2010 to 2012, while GlaxoSmithKline's drugs declined by about 15%.

In 2012 Shionogi, a Japanese pharmaceutical company, that also produces HIV medicines, joined ViiV Healthcare. It received a 10% share, and GlaxoSmithKline's and Pfizer's shares in ViiV Healthcare decreased to 76.5 and 13.5%, respectively. Shionogi and GlaxoSmithKline have developed the HIV medicine marketed as Tivicay in a joint venture since 2001. In order to use the existing distribution network the integration of Shionogi was a suitable solution.

ViiV Healthcare demonstrates how two competitors may cooperate in a selected business segment and how intelligent reward schemes (here: equity shares) yield the appropriate incentives for the partners in order to avoid shirking behaviour.

Table 4.1 Sales from HIV drugs.

Sales in £m	2010	2011	2012	2013	2014
GlaxoSmithKline	**1332**	**1317**	**1127**	**1089**	**950**
Combivir	363	322	179	116	59
Epivir	115	110	49	—	—
Epzicom/Kivexa	555	617	665	763	768
Lexiva/Agenerase	155	142	127	113	87
Trizivir	144	126	107	97	36
GlaxoSmithKline/Shionogi	—	—	—	**19**	**282**
Tivicay	—	—	—	19	282
Pfizer	**80**	**110**	**128**	**143**	**136**
Selzentry	80	110	128	143	136
Others	**154**	**142**	**119**	**135**	**130**
Total	**1566**	**1569**	**1374**	**1386**	**1498**

Source: [16]

does not involve core competences of the cooperating companies, since these are decisive for differentiation in the market. This requires continuous observation and decisions on what the essential capabilities of a company are. Decades ago automotive producers would have (rightly) assumed the production of engines is one of their core capabilities that should not be shared with competitors. Today, we observe alliances of automotive manufacturers that produce engines together because this view has changed.

Owing to the highly sensitive relationship with a competitor, extensive efforts at stabilisation are required. Most strategic alliances have contractual fundaments clearly stating the extent of cooperation, the rules of the cooperation and a time line. A higher degree of stabilisation can be accomplished by establishing a joint venture, but not all strategic alliances are well suited for this stabilisation mechanism, since founding a new legal entity implicates the transfer of rights, assets and – possibly – staff to the joint venture, which is much more complex and burdensome.

4.3.4 Value Chain Cooperation

Value chain cooperation involves two or more companies operating in the same value chain, but for the purpose of the cooperation they contribute inputs from different steps of the value chain. Typical examples of value chain cooperation are outsourcing projects, which helps companies focus on certain parts of the value chain, while partner companies service other parts. Further widespread examples are partnership programmes in companies' procurements, where one (larger) company standardises its vertical relationships to partner companies procuring pre-products.

Owing to the value chain character of this cooperation its objective is mainly generating access to other companies' superior technologies and specialised know-how. In addition, these cooperating partners are providing similar services or products to several other customers and as a consequence are able to reap the benefits of economies of scale.

The main management challenge arises from the vertical relationship of a value chain cooperation, creating unilateral or mutual dependencies between the cooperating companies. These vertical dependencies are frequently related to specific investments, opening the way for exploitation that can seriously damage the success of the transaction; see for example Box 4.4. Value chain cooperation employs numerous instruments to cope with this dependency:

- Contracts can be used to protect the specific investment of one of the cooperation partners. Although a rather obvious solution, the specific formulation of the contents of the contract might lead to fairly voluminous stipulations, which prove to be costly.
- Organisational solutions can be used to reduce the asymmetric dependency in a value chain cooperation due to specific investments. Two examples illustrate possible solutions. Firstly, unilateral dependency can be turned to a mutual dependency by creating similar specific investments (or disinvestments) in the partner company. Establishing research and development know-how in supplier companies in the automotive industry is such a mechanism. On the one hand the supplier has to invest specifically in order to carry out the transaction with the automotive producer; the automotive producer, on the other hand, has to prepare and specify relevant information and input for the supplier that it cannot use in a relationship with another supplying company. Secondly, in some cases the cooperating companies "overlap" each other. Since the supplying partner could refuse to invest in specific machines to manufacture the products for the receiving company, the receiving company could offer to buy the machines and locate them within a plant of the supplying company, which could use these machines, although it does not own them.
- In the later phases of a cooperation the accumulation of mutual trust and building a reputation of faithfulness and loyalty to the cooperation helps to further stabilise the cooperation.

4.3.5 Project Cooperation

A project cooperation connects multiple partners with specific competences for the realisation of a pre-defined project. Because of the character of the project, the cooperation is usually limited in time. Such projects are, for

Box 4.4 Linde and Shell cooperate to build petrochemical production facilities [17]

Linde is a German producer of industrial gases, a provider of engineering solutions for the production of hydrogen and synthesis gas, oxygen and olefins and of plants for natural gas treatment. Its capabilities in providing comprehensive solutions in natural gas treatment opened the way for a cooperation with Shell, which explores and drills for oil and natural gas and produces oil products in its refineries.

In March 2014, Shell and Linde agreed that Linde will provide engineering solutions for Shell's ethane cracking units. The cooperation covers licencing, engineering, procurement and construction services, as well as the supply of proprietary equipment for ethane cracking units. Owing to the complex nature of the technology, the construction of the production facilities has to be carried out in close collaboration with Shell and its requirements. However, Shell falls back on Linde's expertise in building natural gas treatment facilities. The access to another party's know-how or technology is quite typical of value chain cooperation. The main challenge to this type of cooperation is to thoroughly define Shell's requirements for its production and to identify and manage the interfaces to the supplying partner Linde, namely, appropriate communication and information channels have to be developed and implemented.

Source: Handelsblatt. 2014. *Linde und Shell wollen Petrochemie-Anlagen bauen.* http://www.handelsblatt.com/unternehmen/industrie/gemeinsames-projekt-linde-und-shell-wollen-petrochemie-anlagen-bauen/9699002.html

instance, large construction projects, the organisation of large events, a (interdisciplinary) research project or a software development project. Since many partners are involved in a project cooperation, all of whom have to be coordinated in time and location, a project cooperation is headed by a governing body. This could be a specialised project leader or a large partner from the project cooperation, who takes responsibility for the project management.

A precondition for a project cooperation is a very exact project description, including a precise project objective, a time line, a detailed outline of required competences and services and inputs for the project. Since the partners contributing their competences to the project can be chosen freely according to the needs of the project, project cooperation is very flexible in the set-up phase.

Management challenges stem from the flexible design of a project cooperation and its significant information asymmetries. The success of a project cooperation crucially depends on, firstly, the capabilities of the project head to develop the project description and planning, secondly, finding the appropriate partners and, thirdly, handling the problems arising from the information asymmetries between the project participants.

Finding the appropriate partners requires a clear description of the required competences, a method to evaluate possible partners and developing supplementary criteria, such as a corporate culture for providing a positive and stabilising cooperation atmosphere. (Detailed partner search mechanisms will be provided in Section 4.4.)

In order to cope with the information asymmetries in the operational phase of the cooperation, contractual stipulations are a necessary management instrument determining the services to be provided, quality standards, rights and duties and sanctions, and whether the output does not comply with the contracted services or violates the agreed quality standards. Unfortunately, attributing malfunctioning and insufficient performances to specific project participants is often difficult or impossible due to the information asymmetries and to the complementary character of some parts of the project. Thus, implementing information and communication structures that work well is a necessary instrument for managing a project cooperation. The project management has to identify information requirements (who is working together and who is doing subsequent work) and to establish appropriate information channels between these partners and to the project management in order to monitor the project's progress and the performances of the participating companies.

4.3.6 Networks and Virtual Enterprises

Cooperation within networks is widespread and exhibits various forms and configurations, making a precise definition of networks difficult. In general, networks are characterised as rather loose cooperations having a low degree of formalisation that leads to high flexibility and adaptability to environmental changes; see for example Box 4.5. Typically, a network consists of numerous cooperation partners. The objectives of networks are similarly diverse. Some networks aim to achieve economies of scale by joining their businesses, other networks focus on linking different competences of partners. Nevertheless, some networks are more formalised. This is observed in networks in production or in development that have survived a longer period and that have successively increased their degree of formalisation. This formalisation can even lead to the establishment of a legal entity for the network, which increases the network's stability but reduces the flexibility to adapt by changing the structure of the participating companies. A network type that is closely related to a project cooperation is a virtual enterprise. Virtual enterprises emerge from a base network of several companies. Appropriate companies with their special competences are selected for a project from the base network and form a cooperation (a virtual enterprise) tackling the different tasks that are allocated in this project. After the successful completion of the project, the virtual enterprise dissolves and network partners may start new virtual enterprises. Virtual enterprises can be distinguished from one-off project cooperation by their repeated project activities, though the participants in the projects, that is, in the virtual enterprise, may differ from project to project. Because of this

Box 4.5 Google initiates the Open Handset Alliance to develop the Android Operating System [18]

After acquiring Android in 2005, Google initiated the Open Handset Alliance in 2007 in order to develop an open software platform for use in mobile devices such as smartphones or tablets. The founding members of this network included 34 companies. In 2015 the number of members had risen to 87. The development of a mobile operating system must fit with many stakeholders' requirements. Integrating these special requirements and the associated specialized know-how of these stakeholders meant the network idea was a suitable solution for the development of this operating system. The Open Handset Alliance grouped the participating companies according to their particular know-how that they could contribute to the cooperation:

Mobile phone producers: The operating system has to work on mobile phones. Therefore, the mobile phone producers have to provide input for technical interfaces between the mobile phone and the operating system. In addition, they are needed as partners for the swift distribution of the new operating system. (Examples: Samsung, HTC, Huawei)

Mobile network operators: Besides voice information, the mobile devices transmit data and allow to access to the internet. Therefore, mobile network operators have to provide the technical infrastructure to transport the new data, namely, interfaces to mobile networks have to be part of the operating system. (Examples: Vodafone, Telefonica, T-Mobile)

Semiconductor companies: Owing to the extended use of data in these mobile devices, new central processing units (CPUs) have to be developed and the operating system has to run seamlessly on these CPUs. Therefore, producers of computer chips are part of the alliance and contribute information on the CPU interface to the alliance. (Examples: ARM, Intel, Qualcomm)

Software companies: The operating system is used by software that is produced by software companies. They generate the data that are received by or transmitted from the mobile devices and therefore need the capacities provided by the mobile network operators. (Examples: Ebay, Google)

Commercialisation companies: These companies provide input for the distribution of the new operating system. (Examples: Accenture, Teleca)

The Open Handset Alliance and its product, Android, exemplify typical network characteristics. It comprises different competences that are needed to produce the new operating system and the network as a whole is orchestrated by one dominating partner (here: Google). Notice that within the groups the normally fierce competitors cooperate. The willingness to cooperate with competitors in such a project is due to the nature of the operating system as a basis for the technologies and products that the competitors offer. It is not part of the competition but a basis for their competition.

Source: Open Handset Alliance. 2007. *Industry Leaders Announce Open Platform for Mobile Devices*, Press Release. http://www.openhandsetalliance.com/press_110507.html

repetition, the participants in the base network agree to some rules on how they form these virtual enterprises, facilitating the set-up of a project and thus decreasing the costs for a project cooperation.

One of the main challenges in network cooperation is to organise the coordination of the network. Again, different approaches can be observed for addressing this challenge. While most networks have implemented a separate network management to coordinate and supervise the network activities, others rely on defined rules and on joint decision making. The more formalised and stable a network cooperation is, the more it will implement a separate network management. Fairly loose and smaller networks often abstain from implementing such a management. This decision goes along with the transaction structure of the network. Formalised and stable networks are observed to have very frequent and intensive interactions, while in loose networks interaction intensity is rather low. Thus, the costs of establishing a separate network management would be quite high with respect to the interaction. However, joint decision making associated with higher negotiation and decision making costs may be preferable for the loose networks.

4.3.7 Cooperative

A cooperative is a special form of horizontal cooperation, characterised by its legally binding governance elements and mechanism. The objective is generally the achievement of cost advantages from economies of scale. It is frequently observed in procurement and sales activities of companies. The cooperating companies establish a new legal entity, the cooperative, where they allocate certain parts of their business. Again a precise knowledge of the company's value chain facilitates this decision. This legal entity is subject to the stipulations of the law governing cooperatives, which varies in different countries. Some joint features that can be observed for most cooperatives are:

- There are a minimum number of members (usually three).
- The members maintain economic relationships with their cooperative; they consider the cooperative as a vehicle to improve their own business.
- Entry and exit to the cooperative is easy in comparison with other legal forms, such as a limited liability company or a joint stock company.
- Voting rights are on a per-head basis and are not derived from the capital share of the member.

Typical management challenges in cooperatives have their origin in the structural features: the relatively large number of members and the heterogeneity of members that jointly own the cooperative present a management challenge. The management of the cooperative has to balance the joint interests of the members (i.e. the reason for establishing the cooperative), which materialises in the scale effects on a central level and the individual interests of every single member company. For instance, the cooperative and its members have to decide which parts of their joint value chain are executed within the member companies and which parts are the responsibilities of the cooperative. This would be easy to manage if the member companies were symmetrical and homogeneous (i.e. similar size, similar competences, similar strategies, etc.). But most cooperatives show somewhat heterogeneous member structures. These diverse companies deliver different contributions to the cooperative and have differing requirements for the services they expect from their cooperative. Thus, a differentiated member management taking into account these heterogeneous needs has to be implemented. But differentiation results in increasing costs through reduced economies of scale. The management therefore has to consider the decreased benefits because of the lower economies of scale, the costs of implementing differentiated solutions and the additional benefits that can be achieved within the member companies on a local level.

4.3.8 Joint Venture

A joint venture is actually not a stand-alone cooperation type but a special institutionalisation that can also be part of the cooperation types mentioned previously. A cooperative, for instance, is always a joint venture and strategic alliances can be accompanied by establishing a joint venture. The cooperating partners found a new legal entity, the joint venture, as a vehicle for their cooperation, frequently taking the legal form of a limited liability company or joint stock company. All of the cooperating companies (with the exception of possible financing partners) maintain economic transactions with the joint venture, that is, they provide the capital for the company and are customers or suppliers of the joint venture. The founding of the joint venture is often accompanied by a framework agreement or a joint venture statute that determines the objective of the joint venture, the rights and duties of the cooperating partners (e.g. restrictions on the use of jointly manufactured goods) and the contributions of the partners (capital and physical assets provided to the new joint venture or assignment of staff).

4.4 Management of Business Cooperation: A Process Model

4.4.1 The Management Process

The process for managing business cooperation develops five necessary management steps: firstly, the analysis of a company's current competitive position; secondly, the derivation of cooperation options; thirdly, the preparation and institutionalisation; fourthly, the operational management of the cooperation;

and fifthly, monitoring the success of the cooperation (Figure 4.2). Strategic positioning analyses a company's strategic position and explains where and how cooperation might contribute to the improvement of the company's competitiveness. If cooperation is considered to be a suitable solution, the company has to develop its cooperation capabilities and has to search for appropriate cooperating partners. In addition, it has to verify whether the cooperation complies with the competition law, since cooperation always restricts competition and is a concern for antitrust authorities. After the preparation, the institutionalisation of the cooperation has to be developed, namely, it has to be decided how flexible or stable the cooperation should be and how these requirements can be implemented. Moreover, the relationship of the cooperation partners and/or the relations between the cooperation partners and their joint venture has to be structured. After the establishment of the cooperation, the cooperation management focuses on operation of the cooperation. On the one hand the idea that the management moves to "coordinate and cultivate" management becomes relevant, on the other hand the quality of preparation and institutionalisation creates a framework for managing the cooperation operationally. All these steps apply the findings on cooperation from the previous discussions and adapt management methods to the specific conditions of cooperation [19, 20].

As mentioned in Section 4.1, cooperation is a type of governance that is positioned between market and hierarchical governance and therefore inherits selected characteristics of both types of governance. It combines stability from the hierarchical organisation from a firm with flexibility from the market governance. The management process adds to this mix of procedures that allows for stability and flexibility. In the final step of the management process, accomplishment of the cooperation's objectives is monitored. According to the results of this evaluation, various options are available to the management. If the cooperation meets all objectives for all participants, it continues unchanged. But it may turn out that the expected results are not met, which can be caused by misconstructions in every step of the cooperation process. The nature and the extent of underperformance may indicate the origin of the problems in the cooperation and therefore suggest solutions for improving the cooperation performance. Minor problems in monitoring the joint processes or communicating with the partners will imply slight adaptations to the operational cooperation management. If the underperformance is identified to originate from ill-devised rules for the cooperation, these rules have to be reformulated. The cooperation can continue with the same partner, but needs some significant and more complex changes in its framework. With these two types of adaptation the cooperation continues with identical partners and therefore it is stabilised. If the underperformance is caused by insufficient contributions from one of the partners or by malfunctioning processes that can be tracked back to one of the partners, a feedback loop leads back to the preparation of the

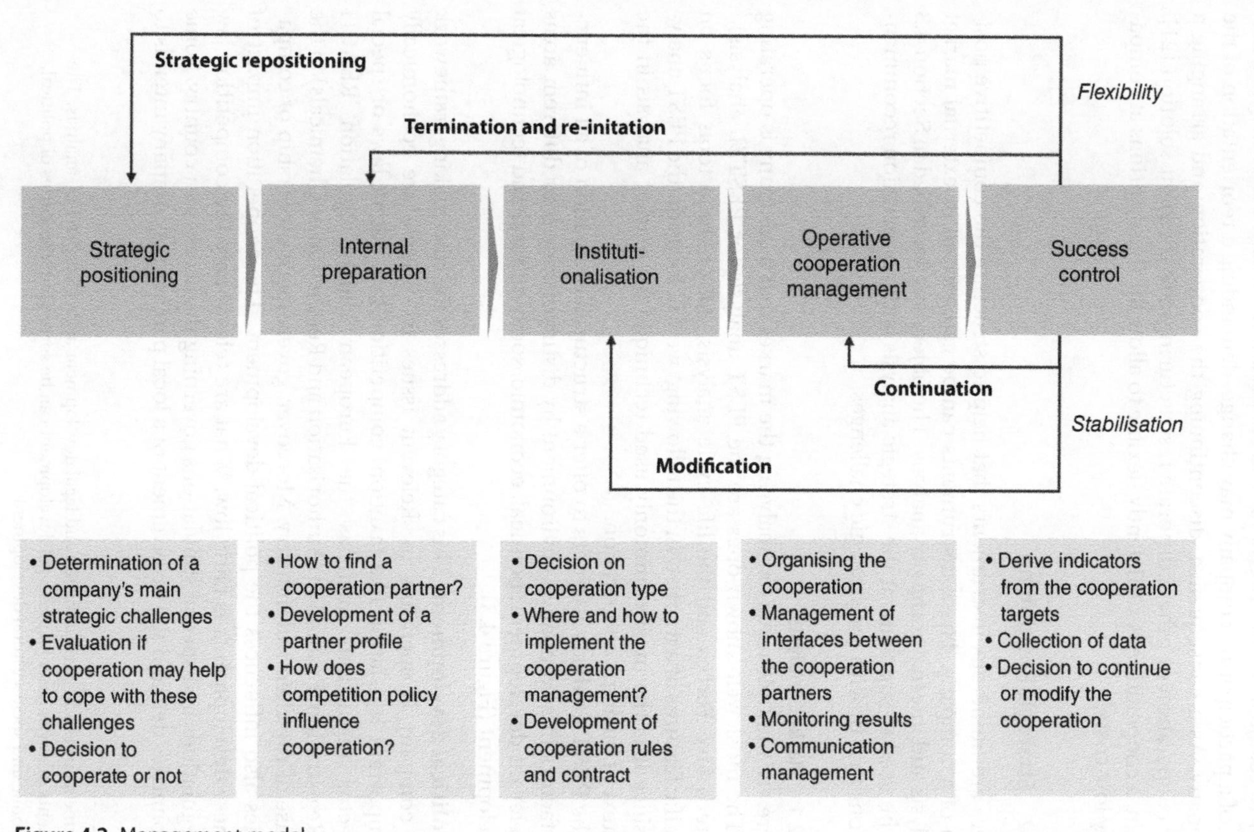

Figure 4.2 Management model

cooperation and the partner selection. An option could be to dissolve the cooperation and start a cooperation with a new partner who displays a better fit or to readjust the fit criteria for the cooperation. Finally, the strategic situation of a participating company may change, demanding a reorientation of the company. As a consequence, discontinuing the cooperation and arranging a new cooperation or other corporate restructurings are the right solution [21]. Again, a cooperation is sufficiently flexible to allow for these options at comparably low costs.

4.4.2 Strategic Positioning

There are numerous instruments that help to structure the competitive position of a company. These instruments can be categorised in external market analysis and internal company analysis. The objectives described in Section 4.3 can be used to supplement the strategic analysis with cooperation recommendations that fit with the strategic challenges.

4.4.2.1 Market Analysis

There are multiple tools for analysing the market that a company is operating in. The most well-known ones are the PEST analysis (or PESTEL analysis),[1] Porter's Five Forces and the life cycle analysis. All of these tools focus on specific features of a market. In the following, we will focus on the PEST analysis since it is the most commonly used technique for market analysis in the context of business cooperation.

The **PEST analysis** attempts to offer a structured evaluation of all influencing factors in a company's environment by distinguishing four different areas of relevant development: political, economic, socio-cultural and technological development (Figure 4.3).

- **Political development:** This category addresses the rule-making framework a company is operating in. Relevant issues to observe are economically important legislation like taxation, competition law, patent laws or special sector regulations such as the European Union regulation REACH (Registration, Evaluation, Authorisation and Restriction of Chemicals) in the case of the chemical industry. Moreover, government ownership of companies also influences the political development. The cooperation impact of these developments is fairly low. What are relevant are the competition law regulations and special stipulations governing the access to a country. Some countries demand the involvement of a local partner if a company intends to

1 Some sources add environmental and legal development and call it PESTEL analysis. The elements of environmental and legal development can be sorted into categories of political, economic and socio-cultural development.

Political development	Economic development	Socio-cultural development	Technological development
• Political parties and their alignments • Legislation, e.g. taxation special sector regulations • Competition law • Goverment ownership of companies	• GOP per head/growth • Sector developments • Disposable income • Currency fluctuations • Investment • Unemployment • Energy costs • Labour costs • Special economic sector conditions • Local economic clusters	• Shifts in values and culture • Lifestyle • Green issues, attitudes towards environmental protection • Demographic change • Income inequality • Attitudes to work/leisure	• New technologies • Speed of technological change • R&D investments • Attitudes towards new technologies • Government investment policies
• Low cooperation impact • Joint lobbying • Eventually size requirements	• Fast growing regional/ sectoral markets → Market access • Cooperation to reduce costs	• Cooperation for access to → Knowledge about customers → Customers → Technologies	• Cooperation to get access to (new) technologies • Economies of scale (e.g. in expensive R&D) • Innovation cooperation

Figure 4.3 PEST analysis and its implications for cooperation management

start a business in that country. This form of involvement is usually by founding a joint venture. Competition law may restrict the companies' organisational decisions by prohibiting special forms of cooperation. Thus, political developments do not imply economic reasons for starting a cooperation, but they either restrict cooperation or make it legally necessary.

- **Economic development:** The economic development captures all economic indicators describing a local market. Some indicators can also be applied to customer-segmented markets. These indicators are, for example, per capita GDP (gross domestic product) and per capita income, exchange rate variations, investment development, energy and labour costs, special local economic conditions, and so on. The analysis of the economic environment can help to identify rapidly growing local or regional markets and a cooperation can be employed to gain faster access to these markets. If a company collects price information (e.g. prices for oil, copper or iron) and anticipates price increases from its analysis, it may decide that substitutes or technologies which reduce the consumption of raw materials will become relevant to reduce the input costs. A cooperation can open the way to these desired technologies. In some rare cases a cooperation can be applied to protect a company's access to rare but strategically important resources (e.g. rare earth metals).
- **Socio-cultural development:** The socio-cultural development attempts to analyse fairly soft developments like shifts in value and lifestyle or people's attitudes (e.g. towards "green" issues or towards leisure), but demographic developments are also part of this development. The analysis of socio-cultural development results in forecasts on new or changing demands and new products that could meet these new demands. Thus, it is a very customer-centred view that is derived from the analysis. Cooperations can provide access to information on these customers and their new attitudes, they can help to get access to these customers (e.g. if their number is increasing and they become a profitable customer segment) or they can offer access to technologies that help to produce these products that satisfy the new demands (e.g. fuel efficient cars).
- **Technological developments:** Analysis of the technological developments considers the general technological developments such as R&D investments or government innovation policies and – on a micro level – new pathbreaking technological developments that are apt to improve or substitute the company's products. In addition, the analysis aims to figure out what future technological developments could reasonably extend the company's existing technologies. Cooperation can be a useful instrument to gain access to these new technologies, particularly if they are new developments that threaten to substitute the company's established technologies. If the analysis outlines necessary technological developments, a company could ally with other similar minded companies to develop these technologies. This applies especially to the research and development of basic technologies and to very expensive technologies. Cooperating companies could achieve economies of scale and share the risk of the development.

In a similar way other strategic analysis tools like Porter's Five Forces or the life cycle analysis can be extended by deriving the corresponding cooperation options. It is necessary to stress that cooperation *can* provide a solution, but this does not implicate that cooperation *is* the best solution. This decision depends on other developments and company characteristics. Thus, a further more detailed analysis is necessary.

4.4.2.2 Company Analysis

For the analysis of the company, three instruments will be introduced: SWOT analysis, value chain analysis and core competence analysis.[2]

SWOT analysis is actually a combination of the analysis of external and internal factors. The external factors are the opportunities (O) and the threats (T) that are beyond the boundaries of the firm. These two categories can easily be analysed by applying the techniques from the preceding section. Opportunities could be market growth in some local markets or market segments, new products and technologies or new demands. Legal changes opening new markets can also be considered an opportunity. Many opportunities come from a thorough PEST analysis. Threats may originate from existing competitors or new entrants or from new technologies that threaten to substitute existing technologies. These threats can, for instance, be derived from Porter's Five Forces analysis. In addition, low growth and stagnating demand or political changes that threaten to foreclose markets pressurise a companies' success. These items are also part of the preceding section with its cooperation recommendations.

The internal perspective addresses the strengths (S) and weaknesses (W) of a company. Strengths could be a company's market position, its specialised product portfolio, its superior technologies or its innovativeness, cost advantages or certain quality standards. A company's weaknesses are costly production technologies, lacking products that fit with market demands, low innovativeness or innovation capacities that do not fit to new developments and poor quality products. A company can use cooperation to tackle its weaknesses, in particular by getting access to other companies' technologies or know-how. Clearly other companies will be less inclined to cooperate and offer their superior technologies to a weak company. Thus, there must be attractive offerings by the weak company that could create a win–win situation for both companies. This is where the analysis of the company's strengths comes into play. The strengths could meet the weaknesses of a prospective cooperation partner and in a cooperation they can mutually exchange their strengths for their individual benefits making both of them better off.

The SWOT analysis bears some similarities to the analysis of a company's **core competences**, since a company's core competences should also be part of

---

2 For a detailed description of the SWOT analysis see Chapter 2.

its strengths. Although the concept of core competences is highly disputed, it may nevertheless help to identify parts of a company where cooperation could be beneficial.

Three different types of competences can be distinguished. Core competences are the most relevant competences for a company, although it is hard to precisely identify these competences in detail. Three characteristics describe core competences: firstly, they should significantly contribute to customer benefit; secondly, they should be hard to imitate; and thirdly, they should be applicable to different markets. Parts of a company that exhibit these characteristics are recommended to be kept within the boundaries of the company due to their strategic relevance. Complementary competences support the core competences, but can be imitated and do not have an immediate impact on the quality, innovativeness or other basic characteristics of a product. They are "further away" from the product. Having these competences would not enable another company to manufacture the products with a similar quality and at similar costs. Complementary competences can be supplied by cooperation. Owing to their support of the core competences they are still relevant for the company, but not so relevant to have them within the firm.

Peripheral competences do not directly affect the production of the core products and are of low strategic importance. Typical examples are competences that are needed to produce standardised products that can be easily substituted. Peripheral competences should be obtained from the market.

In many cases, one company's core competences are another company's complementary competences and vice versa, or to be more precise: this configuration allows the start of a cooperation that mutually exchanges the core products which serve as complementary products in the other company. Thus the cooperation increases each company's benefit.

Finally, and most importantly, the value chain analysis is highly relevant for a decision to cooperate. Only precise knowledge of the value chain enables the management to identify areas where cooperation could be applied and, moreover, is fundamental for arranging the appropriate cooperation structure. Knowledge of the value chain also includes information of the adjacent parts of the value that are covered by other companies who could become a partner in a cooperation.

The analysis of the value chain goes far beyond a rough classification of preproduction processes, production, sales and supporting processes. These very comprehensive steps have to be refined into smaller steps. By refining the analysis, the company is able to identify more aspects in its processes that could be part of a cooperation. A more detailed look at the processes also enables the management to identify products and services that are part of the entire production process but that could also be offered to other companies without constraining the production activities of the firm. After refining the value chain, with a detailed process map at hand, the company has to carry out the subsequent steps:

- **Analysis of the process steps**: Bearing in mind the typical purposes of cooperation, every single process step must be checked for the following criteria:
  - Do the process steps significantly contribute to the value of the products? Could these steps be part of the company's core competence?
  - Could it be possible to reduce costs by increasing production volumes, that is, could economies of scale be realised in this process step?
  - Which investments are necessary in this process step? Are these investments specific with respect to the following steps in the value chain?

Answering these questions gives an indication of whether the step is cooperation appropriate or not. If it is considered as part of the core competences of a company or if there are large specific investments in this step, cooperation is not an option and integration is the appropriate governance mode. Low specific investments or the option for economies of scale clearly indicate a cooperation option.

- **Analysis of interfaces:** In addition to considering the content of the value chain steps, the interfaces between the steps have to be assessed. If one of these steps is actually chosen for cooperation, how this step is connected to the other process parts that remain within the company will be important. Three aspects have to be considered:
  - *Products*: The products passing the interfaces have to be identified. Moreover, how complex these products are must be verified and whether extensive quality controls are necessary, which would increase in importance if the process steps were to be carried out by a cooperating partner.
  - *Services*: Similar to the exchange of products, services passing the interface have to be identified and classified.
  - *Information*: Most crucially, the information passing the interfaces of the process have to be assessed and evaluated. Information is transferred in different shapes. Part of the production knowledge is embedded into products that pass the interface, so that it is not accessible in later steps. To access this "coagulated" information one has to turn back to the production step, which becomes much more difficult if it is decided that this process step is to be carried out in cooperation. Other information is carried by employees working in different process steps. Cutting the process in order to cooperate in a particular process step, and subsequently assigning the employee to the process step that remains within the firm, implies a loss of (necessary) information in the process step carried out in the cooperation, causing problems in this step due to a lack of information.

The interface analysis actually prepares later steps of the cooperation management by raising awareness of the relevance of exchanged goods, services and information that have to be taken into account in institutionalising the cooperation.

4.4.3 Preparation

If the management has decided that cooperation could be a solution to one of its strategic challenges, the company has to prepare for the cooperation. Three questions have to be answered in this step of the cooperation management:

- Who are prospective cooperation partners?
- Is the cooperation subject to constraints from competition law?
- Which cooperation mode should be chosen?

The third question can be easily answered from earlier discussions in Section 4.4, so we will focus on the partner choice and the competition law implications for the cooperation.

4.4.3.1 Partner Choice

There is a generic three-step procedure for structuring the partner choice. In step 1 (partner screening) a partner profile is developed. This partner profile is derived from the analysis of the strategic positioning. The profile will thus contain a description of the prospective partner's position in the value chain, an outline of expected competences, technologies or know-how the partner should have, depending on what has been identified in the strategic analysis. Moreover, a rough sketch of other partner characteristics like size, national origin, infrastructure requirements (e.g. requirements for information systems or certain technologies that have to be used by the partner) or partner's strategy could be added to the profile. Since every cooperation aims for a mutual benefit, a similar profile should be developed for the company itself, so that their own contributions to a cooperation are clear. The result of this first step will be a rather long list of prospective partners fulfilling the characteristics given by the partner profile. Step 2 (analysis of prospective partners) will reduce the long list of prospective partners to a short list, by developing a detailed set of quantitative and qualitative criteria (e.g. for qualitative criteria – quality orientation, innovation orientation; for quantitative criteria– size requirements, financial performance). Elements of such a criteria catalogue will be introduced later. In step 3 (negotiation to select one partner) preliminary negotiations with prospective partners from the short list begin. In these negotiations the desired solutions are discussed and the offers from the company to the prospective partner company are negotiated. At the end of this process the partner for the cooperation will be selected [22].

The success of the partner search process is determined by conceiving a comprehensive partner profile that helps to find possible partners and to evaluate the partner's fit to the company's requirements. In order to structure the criteria for finding the appropriate partner, three dimensions of fit can be distinguished: strategic fit, fundamental fit and cultural fit [21].

Strategic fit Strategic fit refers to three sub-dimensions: objectives, strategies and timing. First, the **objectives** for joining the cooperation have to be compared. For a cooperation seeking economies of scale these objectives should be identical. In contrast, if access to technologies or markets is the predominant reason for a cooperation, reconciling the companies' objectives may be more difficult. As outlined earlier, companies then offer complementary assets to generate a mutual benefit. Consequently, their objectives in the cooperation may differ and whether each company receives a fair share from the cooperation benefit has to be evaluated. The second component of strategic fit is concerned with the companies' **strategies**. For instance, one partner could pursue a strategy of cost leadership, while the other partner attempts to be a differentiator. Different strategies do not necessarily imply that the cooperation will not work but various strategies must be thoroughly taken into account when planning the cooperation. Furthermore, the relevance of the cooperation for the companies' overall success has to be evaluated, too. If the cooperation is very important for one partner but not very relevant for the other one, it creates an asymmetry that is very hard to manage. Third, the planned time line of the cooperation must fit with both companies' needs [17]. The companies have to agree whether the cooperation is established for a fixed time period or whether it is to be created for an indefinite time.

Fundamental fit The fundamental fit evaluates if the cooperation is able to create a win–win situation, making the cooperation a benefit to all participants. Therefore, companies willing to cooperate have to account for all benefits and costs that might arise from the cooperation. Every cooperation results in intended changes of revenues and costs, which may decrease or increase. In order to estimate these effects it is advisable to distinguish between increases and decreases in costs and revenues, which may occur directly or indirectly. Most relevant are the direct revenue increases or cost decreases that are linked to the intended objectives of the cooperation. For example, a given horizontal cooperation may target reducing costs through economies of scale, or a cooperation aiming to gain access to certain markets expects revenue increases through sales in the new market. New technologies provided by a partner result in new products that increase revenue or new technologies can be used for cheaper production and decreased costs. These expected effects may vary with the different cooperation partners that are analysed in the partner choice procedure. Hence, it is necessary to derive the size of the positive expected effects for the planned cooperation objective. In this example, different cooperation partners may have different networks and experience in promoting sales in the new markets, leading to widely different revenue increases.

The same organisational change that is necessary to implement the cooperation may also have other positive and negative effects, which mainly occur as cost increases that are needed to set up and run the cooperation, for example, contracting costs, costs for adapting infrastructure like IT, costs for monitoring the interface to the cooperation, cooperation partner or costs for protecting specific investors or positive learning effects in other company areas. Listing possible indirect effects illustrates that these effects are predominantly negative. Similar to the intended positive impacts of cooperation, these cost effects vary with different partners. If two companies have similar infrastructures (e.g. information systems), the costs will be much lower than for a cooperation where significant adaptations are necessary.

Effects not only occur in parts of the company that are directly involved in the cooperation, but also in other parts of the firm. For instance, if changes in data format or software are needed to manage the interfaces between the cooperating parts of the partner companies, other areas using these data formats also have to adapt to these changes. Figure 4.4 shows a summary of possible effects that should be checked in the analysis of the fundamental fit.

Cultural fit Finally, companies may differ with respect to their national or organisational culture. Since partners remain independent companies within the cooperation, the coordination of the partner companies and the contacts of their employees at the cooperation interfaces are highly relevant. A similar cultural mind-set promotes a joint understanding of a problem, while significant cultural differences may trigger misunderstandings. For example, in a cooperation of a small and young start-up company with a large and established strategy, the organisational culture will differ significantly. On the one hand, the start-up will exhibit a culture of quick decisions and flat hierarchies, while the large company will have established decision procedures that will be more time consuming due to a hierarchical organisation. If the start-up does not receive a quick reply to a request it will be interpreted as an unwillingness to reply, while the real reason is the large company's elaborate ways of decision making. Therefore, communication becomes more complicated with increasing cultural distance, because different cultural mind-sets result in more needs to explain certain behaviour in order to avoid the wrong conclusions. In particular, cooperations with foreign partners face the challenge of different cultural settings. In contrast, similar cultural backgrounds facilitate cooperation since it reduces the need to communicate. These differences not only refer to different national cultures but also to organisational cultures. The understanding of how business should be done can be different in family-owned companies in comparison with capital-markets-focussed joint

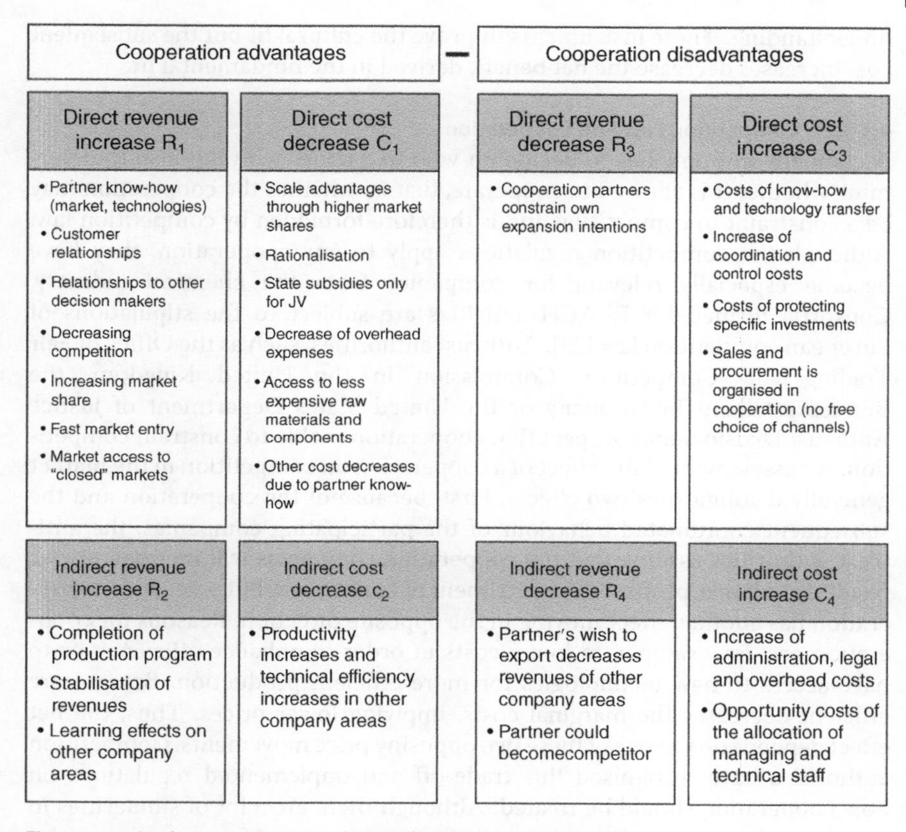

Cooperation advantages
—
Cooperation disadvantages
Direct revenue increase R_1
• Partner know-how (market, technologies)
• Customer relationships
• Relationships to other decision makers
• Decreasing competition
• Increasing market share
• Fast market entry
• Market access to "closed" markets
Direct cost decrease C_1
• Scale advantages through higher market shares
• Rationalisation
• State subsidies only for JV
• Decrease of overhead expenses
• Access to less expensive raw materials and components
• Other cost decreases due to partner know-how
Direct revenue decrease R_3
• Cooperation partners constrain own expansion intentions
Direct cost increase C_3
• Costs of know-how and technology transfer
• Increase of coordination and control costs
• Costs of protecting specific investments
• Sales and procurement is organised in cooperation (no free choice of channels)
Indirect revenue increase R_2
• Completion of production program
• Stabilisation of revenues
• Learning effects on other company areas
Indirect cost decrease c_2
• Productivity increases and technical efficiency increases in other company areas
Indirect revenue decrease R_4
• Partner's wish to export decreases revenues of other company areas
• Partner could becomecompetitor
Indirect cost increase C_4
• Increase of administration, legal and overhead costs
• Opportunity costs of the allocation of managing and technical staff

Figure 4.4 Analysing revenue and cost effects of a cooperation

stock companies. A new technology-focussed start-up has completely different leadership and decision making structures compared with a large, established company. To evaluate cultural fit, the following criteria should be analysed:

- cultural distance
- degree of internationalisation
- strategic orientation (e.g. customer orientation, cost orientation)
- innovativeness
- quality orientation (e.g. reliability, minor error range)
- employer–employee relationships [10, 23, 24].

Although similar organisational cultures facilitate cooperation, different cultures do not necessarily prevent cooperation. Following an analysis of criteria of cultural fit, cooperation partners are able to address the cultural diversity in their cooperation and react to these dissimilarities by allowing for more extensive communication, establishing workshops and promoting a mutual understanding. These instruments improve the cultural fit but the subsequent cost increases decrease the net benefit derived in the fundamental fit.

4.4.3.2 Competition Law and Cooperation

A company not only has to decide on who to partner with but also to determine whether it is allowed to cooperate, that is, whether the cooperation may be a constraint to competition and is therefore forbidden by competition law. Although the competition regulations apply to any cooperation, they have become especially relevant for companies from the chemical industry. Consortia formed for REACH activities are subject to the stipulations of European competition law [25]. Antitrust authorities, such as the Office of Fair Trading and Competition Commission in the United Kingdom, the Bundeskartellamt in Germany or the United States Department of Justice Antitrust Division, may suspect that cooperation is able to constrain competition. An assessment of the effect of a cooperation on competition in the market generally distinguishes two effects. First, because of the cooperation and the subsequent coordinated behaviour of the participating companies, the antitrust authorities assume that the cooperating companies will increase prices, resulting in larger profits to the detriment of consumers. But – second – cooperation has another effect moving in the opposite direction. Reasons for cooperation are, for example, to lower costs in order to enhance efficiency or to have access to new technologies for more efficient production. Being more efficient decreases the marginal costs, implying lower prices. Thus, the net effect depends on the size of these two opposing price movements. Competition authorities have recognised this trade-off and implemented regulations on how cooperation should be treated. Although there are a lot of similarities in various national competition laws, different treatments of cooperation under these laws still exist and, in particular, in the procedures for dealing with these cases. Owing to the lack of space we will focus on the European competition rules for horizontal and vertical cooperation.

Art. 101 par. 1 TFEU (Treaty on the Functioning of the European Union) forbids horizontal cooperations, since they directly or indirectly influence prices and/or limit or control production, markets or technical development. But in order to allow for the positive efficiency effects mentioned previously, Art. 101 par. 3 TFEU defines conditions for exemptions from this general prohibition. Cooperation agreements can be exempted from this ban if they cumulatively meet four conditions:

- The cooperation must improve the production or distribution of goods or must promote technical or economic progress.
- Consumers must receive a fair share of these benefits.
- The cooperation must be indispensable to attain these objectives.
- The cooperation must not afford the parties the possibility of eliminating competition.

For the application of the exemption regulations, two procedures exist. First, for some horizontal cooperation agreements the European Commission has decided to grant block exemption, that is, every cooperation that is covered by a block exemption can be implemented. A block exemption exists, for example, for horizontal R&D cooperation agreements. Second, the four criteria cited can be evaluated on an individual basis and an individual block exemption applies.

European regulations on horizontal cooperation stipulate that every company starting a horizontal cooperation has to assess the provisions of Art. 101 TFEU itself. A horizontal cooperation does not have to be filed at the Commission and it is not possible to ask the Commission for a binding judgement on a cooperation case before the cooperation is implemented and operational. This adds to the legal insecurity for horizontal cooperation and increases the (legal) risks for the cooperation, because the Commission may decide to pick up the case at any time if it suspects that the cooperation is illegal. If the European Commission comes to the conclusion that the self-assessment was wrong and the cooperation is actually illegal, the cooperating partners will be fined [26].

For vertical cooperation the self-assessment is easier, since the European Commission has decided to exclude vertical agreements from the ban of Art. 101 par. 1 TFEU for market shares below 30% by issuing a block exemption for vertical agreements. If the market share exceeds 30%, an individual exemption is possible if the provisions of Art. 101 par. 3 TFEU apply [26].

4.5 Institutionalisation

After finding the appropriate partner, the cooperation partners now have to agree on the shape of their cooperation. Three issues have to be considered to shape the cooperation: the cooperation management must be organised and allocated, the parties have to agree on the rules for their cooperation and the terms for exchanging goods and services must be defined.

4.5.1 Institutionalisation of Cooperation Management

For the institutionalisation of the cooperation management, different types of cooperation must be distinguished. If as in the case of a joint venture a new legal entity is founded, then this new cooperation company will have its own management. Nevertheless, the companies owning this joint venture have to assign cooperation competences internally, that is, responsibilities for the relationships to the joint venture have to be allocated to managers. If no new legal entity is established, institutionalisation is focussed on creating cooperation management structures within each company and constituting joint managing structures necessary to take responsibility for the joint operations. These different types of cooperation management can be characterised as internal or external and symmetric or asymmetric.

For every cooperation (with and without founding a new entity) an internal cooperation management has to be established to deal with the relationships with the cooperation partner and/or the jointly owned company. For this purpose, different implementation options are available. The cooperation management can be allocated to a dedicated organisational unit for managing the cooperation (e.g. adjunct to the management board) or it could be assigned to units within the company that deal with the processes adjacent to the cooperation, where the cooperation management would be added to their other tasks, that is, it is a part-time position for the cooperation management. Both approaches have advantages and disadvantages. Having a separate unit facilitates the coordination of the operating units of the company. This could be important if several units are involved in the cooperation or if the company maintains several cooperation projects that have to be orchestrated. Thus, a separate unit is useful if the cooperation is complex and there are multiple interfaces to the cooperating partner that have to be coordinated. However, a separate cooperation management unit has disadvantages in generating and processing relevant information originating at the interfaces to the cooperation partner. Here a cooperation management allocation to operational units that maintain direct contacts with the cooperating partner is advisable. In particular, if the interfaces are well defined, namely, if the products or services are fairly standardised and the cooperation impact to other process steps and business units of the company is somewhat negligible, the operational unit is better suited for managing the cooperation. Moreover, the size and relevance of the cooperation for the company co-determine the allocation of the management. Large and relevant cooperation projects will have the cooperation management in separate units or even at the board level, while smaller cooperation projects can be managed by unit managers [19].

Concerning the management of the jointly operated processes, the implementation of the cooperation management depends on the nature of the cooperation. If intensive coordination is necessary, that is, when the processes of the two companies are interwoven or if operational decisions have to be made frequently, a standing steering board for the cooperation, consisting of delegated managers from the partner companies, is advisable. With a decreasing intensity of necessary contacts, a steering committee composed of members of the participating companies' internal cooperative management is another organisational option.

Depending on the cooperating partners' involvement in the cooperation, the cooperation management may take different forms. In particular, the capital shares in newly established joint ventures can result in corresponding management involvement in the joint venture. Thus, symmetric or asymmetric cooperation management structures are observed. The more involved in assets or capital one cooperation partner is, the more the cooperation will tend to be asymmetric, with one partner dominating the cooperation. For a cooperation that is not implemented by establishing a new legal unit (such as a joint venture), the symmetry or asymmetry in the management usually follows the companies' economic involvement in, or contribution to, the cooperation.

4.5.2 Rules and Rights

Describing rules and rights in a cooperation is an essential part of the institutionalisation, since the normal command and control rules that work within a firm (see Section 4.1) cannot usually be applied to a cooperation, and governance by means of market prices are equally not sufficient for a cooperation. Rules and rights for the cooperation are part of the partial stabilisation of the relationship. Moreover, they co-determine the mutual influence of the partner companies. Keeping in mind that the "command and control" paradigm cannot be applied in a cooperation, rules and regulations form the basis for influencing the partner by means of "coordination and cultivation". Decisions on a cooperation's rules must address the issue of relevant areas to be regulated and to what extent rules have to be negotiated and agreed on. Since a cooperation may have different objectives, operates in different parts of the value chain and has changing degrees of flexibility and stability, there is no general rule for these decisions, but there are guidelines that can be followed.

One could suppose that extensive rule books and long cooperation contracts are a useful framework for cooperation with another company, but defining rules is subject to an inherent trade-off. On the one hand rules are appropriate to protect the individual partner's investments in the cooperation and to assign the outputs of the cooperation to the partners. Clearly assigned property rights in a cooperation will reduce the costs of resolving any conflicts, but in complex cooperative arrangements not all future contingencies can be foreseen, and it could be unclear what the output will be (particularly in R&D cooperation). Hence, even if partners were to create a long rule book, unforeseeable events can still occur and require supplementary mechanisms. Evidently, the larger a cooperation is and the more relevant it is for the cooperating companies, the more elaborate the regulations and rules will be, since large investments in the cooperation have to be protected. On the other hand, long contracts containing countless duties and rights may have negative incentives for the cooperative behaviour of the partners. In order to comply with the extensive rules, the cooperating partner may only check with all the rules and duties that they have agreed on. This "checking behaviour" may restrict the partners' activities and even worse may hamper the partners' creativity. One of the reasons for a cooperation is to have adaptability and to leave some room so that transaction partners can react to incentives from the market and to provide suitable new solutions.

Hence, reduced sets of rules may increase the flexibility of a cooperation, which is valuable in fast-changing markets. Substituting for the extensive rule books, efficient conflict resolution mechanisms should be implemented that stabilise the cooperation but leave enough freedom for individual activities of the partners.

The contents of the rules and regulations can be categorised into: (1) regulations on contributions, (2) outputs and compensation, (3) organisational and infrastructure regulation and (4) regulations on behaviour. The contribution part determines which assets (machines, capital, etc.) the partners have to provide to the cooperation and if monetary compensation has to be paid in the case of asymmetric asset contributions. Additionally, there should be stipulations on how many staff have to be dedicated to work for the cooperation or are transferred to a joint venture. Infrastructure rules concern the precise assignment of competences for the cooperation within the partner companies, so that requests can be answered swiftly and competently. Corresponding to the assignment of competences for the cooperation within the companies (and the cooperation), communication structures have to be defined precisely and communication infrastructures have to be provided to the cooperation. Behaviour rules are important but sensitive, since too much regulation of the partners' (and their employees') behaviour restricts their creativity (see later) and may be in contradiction of the rules of the company. Behaviour rules substitute for the "command and control" structures in hierarchies, they are general "commands" to all participants. Nevertheless, some rules that have proved to be highly relevant are proposal rights and stipulations on how to behave in situations lacking clear guidelines from the cooperation agreement or in the case of external shocks. Conflict resolution mechanisms are a necessary inclusion in cooperation rule books [21, 27].

4.5.3 "Cooperative Transfer Prices"

Between the cooperation partners, or between the cooperation partners and a joint venture that they own, goods and services are exchanged and therefore have to be priced, because these are transactions between entities that are legally still independent. One could assume that this is not a problem as long as comparable market prices are readily available. But a cooperation is *not* based on market transactions, so choosing market prices would mean applying a governance mechanism that exhibits different transaction cost structures which would lead to the wrong prices within the governance mechanism of the cooperation. Similar pricing problems are well-known for exchanging goods and services within a firm, where transfer prices are used to solve the problem. Accordingly, for a cooperation "cooperative transfer prices" have to be defined. Transfer prices have different functions. Within a company the accounting and measuring of a company's performance are main functions. In a cooperation the coordination function becomes predominant. These prices for goods and services that cross the interface between the cooperation partners are able to influence the behaviour of the partner and are therefore essential for coordinating the cooperation.

If market prices were chosen as "cooperative transfer prices", they would have the advantage that they are hard to manipulate, leading to broad acceptance in the cooperating companies. But they do not reflect the special cost characteristics of a cooperation, a cooperation transaction is not a market transaction. Thus, applying market prices in a cooperation relationship could imply misleading incentives to the partners. Cost-based transfer prices are another mechanism for solving the pricing problem. But cost-based prices are accompanied by significant problems too. It has to be decided whether marginal or average costs should be chosen. Marginal costs (i.e. the costs that are necessary to produce one small (marginal) additional unit) are superior in setting appropriate incentives, but then prices would not cover expenditures for fixed costs and associated overheads. Choosing average costs would solve the problem of covering total costs but the incentives would be discouraged and companies could be inclined to transfer their overhead costs to the cooperation partner. In a cooperation, identifying marginal or average costs is more complicated due to the information asymmetries. It is especially hard to gauge the "right" overhead costs that are associated with the cooperation, resulting in somewhat rough estimates of these costs.

It becomes evident that no pricing mechanism is superior. If there is sufficient transparency between the cooperation partners, cost-based prices on the basis of marginal costs with some mark-up could be a solution. If retrieving the cost information is difficult in a cooperation or if fixed costs are particularly relevant, then cost-based prices may show inferior results [28].

4.6 Operational Management of a Cooperation

After defining the shape and organisation, the cooperation begins to operate. There are numerous tasks that have to be fulfilled in the operational cooperation management. In the following, we will focus on the implementation of coordination tasks in the cooperation. In general, the operational cooperation management will be significantly facilitated if the preceding steps of partner selection have been carried out thoroughly, for example, heterogeneous partners implicate corresponding provisions in the institutionalisation, which is the basis for the operational cooperation management.

4.6.1 Monitoring

In order to coordinate the operations of the cooperating partners, their activities have to be monitored. Information can be detected at the interfaces between the cooperation partners or between the joint venture and the cooperation partner, namely, all information that is transmitted or that can be derived from products or services crossing these boundaries can be evaluated. The cooperation agreement defines each cooperation partner's rights of information acquisition. The cooperation partners have access to all the information that they can record at the cooperation interfaces. In some cases, the cooperation rules allow deeper inspection into the joint venture or into the partner company. This inspection may show opposite effects. On the one hand it can promote the success of the cooperation. Better information that is gathered within the cooperation but outside the company can improve a company's contribution to the cooperation by adapting its production or services. Getting access to information beyond a company's boundaries aims to imitate common intra-company behaviour, whereas information collection within the company is always available and clearly betters a company's performance. On the other hand, a company could use information from the cooperating partner to the detriment of the partner company. Particularly in a horizontal cooperation with cooperating competitors or in a cooperation where getting access to the other company's technology or know-how is an objective, additional information could be absorbed without using the information to the benefit of the cooperation [19].

4.6.2 Influence and Communication

Since direct intervention in another company beyond a company's boundaries is not possible, installing reliable information channels and understanding mechanisms for influencing the partner company become essential in cooperation. Providing information is the basis for the management paradigm of "coordination and cultivation".

Planning the communication requirements and determining appropriate information techniques and assigning them to the communication requirements should be part of the institutionalisation. Nevertheless, they are also used and adapted in the operation of the cooperation. Typical communication techniques that can be used in cooperations are:

- establishing communication norms (e.g., accessibility, reply deadlines)
- creating a directory and assigning competences
- creating direct contacts between employees with similar tasks (especially for parts of a cooperation where intensive exchange of information is expected and crucial)
- periodic workshops, meetings, or phone calls
- creating a wiki or handbook for the cooperation (collecting widely distributed information, especially suited to information that is needed infrequently).

These communication tools can be used to influence the partner company and its decisions. For this purpose, not only is collecting and evaluating information necessary (as described earlier), but it also intelligently provides information to the partner. From the knowledge of the value chain and from communication with the partner, their information needs can be derived. Knowing these information requirements, a company can provide the appropriate information and make assumptions on the effects of the information, providing an instrument for influencing decisions. A company could be tempted to provide wrong or manipulated information or data to the partner, which would violate a trusting relationship and also destabilise the cooperation. Providing the right information at the right time is the "cultivation" in the cooperation management process. Providing information provision works like a fertilizer: the partner company decides on which information it uses but it helps to make the cooperation grow.

Besides this informal but highly relevant mechanism for coordinating a cooperation, there are the institutionalised coordination mechanisms that are used in the operational cooperation management. The cooperation agreement usually provides stipulations for management meetings or other means of coordination. Moreover, the transfer prices for exchanging or providing goods are another way of coordinating actions in the cooperation.

4.7 Monitoring Cooperation Success

Cooperation is not an end in itself but serves to increase the profits of a company. Hence, measuring the success of a cooperation and how it contributes to a company's profits is a necessary final step for the cooperation management. Measuring the cooperation's success is again an information problem that has to be structured and solved [29].

From the objectives that have been defined for the strategic positioning, some benchmarks for measuring the cooperation's success can be derived. Moreover, in the analysis of the fundamental fit, some revenue and cost effects have already been identified and have to be refined to measure the success of the cooperation for the company. Unfortunately, not all the necessary data are readily available for calculating the net profit of a cooperation. Data availability depends on the cooperation type and the institutionalisation of the cooperation. In rather loose cooperation arrangements, collecting relevant data is very difficult, while intensive cooperation relations with interwoven relationships and precise regulations of data provision facilitate the collection of relevant data for measuring the cooperation performance. Moreover, isolating the effect of the cooperation from other influencing factors is another protracted task. Even if data are available, it can be hard to discern the effect of the cooperation from the internal efforts of a company to manufacture a product. Complementing quantitative (accounting) data, qualitative data can be used to measure soft factors that contribute to the success of a cooperation. Questionnaires to people working together with the partner company are a suitable instrument for evaluating soft factors and for early detection of malfunctions of the cooperation.

The data collection for monitoring the performance and success of the cooperation involves the following steps:

- **Deciding on the purpose of cooperation performance measurement:** There are different purposes for monitoring cooperation success. If performance measurement is used in a company's accounting system, quantitative data are necessary. Alternatively, the performance measurement could serve the operational cooperation management. Then quantitative and qualitative data (e.g. from questionnaires) can be used. Not all information is readily available especially if it is created across company borders. This applies especially to quantitative data. Questionnaires may help to get an indication for how well the cooperation is running. Typical questions ask for the perceived success of the cooperation, the impact on the company, the degree of change that was necessary for the establishing of the cooperation, etc.
- **Deciding which information is needed:** From the cooperation objectives, indicators are derived that could measure the accomplishment of the objective. A selection of indicators that are assigned to cooperation objectives is presented in Figure 4.5. Indicators have to fulfil certain criteria, for example, they should be closely related to cooperation activities and should be dissected from a company's own efforts. In addition, indicators can be derived from the fundamental fit analysis.

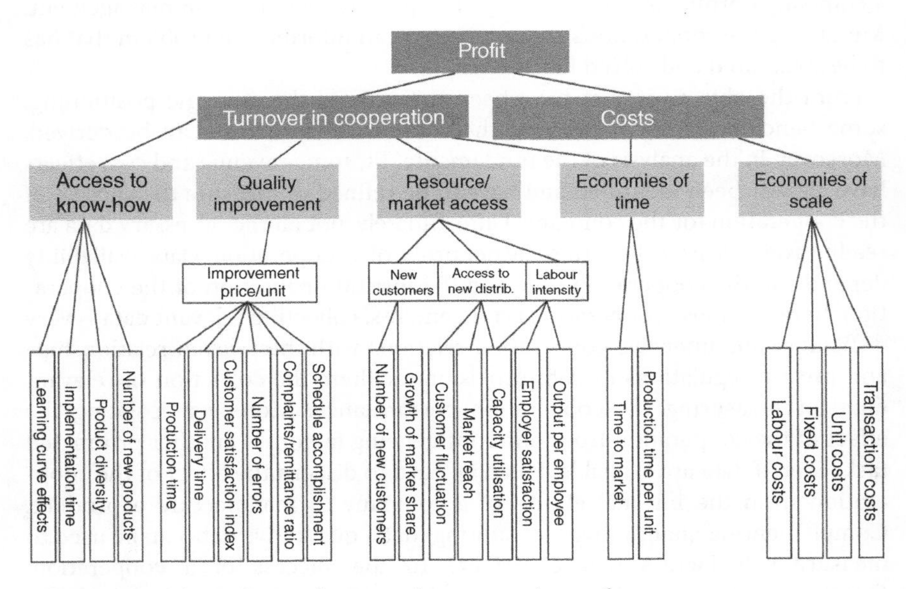

Figure 4.5 Deriving indicators for measuring cooperation performance

- **Identifying which information is available:** Not all requested data will be available. If data are unavailable, these loopholes can be filled by constructing proxies for the missing information. While the construction of information proxies may yield relevant information, these proxies do not have the same quality as the wished-for data that are not available. Thus, by using proxies the margin of error is increased and this uncertainty of the proxy data should be taken into account for evaluating this type of data.
- **Integration of data to internal accounting systems:** In the case of using measured data for accounting purposes, the collected data have to be integrated into the accounting system to allow an integrated control of the cooperation within the firm's control systems.

4.8 Summary

- **Cooperations have become an important part of companies' management.** They allow companies to combine elements of control from the internal organisation with the advantages of market-like, high-powered incentives. They help companies to create cost advantages and to gain access to new technologies and markets.
- **Cooperation management differs from the usual corporate management**. While corporate management can rely on the principles of command and control, these cannot be applied in a cooperation. Instead, methods of coordination and cultivation have to be used as incentives for a cooperation partner.
- **Every cooperation is the result of a strategic analysis**. Cooperation is not an end in itself, but the result of a strategic analysis. Why and how a cooperation may help a company and why it is a superior organisational solution has to be analysed.
- **A reasonable cooperation management is organised by a five-step process**. After the strategic analysis, the company has to prepare for the cooperation, that is, it has to find the right partner. It has to develop the appropriate institutionalisation (rules and rights) and it has to carry out the operational management of the cooperation while the cooperation is working. Finally, a system for monitoring the cooperation success has to be introduced.

References

1 PwC. 2015. *18th CEO Survey – Industry Focus – Chemicals*. http://www.pwc.com/gx/en/ceo-agenda/ceosurvey/2015/industry/chemicals.html (accessed 17 March 2016).

2 Coase R. 1937. The nature of the firm. *Economica*, 4(16): 386–405.

3 Williamson OE. 1975. *Markets and Hierarchies*. Free Press: New York, pp. 26–30.
4 Williamson OE. 1985. *The Economic Institutions of Capitalism*. Free Press: New York, pp. 52–63.
5 Williamson OE. 1991. Comparative economic organization: The analysis of discrete structural alternatives. *Administrative Science Quarterly*, 269–296.
6 Williamson O. 2005. Networks – organizational solutions for future challenges, in *Economics of Interfirm Networks* (ed. T Theurl). Mohr Siebeck: Tübingen, pp. 3–27.
7 Theurl T. 2005. From corporate to cooperative governance, in *Economics of Interfirm Networks* (ed. T Theurl). Mohr Siebeck: Tübingen, pp. 162–165.
8 Besanko D, Dranove D, Shanley M, and Schaefer S. 2009. *Economics of Strategy*. John Wiley & Sons Ltd: Chichester.
9 Malone TW. 2004. *The Future of Work – How the New Order of Business Will Shape Your Organization, Your Management Style, and Your Life*. Harvard Business School Press: Boston, MA.
10 Child J and Faulkner D. 1998. *Strategies of Cooperation – Managing Alliances, Networks, and Joint Ventures*. Oxford University Press: Oxford.
11 BASF, EnerG2. 2014. *EnerG2 and BASF announce multifaceted partnership*, Joint Press Release. https://www.basf.com/documents/corp/en/news-and-media/news-releases/2014/11/P401e_BASF_EnerG2_Release.pdf (accessed 12 December 2015).
12 KPMG. 2010. *The Future of the European Chemical Industry*. http://www.kpmg.com/Global/en/IssuesAndInsights/ArticlesPublications/Lists/Expired/The-Future-of-the-European-Chemical-Industry.pdf. (accessed 12 December 2015).
13 Lanxess. 2013. *LANXESS appoints Mito Polimeri as Strategic Partner*, Press Release. http://lanxess.com/en/media-download/2013-00116e-pdf_en/ (accessed 20 December 2015).
14 ViiV Healthcare. 2009. *GlaxoSmithKline and Pfizer announce innovative agreement to create a new world-leading*, specialist HIV company, Press Release. https://www.viivhealthcare.com/media/press-releases/2009/april/glaxosmithkline-and-pfizer-announce-innovative-agreement-to-create-a-new-world-leading-specialist-hiv-company.aspx (accessed 18 December 2015).
15 ViiV Healthcare. 2012. *Shionogi and ViiV Healthcare announce new agreement to commercialise and develop integrase inhibitor portfolio*, Press Release. https://www.viivhealthcare.com/media/press-releases/2012/october/shionogi-and-viiv-healthcare-announce-new-agreement-to-commercialise-and-develop-integrase-inhibitor-portfolio.aspx (accessed 18 December 2015).
16 Data collected from GlaxoSmithKline's annual reports (available at: http://www.gsk.com/en-gb/investors/corporate-reporting/corporate-reporting-archive/) (accessed 12 June 2017).
17 Handelsblatt. 2014. *Linde und Shell wollen Petrochemie-Anlagen bauen*. http://www.handelsblatt.com/unternehmen/industrie/gemeinsames-projekt-linde-und-shell-wollen-petrochemie-anlagen-bauen/9699002.html (accessed 21 July 2016).

18 Open Handset Alliance. 2007. *Industry leaders announce open platform for mobile devices,* Press Release. http://www.openhandsetalliance.com/press_110507.html (accessed 12 December 2015).
19 Yoshino MY and Rangan S. 1995. *Strategic Alliances: An Entrepreneurial Approach to Globalization.* Harvard Business School: Boston, MA.
20 Bardin L, Bardin R, and Bardin G. 2014. *Strategic Partnering – Remove Chance and Deliver Constant Success.* Kogan Press: London.
21 Doz, YL and Hamel G. 1998. *Alliance Advantage: The Art of Creating Value Through Partnering.* Harvard Business Press: Boston, MA, pp. 169–193.
22 Lendrum T. 2000. *The Strategic Partnering Handbook: The Practitioners' Guide to Partnerships and Alliances.* McGraw-Hill: Sydney.
23 Bronder C and Pritzl R. 1992. Ein konzeptioneller Ansatz zur Gestaltung und Entwicklung Strategischer Allianzen, in *Wegweiser für Strategische Allianzen.* Gabler Verlag: Wiesbaden, pp. 16–44.
24 Bleicher K. 1992. Der Strategie-, Struktur-und Kulturfit Strategischer Allianzen als Erfolgsfaktor, in *Wegweiser für Strategische Allianzen.* Gabler Verlag: Wiesbaden, pp. 266–292.
25 Chemicalwatch. 2009. *REACH consortia without breaching competition law.* https://chemicalwatch.com/1793/legal-spotlight-reach-consortia-without-breaching-competition-law (accessed 26 November 2016).
26 Jones A and Sufrin B. 2011. EU Competition Law, Text, Cases and Materials. *World Competition,* **34**(4): 716–717.
27 Spekman RE, Isabella LA, and MacAvoy TC. 2000. *Alliance Competence: Maximizing the Value of Your Partnerships.* John Wiley & Sons, Inc: New York.
28 Theurl T and Meyer EC. 2003. *Verrechnungspreise in Unternehmenskooperationen – Eine Einführung [Transfer Prices in Business Cooperation – An Introduction], in Verrechnungspreise in Unternehmenskooperationen: Theorie – Strategie – Anwendung* (eds T Theurl and A Crüger). Verlag für Wirtschaftskommunikation: Berlin.
29 Segil L. 2004. *Measuring the Value of Partnering – How to Use Metrics to Plan, Develop, and Implement Successful Alliances.* AMACOM: New York.

Part II

Innovation

5

Principles of Research, Technology, and Innovation

Jens Leker[1], *Thibaut Lenormant*[1], *and Gerald Kirchner*[2]

[1] *University of Münster, Department of Chemistry and Pharmacy*
[2] *ALTANA AG, Corporate Environment, Health, and Safety*

> *Anything that won't sell, I don't want to invent. Its sale is proof of utility, and utility is success.*
>
> Thomas A. Edison (1847–1931), American inventor and businessman

In this chapter the key dimensions of innovation are explained, before sources of innovation are elucidated from a theoretical as well as from a functional perspective. How to structure innovation activities within a company will subsequently be explained, focusing on either a centralized or decentralized R&D department and the advantages of the open innovation approach discussed. In the last part of the chapter the Stage-Gate® model to manage the innovation process will be described in detail, providing practitioners with a tool for a structured approach to innovations.

5.1 What Is Innovation and Why Do You Need It?

On a certain evening in June 1878, Russian chemist Constantin Fahlberg suddenly realized that he missed dinnertime. He had been working all day at the Johns Hopkins University laboratory of Professor Ira Remsen, where he had begun his own research one year earlier. The coal tar derivatives with which he had been experimenting had occupied much of his attention and he didn't notice how late it was. Thereupon, he rushed off for a meal. As he bit into a piece of bread, it tasted surprisingly sweet. Maybe this was in fact not bread, but some sort of cake or confectionery. Yet, his napkin, and the water too, had

Business Chemistry: How to Build and Sustain Thriving Businesses in the Chemical Industry, First Edition. Edited by Jens Leker, Carsten Gelhard, and Stephan von Delft.

this remarkable sweet taste. There could be only one reason. He must have spilled some of the product from his experiments on his hands and had then forgotten to wash them before dinner. Thrilled by this insight, he ran back to the laboratory, searching for the source. Frantically, he tasted every beaker, dish or vial he had used for his experiments that day until he found it: a beaker, in which he had reacted ortho-sulfobenzoic acid with phosphorus pentachloride and ammonia. In the months following his discovery, Fahlberg worked together with Remsen on determining the substance's chemical composition, its characteristics, and the optimum synthesis route [1]. They reported on their results in a joint publication in February 1879. Indeed, they had discovered a brand new substance, "even sweeter than cane sugar": benzoic sulfimide, later to become known as saccharin [2].

The story of saccharin illustrates the first element of any innovative activity, that is, ***invention***. An invention is an act of creation, like the synthesis of a new molecule. It may also involve the development of a new idea. Think, for example, of Edison experimenting at his Menlo Park laboratory and trying to figure out a solution to the problem of designing a practical incandescent light bulb. As long as activities such as these remain in the realm of ideas, of pure science, they remain inventions. But the resolution of a scientific puzzle or the development of a laboratory prototype makes no direct economic contribution. In this regard, Ira Remsen considered himself as a man of pure science and ignored industrial chemistry. Fahlberg, in contrast, rapidly sensed the potential industrial applications of saccharin. After leaving Remsen's laboratory, he applied for patents in Germany and the United States and claimed rights on both the molecule and the production process. Then, he founded a company to manufacture saccharin and encountered great commercial success on both sides of the Atlantic [1]. Fahlberg had turned a scientific discovery into a successful innovation. It is important to note that merely applying for patents, or having these patents granted, was not sufficient to qualify saccharin as an innovation. A patent provides only a particular legal protection, which prevents potential competitors from marketing similar products in a given geographical area, and for a certain period of time. The decisive move that made saccharin an innovation was its commercialization and subsequent adoption by customers. Accordingly, an "innovation comprises the development, production, and market commercialization of an invention as well as product diffusion and adoption by customers" [3: 1066]. Figure 5.1 summarizes this definition.

That being said, we need to dive deeper into the concept of innovation so as to better grasp its multiple facets. Innovation is always about doing something

Innovation = Invention + Economic exploitation

Figure 5.1 Definition of innovation [4]. Adapted from: Bröring S. 2005. *The Front End of Innovation in Converging Industries*. Springer Verlag: Berlin

new, and novelty is a relative concept. What is new today, will no longer be so tomorrow. What is new to one firm, may already exist in another. How novelty is perceived may well change when considering the firm/manufacturer's or the customer/user's perspective. The same is true for innovation, which makes it a many-sided concept. Drawing upon Hauschildt and Salomo (2011), we will consider five key dimensions of innovations, that is: (1) temporality, (2) content, (3) subjectivity, (4) intensity, and (5) normativity [5]. These are not just fancy scientific refinements prescribed by scholars eager to develop precise but abstract theoretical definitions. The distinction between these different dimensions has crucial practical consequences for the management of innovations in the chemical industry.

5.1.1 Temporality

The temporal dimension refers to the relationship between innovation and time. Firstly, innovation is not a one-off event. From that day in 1878 when Constantin Fahlberg invented saccharin in his laboratory, it took several years for him to develop, step-by-step, a profitable business. After his fundamental research on sulfobenzoic acids, he went on to focus more on applied research when developing an efficient manufacturing process. He also had to uncover the correct formulation for the saccharin, manage its production, plan distribution channels, advertise, handle customer requests, and, eventually, adapt and upgrade its saccharin product for new applications. Here, it is crucial to emphasize that innovation is not only an outcome, but principally a process. To this extent, consider that, "while innovation is defined as the (commercial) introduction of a new idea, the process of innovation refers to the temporal sequence of events that occur as people [...] develop and implement their innovation ideas within an institutional context" [6:32]. This process view also implies that innovations go through several steps from invention to commercialization. Despite the many approaches to the innovation process, these steps commonly include idea generation and screening, technological research, business and market opportunity analysis, technical development, testing, and launch [7]. We will return to the issue of managing the innovation process in Section 5.4, where we present what is probably the most widely applied innovation process framework: the Stage-Gate® model.

Secondly, innovation is perishable. An innovation doesn't retain its status indefinitely. Even saccharin became a common consumer good in the United States after sugar prices skyrocketed during World War I. Besides, innovation in the field of artificial sweeteners didn't stop with saccharin: aspartame, sucralose, and, more recently, neotame (up to 40 times sweeter than saccharin) have been marketed since then [1]. Hence, from an overall perspective, innovation as a process is relentless, iterative work, which also includes inventing, developing, and introducing further improved innovations.

5.1.2 Content

The content dimension addresses the question: What is new? Theoretically, there is no limit to what you can innovate, as long as it leads to a useful economic application, namely, to its adoption by a group of customers. Research has produced an impressive amount of systematic typologies to organize this diversity and it is far beyond the scope of this chapter to cover them all.[1] This chapter will focus on the most common forms of innovations in the chemical industry: technological innovations, in particular products[2] and processes. Technological innovations refer to those innovations originating from different scientific focus areas, such as engineering, applied and/or pure natural sciences [9]. While it seems obvious that most of innovations in the field of chemistry rely on the contribution of natural sciences, the distinction between (i) product and (ii) process innovation (not to be confused with the innovation process) deserves more attention.

A useful representation of a product is that of a bundle of attributes, where these attributes fulfill customers' needs and constitute technical specifications. Namely, a product is composed of two interdependent dimensions: technology and market, whereby the former offers a solution to satisfy the needs of the latter. Hence, product innovations aim at satisfying new needs, or the fulfillment of existing needs, in a completely new fashion.

By contrast, "a production process is the system of process equipment, work force, task specifications, material inputs, work and information flows, etc. that are employed to produce a product or service" [10: 641]. Process innovations consist of new combinations of factors aimed at improving efficiency. The distinction between product and process innovations is all the more important since, at an industry level, their development follows a systematic pattern of evolution; see for example Box 5.1.

5.1.3 Subjectivity

The subjective dimension of innovation relates to the question: From whose perspective is it new? As mentioned earlier, innovation is a relative concept. Therefore, in order to characterize innovation, taking into account by whom the novelty is perceived is just as important as what is new, because different players may evaluate novelty with different criteria. In particular, we distinguish here between: (i) the perspective of the firm within which innovations develop, (ii) the customer perspective, (iii) the viewpoint of the industry, and (iv) that of the firm's executives. Box 5.2 provides an example of different perceptions of innovation.

---

1 For a more detailed presentation of innovations types, please refer, for example, to the textbook of Tidd and Bessant (2009) [8].

2 Comprises goods and services.

Box 5.1 Development of innovations: The case of ethylene manufacturing

This evolutionary pattern of development is well exemplified by the case of ethylene manufacturing as related by Hutcheson, Pearson, and Ball (1995) [11]. In 1923, when Union Carbide Corporation (UCC) first succeeded in producing large volumes of ethylene by thermal cracking of hydrocarbon feedstocks, it advanced the development of numerous ethylene-based product innovations. UCC and The Standard Oil Company spearheaded the manufacture of ethylene glycol, ethylene chloride, and ethylene oxide, which played a key role in the production of gasoline for the burgeoning automobile industry. The large amounts of ethylene that were available also facilitated the development of styrene, whose use literally boomed with the surge in demand for synthetic rubber caused by World War II. The discovery of polyethylene by Imperial Chemical Industries in the late 1930s was another major product innovation originating from ethylene. This first phase is called the fluid phase: product innovation dominates innovative activities as markets are still in formation, product designs are non-standard, and product processes remain rudimentary and adaptable [12].

In the second phase – the transitional phase – process innovations come to the foreground, mainly driven by reduced market uncertainty, heightened competition, and the emergence of dominant designs [12]. These factors encourage firms to increase outputs and, in turn, foster major innovations in production processes, which become more specialized, integrated, and efficient. Indeed, from the mid-1940s to the early 1970s, the scope of innovations in the ethylene industry shifted towards the improvement of the production process and economies of scale, in particular through increase in plant capacity. In about 20 years, the typical production from ethylene plants rose from between 25 000 and 50 000 tons per annum to capacities of between 300 000 and 500 000 tons per annum. This was accomplished thanks to the generalization of high severity/short residence time cracking technology, and several ameliorations in equipment, such as the development of significantly larger furnaces.

In the third phase – the specific phase – the innovation behavior becomes less flexible, due to product standardization and tight process integration [12]. Product and process become highly interrelated, turning every modification into a challenging and risky endeavor. Efficiency and economies of scale drive innovation in a cost-based competitive environment, which then occurs at a much slower rate and tends to focus on incremental improvements. The ethylene industry didn't escape this fate. From the early 1970s, it was beginning to reach maturity. The rate of increase in plant capacity leveled off as state-of-the-art process innovations became widespread as standard. The cumulative effects of oil price shocks and market saturation in developed countries spurred on cost-optimization measures. Innovative efforts were then focused on incremental refinements with respect to product quality, energy consumptions, and production yields, such as process design modifications, allowing flexible use of

(*Continued*)

Box 5.1 (Continued)

alternative feedstocks (e.g., naphtha), or the implementation of heat recovery systems. In summary, being aware of this pattern of evolution is fundamental in order to understand the dynamics of the chemical industry from a technological point of view. To conclude the discussion on the content aspect of innovations, it is worth noting that technological innovation is just one possible type among many others. Innovation may also relate to new forms of organization, new procedures, new services, new types of contracts, etc. In that respect, this book presents yet another kind of innovation – business model innovation – in Chapter 7.

Box 5.2 Different perceptions of innovation

The following case demonstrates the importance of consciously delineating these different perceptions. During his career, an employee took part in the development of a new sterically hindered amine for gas-treatment applications. Amine treatment has been used for almost a century by the natural gas industry to remove acid gases such as CO_2 and H_2S prior to pipeline transportation. The project team was well aware of this fact. Hence, from an industry perspective, it was clear that the new amine would actually substitute existing offerings rather than create its own market. The project could also learn a lot from established products, especially for benchmarking purposes. Eventually, the new amine proved to reduce energy consumption at gas treatment plants by more than 20% compared with the next best alternatives already on the market. From the perspective of the firm, natural gas treatment was a brand new market, but amine chemistry was an established area of expertise. Thus, technical developments relied heavily on the knowledge of the in-house amine experts, while significant resources were invested in exploring the gas-treatment industry. Internally, management supported the project, convinced by the technical benefits of the new amine and positive customer feedback. After some time, the team realized that this early feedback came in fact from enthusiastic R&D employees and not from plant operators, who turned out to be key decision-makers with respect to the amine systems on the customer side. These first market contacts had prevented the team from including the point of view of the true customers. Unfortunately, from the perspective of plant operators, the benefits offered by the new amine did not compensate for the perceived risks and, suddenly: the new product apparently no longer had a market outlet. The management team started to cast doubts on the project: How could it recover from such a setback? It took a lot of persuasive effort and pedagogy from the project leader to restore confidence in the project's chances of success. Arguing on the strategic role of this project in renewing the firm's declining activities, he even succeeded in unlocking additional resources to improve customer understanding and devise a new market approach.

Firms tend to assess innovations on the basis of competitor offerings, or by the extent to which new competencies or resources had to be used. Customers are inclined to call on their current mental models and behavioral habits [3]. Failure to consider this difference may lead to dramatic consequences. The firm's perspective is appropriate to deal with management issues, such as the organization of the development of new products. As Garcia and Calantone (2002) [9] comment, using the customer's perspective to address how a firm should approach the development of new products would be like "letting the customer drive the innovative process of the firm." Conversely, the customer perspective is better suited to treating marketing aspects, like defining actions to promote the adoption and diffusion of innovations. Innovation history offers countless cases of failed new products caused by ill-defined marketing strategies, which relied solely on the perspective of the manufacturing firms.

The third perspective, the industry perspective, is equally important because it implies, again, a specific set of criteria against which innovations can be appraised. From the viewpoint of an industry, the relevant factors to evaluate innovations exceed the resources, competencies, or strategy of a single firm; the industry perspective includes those of a group of firms engaged in a similar economic activity, like their competitors. This point is particularly critical for evaluating the degree of novelty of innovations.

Finally, the perception of executives has a somewhat special role when it comes to characterizing innovations. Within firms, executives have the decision-making power to start or stop innovative activities, they devise the innovation strategy, prioritize goals, and allocate the necessary resources. In short, whether an idea or an invention may have the chance of turning into an innovation largely depends on the interpretation of executives [5]. Practical consequences are twofold: (i) a potential innovation only becomes realized once it has been *recognized* as such by the management team; (ii) the individual/team supporting an innovative idea or an invention has to *demonstrate* to executives why this idea/invention deserves such a recognition. Firms are social entities in which decision-making often relates to power games and politics.

5.1.4 Intensity

The intensity dimension concerns the degree of novelty of innovations, that is, the extent to which the company is familiar with the new product/process. It is usually referred to as innovativeness, or, more specifically, as product innovativeness when referring to new products or processes. Being able to precisely answer the question "How much is it new?" is maybe the most decisive competence to develop in order to take the right decisions when managing innovations. However, if there is a common thread among the multitude of contributions addressing this issue, it is the remarkable inconsistency in the terminology employed to characterize product innovativeness. The terms "incremental" and "radical" innovations have established themselves in the managerial discourse, but you may also encounter many others, such as "imitative," "breakthrough," "disruptive," or "revolutionary" innovations. Thus, to cut across this unnecessary complexity, while acknowledging the diversity of innovation types along the incremental–radical continuum, we will retain three categories, namely, in increasing order of innovativeness: incremental, really new, and radical innovations. This categorization refers to the typology of Garcia and Calantone (2002) [9], whose definition of product innovativeness is probably one of the most predominant in the recent innovation literature. Building upon their work, we suggest a simple but systematic heuristic to appraise product innovativeness. Before that, we present their definition of product innovativeness and the characteristics of the three associated innovation categories.

In essence, product innovativeness is a measure of the potential discontinuity an innovation can generate with regard to the technological and marketing components of a product/process. We consider two levels at which these discontinuities can take place: a macro level – the industry – and a micro level – the firm. This distinction is important as it clarifies from whose perspective and to whom a product/process is new. Innovations that are new from the perspective of the industry cause discontinuities on factors that are exogenous to the firm, which obviously requires a lot more innovative power than those merely new to the firm.

On that basis, incremental innovations are innovations that have the potential to cause market and technological discontinuities, but *only* at the level of the firm. Really new innovations correspond to the moderately innovative type. They generate *either* a market *or* a technological discontinuity on an industry-level, but not both, in combination with any firm-level discontinuity. Radical innovations result in market *and* technological discontinuities on *both* an industry and a firm level. Figure 5.2 provides a convenient matrix representation along with examples from the chemical industry.

Based on the definition given previously, there are more mathematically possible combinations than the eight represented here. However, several combinations are simply impossible as a discontinuity at the industry level systematically implies a discontinuity at the firm level – a firm being a subset of the industry. The distribution of innovations across types suggested by the matrix – 37.5% of incremental innovations, 50% of really new innovations, and 12.5% of radical innovations – is consistent with statistical breakdowns reported by several empirical studies [9]. To complement the matrix representation, Figure 5.3 shows a simplified heuristic to allow rapid identification as to which type an innovation belongs. For a given innovation, answering the following questions sequentially leads to the relevant degree of innovativeness.

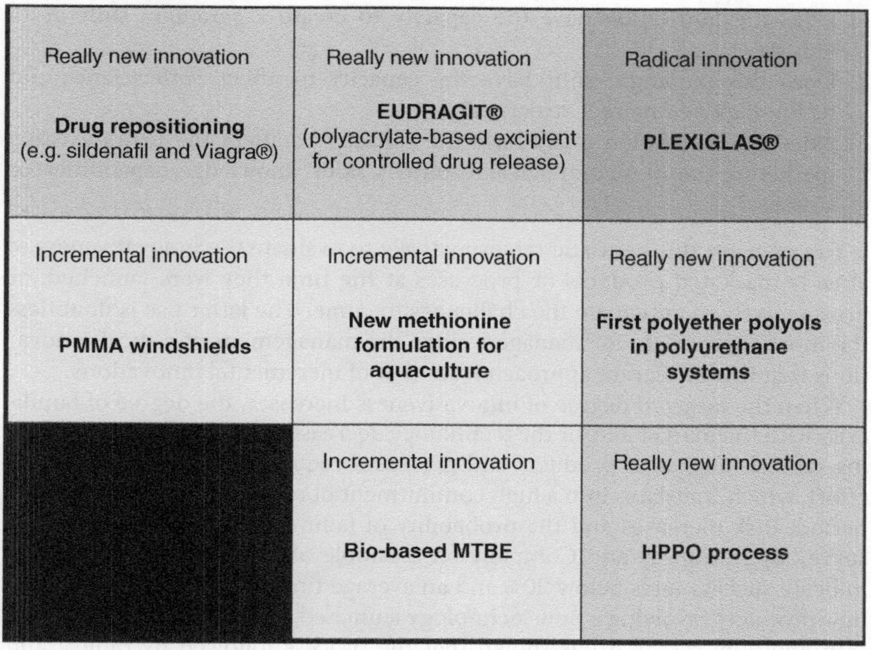

Really new innovation **Drug repositioning** (e.g. sildenafil and Viagra®)	Really new innovation **EUDRAGIT®** (polyacrylate-based excipient for controlled drug release)	Radical innovation **PLEXIGLAS®**
Incremental innovation **PMMA windshields**	Incremental innovation **New methionine formulation for aquaculture**	Really new innovation **First polyether polyols in polyurethane systems**
	Incremental innovation **Bio-based MTBE**	Really new innovation **HPPO process**

Figure 5.2 Innovation typology according to the degree of innovativeness

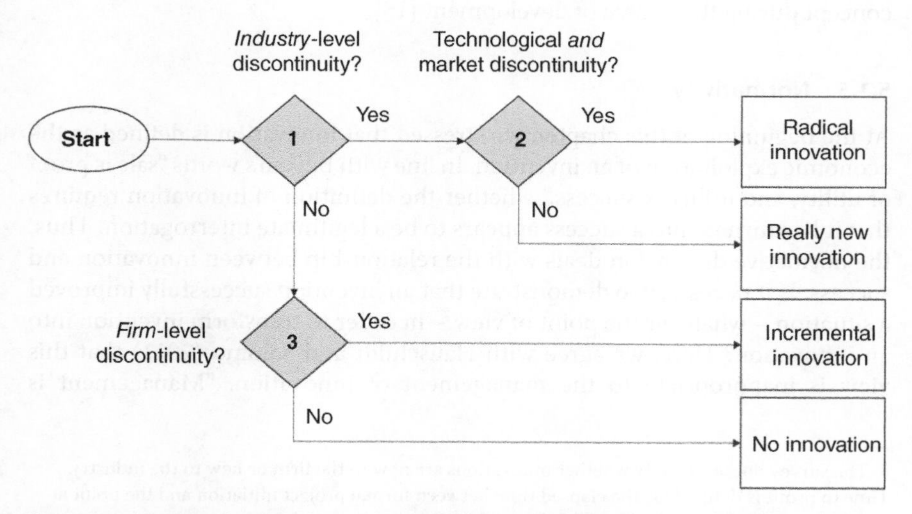

Figure 5.3 Heuristic for identifying innovations depending on their innovativeness

1) Does the innovation have the capacity to create a paradigm shift in an industry?
2) Does this paradigm shift have the capacity to affect *both* science and technology *and* market structure?
3) Does the innovation *solely* have the capacity to affect the firm's existing marketing and/or technological resources, skills, knowledge, capabilities, or strategy?

You may use this heuristic retrospectively to evaluate the innovativeness of already marketed products or processes at the time they were launched, or prospectively to anticipate the challenges to come. The latter use is doubtless the most appropriate for managers since the management of radical innovations requires a different approach from that of incremental innovations.

When the targeted degree of innovativeness increases, the degree of familiarity with the market and/or the technology decreases. Consequently, developing radical innovative products and processes requires a significant learning effort, which translates into a high commitment of resources over an extended period. Risk increases and the probability of failure too. Data compiled in a survey by McKinsey and Company on a sample of 35 chemical innovations indicate success rates below 20% and an average time to profit of 14 years for new products involving a new technology launched on a new market [13].[3]

In addition, research has shown that the process followed by radical and really new innovations diverges markedly from the development of an incremental innovation. Among other things, the development of more innovative products seems less straightforward and tends to experience more feedback loops [14]. This may involve the complete redefinition of the product/process concept during the course of development [15].

5.1.5 Normativity

At the beginning of this chapter, we stressed that innovation is defined as the economic exploitation of an invention. In line with Edison's words "sale is proof of utility, and utility is success," whether the definition of innovation requires that it has turned into a success appears to be a legitimate interrogation. Thus, the normative dimension deals with the relationship between innovation and success. Is it necessary to demonstrate that an invention successfully improved a situation – whatever the point of view – in order to transform invention into an innovation? Here, we agree with Hauschildt and Salomo (2011) that this view is inappropriate to the management of innovation. "Management is

3 The survey doesn't specify whether innovations are new to the firm or new to the industry. Time to profit is defined "as the elapsed time between formal project initiation and the point at which the project's annual sales equal the total R&D investment in it."

future-oriented": making success a decisive criterion in the definition of innovation would contradict the bare reality experienced by innovation managers, who strive to achieve an *anticipated* concrete success [5]. Nevertheless, success remains an essential objective. That is why we will now examine how the decision of managers may influence innovation success and commercial performance.

A recent meta-analysis drawing on 64 studies published between 1970 and 2006 provides an initial compelling result: innovation consistently fosters performance, whether at the product or at the business unit level [3]. Interestingly, this study also shows that innovations perceived as new by customers contribute more to performance than new-to-the-firm innovations. Consequently, developing high performing products or processes requires focusing on the customer perspective rather than on the firm's perspective.

This leads to a second critical insight: pursuing innovation for its own sake of incorporating new technologies into products is a sterile strategy. While product innovativeness improves product advantage against competitor offerings, it negatively affects the familiarity of customers with the new product/process. Therefore, ensuring innovation success requires education and familiarizing customers with the new product/process so as to demonstrate its superiority and reduce perceived uncertainty. "Unless the technology in a new product overcomes customer uncertainty and is perceived to provide an advantage over competitor offerings, it is unlikely to improve new product profitability" [16].

Thirdly, from the perspective of the innovating firm, technological innovativeness is also a double-edged sword regarding commercial success. On the one side, technological innovativeness has the potential to improve product advantage, which, in turn, favors commercial success. On the other side, it generates changes within the firm and its environment (regulation, infrastructure, social norms, etc.), which reduces the likelihood of commercial success [17].

In a nutshell, the more innovative products and processes are the more they contribute to commercial performance, but *indirectly*. On the way to success, managers must acknowledge and address the four following challenges:

- demonstrate advantage over competitive offers
- tackle customer uncertainty with regard to novelty
- manage organizational changes ensuing from innovation
- address the need for transformations in the firm's environment.

As this discussion on the nature of innovations shows, innovation is a complex, multi-sided, and dynamic problem, whose *raison d'être* lies well beyond the mere fascination for science and technology. Innovation contributes directly to the economic performance of firms. Innovation underpins technological change and, as such, determines the ability of firms to adapt and survive in an ever-changing environment. Keep in mind the tragic fate of Eastman

Kodak, which has almost been wiped out of the photography industry after missing the shift from chemical-based film technology to digital. Despite tremendous challenges, harnessing the complexity posed by innovations is not an option. It is a necessity.

5.2 Sources of Innovation

Whereas the economic value of innovations has long been a topic of consensus, identifying the prime forces driving innovations has created much more controversy. Thus, an intense debate emerged in the 1960s and 1970s about the factors determining the direction of innovations, which opposed the technology-push against the demand- or market-pull argument. The core of the dispute can be stated as follows: proponents of the technology-push argument maintained that advances in science and technology drive the direction of innovations. On the contrary, defenders of the demand-pull argument argued that innovations arise from changes in market conditions and the recognition of unmet needs. Debates on this matter were highly polarized, each camp regarding the views of the other as irreconcilably opposed [18]. Even though this debate has now faded in the academic literature, the terms technology-push and demand-pull are still widely used in practice as they continue to be invoked in discussions about the relative importance of market signals over R&D efforts in fostering innovative activities. Thus, before introducing how the dispute was settled, we will begin with a more detailed presentation of the two paradigms followed by an analysis in the context of the chemical industry.

5.2.1 Technology-push Versus Demand-pull

Several authors have been associated with one or the other side of the controversy between technology-push and demand-pull (see for example Box 5.3 in the context of the chemical industry).[4] Yet, the conceptual underpinnings of this debate are frequently depicted as the opposition between the theories of two scholars: Schumpeter for the technology-push tradition, and Schmookler for the market-pull tradition [20]. The pattern suggested by Schumpeter's theory of "creative destruction" can be summarized as follows:

1) Developments in science lead to the birth of major inventions promoted by individual inventors or the R&D activities of large firms.
2) Entrepreneurs (firms or individuals, sometimes the inventors themselves) sense the opportunity behind these inventions. As a result, they are willing to take the risk of developing, producing, and marketing the inventions, turning them into innovations.

4 For an overview of these studies, see, for example, Chidamber and Kon (1994) [19].

3) As the innovation spreads, it disrupts existing market structures. The innovators benefit from a temporary monopoly position, which translates into exceptional growth and profit. Firms that cannot adapt go bankrupt.
4) Entry of secondary innovators progressively undermines this monopoly, while reproducing the initial conditions of equilibrium in the market.

Box 5.3 Technology-push versus demand-pull in the context of the chemical industry

Walsh (1984) [20] investigated the rate and direction of innovative activities on the basis of a large array of evidence in the field of dyestuffs, plastics, and pharmaceuticals over the period of 1830–1980. Here, we report his findings on the early developments of the synthetic dye industry. Most historical accounts on the origin of this industry describe what resembles an archetypal technology-push innovation. In the middle of the nineteenth century, many chemists were focusing on a new discipline, organic chemistry. A major area of research dealt with the transformation of coal-tar, a waste by-product from the burgeoning coal-gas industry, into more useful compounds. In 1856, a research assistant to the organic chemistry pioneer August von Hoffman, the 18-year old William Henry Perkin, was trying to synthesize quinine using coal-tar derivatives. Reacting aniline with potassium dichromate, he obtained a dark precipitate, which obviously wasn't quinine [22]. After rinsing the flask with alcohol, he noticed the purple color of the solution and its ability to dye silk. Perkin had discovered mauveine, the first synthetic dye. Unlike other natural purple dyes, it was very resistant to light and to washing, allowing Perkin to successfully commercialize mauveine as a dyestuff [23]. As a matter of fact, secondary innovators entered the market with new synthetic dyes, and the modern organic chemical industry was born. Thanks to the profits generated by his invention, Perkin retired at the age of 36 and then devoted himself to basic research.

From this account, the synthetic dye industry seems to have emerged directly from organic chemistry research, independent of any potential demand by dyers or customers for improved dyes, confirming the technology-push argument. Perkin's serendipitous discovery, as well as the fact that research into coal-tar chemistry was completely disconnected from the dyers' own research, supports this interpretation. However, statistical data support the exact opposite argument.

Looking at the consumption of cotton in the United Kingdom as an indicator of demand in the textile industry and, as a result, of the demand for dyes, Walsh (1984) comments that demand had already started to increase massively well before 1820. At first, this surge in demand would have been met by relying on existing sources of dyes. This is reflected in the concurrent increase in import volumes of raw material for natural dyes such as madder. Rapidly, existing

(Continued)

Box 5.3 (Continued)

techniques for dye production proved inadequate for the large-scale production, which triggered a wave of inventions. In the 1850s, the dyestuff industry engaged in innovative activities to meet continuously rising demand and the output of dye patents rose dramatically. It is in this economic context, where dyers developed more efficient extraction processes, new natural dyes and of better quality, that Perkin discovered mauveine in 1856. From this perspective, synthetic dyes are only the culmination of a series of innovations in the textile and dyestuff industry. In addition, favorable market conditions probably contributed to ensuring the market success of Perkin's "mauve." Purple was already a very fashionable color and in high demand in the market, but dyes couldn't be applied on silk and only after an expensive process on cotton. Dyers had been rigorously searching for a proper purple dye since the late 1840s.

The development path of the early synthetic dye industry appears to mirror the demand-pull argument as well. It is as if "the growth of the textile industry stimulated increased demand for dyes which in turn stimulated innovation and eventually a new industry to meet that demand" [20].

In summary, there is evidence for both the technology-push and the demand-pull argument in the early development of the synthetic dye industry. Organic chemistry led to new chemical inventions like mauveine, and the entrepreneurial endeavors of people such as Perkin transformed them into dye innovations, turning scientific discoveries into business. More importantly, these innovations were made possible because of an economic context where dyestuff manufacturers were looking for innovative solutions to fulfill pressing customer demand. Hence, it is the *combination* of demand-pull and technology-push that best explains the origins of innovations. The two models are not mutually exclusive, contrary to what the original dispute suggested. It is not even that both supply and demand factors contribute to an explanation, as Walsh (1984) [20] suggests in the conclusion of his study, but rather that market demand and technological advances *interact* to generate innovations [18]. Moreover, another interpretation superseded the simplistic dichotomy of the early debates.

Derived from a series of follow-up empirical studies, the demand-pull approach emerged. Advocates for this approach concluded as to the primacy of market-related factors in determining innovation success and, thus, were perceived to contradict the technology-push approach. Among tenants of the demand-pull paradigm, the work of Schmookler is regarded as a major contribution [20]. His theoretical developments may lead to the following pattern being expected:

1) Market demand rises, driving firms to increase production and investment.
2) Rising demand is first satisfied using existing means, like existing plants.

3) Then, steady market pressure and the related technical challenges foster inventive activities, which translates into an increased rate of inventions, within and outside firms. Growth in the number of patents as well as corporate R&D activities reflect this trend.
4) Inventions are incorporated into new and improved products and processes so as to meet market demand.
5) Subsequent variations in demand are expected to generate a similar variation in the rate of invention after a certain time lag.

In this model, the role of science and technology is subordinate to the strength of demand; people (or firms) develop inventions in response to a market opportunity. The demand-pull theory found a wide audience upon its introduction and "it became fashionable to assume that the debate was over and that it had ended in a clear victory on points, if not a knockout, for the demand school" [21: 207].

While acknowledging the complex feedback structures between market demand and technological developments, Dosi (1982) [24] suggested that the importance of the technology-push and the demand-pull model depends on the degree of innovativeness of the underlying technology. Thus, radical technological innovations, which he terms changes in technological paradigms, result mainly from technology-push efforts, whereas incremental innovations within existing technological paradigms are essentially market-pull.[5] This is not to say that you should consider market signals any less when developing radically new products or processes; analyzing the market is still a major determinant of innovation success. Yet, merely reacting to customer demand will tend to produce more incremental innovations while investing in R&D, or establishing partnerships with universities to stay ahead of technological advances, is a necessary step, but not sufficient, in order to develop radical technological innovations.

In the previous discussion, we pointed out the limits of early debates on the origins and direction of innovations, and emphasized the role of interactions between market signals and technological developments in explaining their emergence. Cutting across the technology–market dichotomy, more recent views address the issue of categorizing the sources of innovations by taking a functional perspective. That is, they distinguish between sources depending on how they generate benefits from a given innovation. Drawing on this perspective, we can distinguish between the following:

- corporation (corporate R&D programs, culture)
- individuals/employees ("Google" culture)
- competitors (e.g., use of knowledge spillover, analysis of patents)
- process demands (e.g., improvement of manufacturing/logistic processes)
- government (e.g., e-mobility)
- suppliers/customers.

5 This approach is also reflected in recent research into disruptive innovation, see Christensen (2013) [25].

In the chemical industry, a significant amount of innovative efforts originates from the manufacturers themselves. In a study, Von Hippel (1988) [26] found that over 90% of innovations in the field of engineering plastics (e.g., polycarbonate) and plastic additives (e.g., butyl benzyl phthalate) had been developed by polymer manufacturers. Even today, internal innovative activities remain a key source of new products and processes. Here, technological forecasting has proven to be very valuable in this endeavor.

Any forecast involves the evaluation of the probability and significance of possible alternative futures, and technological forecasts do just the same. However, let us clarify right at the outset the misunderstandings about the real purpose of this activity. The goal of technological forecasting is not to predict a precise event whereby, at a given date, a technology comes into existence for a certain application. Instead, technological forecasters produce "range forecasts" and "probability statements" so as to anticipate the future characteristics of technologies and their potential consequences. A useful forecast helps to identify opportunities and threats, supports decision-making, and allows managers to act in order to improve the firm's future positioning [27]. "Technological forecast is not a picture of what the future will bring; it is a prediction based on confidence that certain technical developments can occur within a specified time period with a given level of resource allocation" [28: 129]. Chemical companies sometimes go as far as institutionalizing this activity, as Evonik did with its corporate foresight team [29]. Several methods have been developed for technological forecasting. We will focus here on some of the most widespread techniques, namely: (i) environmental scanning, (ii) models, (iii) Delphi, and (iv) extrapolations [30].

5.2.1.1 Environmental Scanning

Environmental scanning relies on the assumption that technological developments follow a sequential path of evolution from scientific work to product commercialization. For a given technological field, evidence of its position on this sequence may be found in various publications, such as scientific articles, patents, or the business literature. Then, by searching in the relevant databases for events that foreshadow future developments, it is possible to identify warning signals that indicate when a technology will probably reach the next stage of development [30]. Patent and bibliometric analysis have proven particularly useful to forecast emerging technologies [31]. For example, Wagner *et al.* [32] conducted a patent analysis relating to lithium-ion battery (LIB) technology using the patent database PatBase®. Given the disproportionate growth in LIB patent application compared with other battery technologies, they conclude that LIBs will continue to have a major impact on future applied research into energy storage. In addition, for each battery component, they were able to specify which technological options have the highest potential to impact future applications.

5.2.1.2 Causal Models

Causal models require identifying the variables underpinning a phenomenon as well as the relationships between those variables, let alone whether it is possible. Causal models also presume that these variables and their relationships can be expressed in mathematical equations. In practice, these explanation-oriented approaches are almost exclusively applied to forecast the diffusion of innovations [30]. Sick, Golembiewski, and Leker (2013) [33] investigated the diffusion of renewable energy technologies by integrating raw material prices into an established diffusion model. Using data from the wind and solar power industry in 18 OECD countries, the expanded model demonstrates the impact of crude oil and natural gas prices on the diffusion of renewable energy technology, represented by the net investment in these technologies. Thus, the authors provide managers in energy-intensive industries, like the chemical industry, with a useful parameter to forecast the evolution of the energy market and plan their investment in energy facilities. A variation of this approach uses probabilistic models, such as the stochastic cellular automata model of diffusion, and employs computer simulation to generate a range of outcomes and the associated probability distribution [30].

5.2.1.3 Delphi

Delphi is a method designed to obtain the opinion of a panel of experts on a particular subject, which has proven popular among technology forecasters.[6] Originally conceived to benefit from the positive effects of groups while reducing their inconveniences, Delphi significantly differs from face-to-face group interactions. In particular: all group interactions take place anonymously through the use of questionnaires; the content of the feedback is controlled by a moderator; and the response of the group is presented in statistical form. Examples of applications of the Delphi method can be found in the forecasting literature [34].

5.2.1.4 Extrapolations

Extrapolations involve using a model to fit historical data of a particular parameter, for instance a performance characteristic of a certain technology, in order to predict future values of this parameter. A fundamental assumption of extrapolations is that series from the past contain sufficient information to derive projections for the future. Accordingly, forecasters start by identifying a pattern in past data (i.e., the appropriate model), which is then extended to the future to infer a forecast. The most common models used by technological forecasters are growth curves – the well-known S-curve – especially the logistic and the Gompertz curves. In view of the popularity of growth curves in technological forecasting, two caveats are worth mentioning. First, when searching for the appropriate model, it is more important to select a model that

6 Please note that the Delphi method can be used in other contexts, such as strategic analysis.

adequately represents the process underlying the data than a model that provides the optimal fit with the historical data. As Martino (2003, p. 728) remarks: "a good forecasting model is one that will fit the future data." Second, growth curves perform rather badly for predicting the upper limit of the parameter they describe – representing, for example, the maximum theoretical performance of a technology – based on data from the early portion of the curve. Hence, the low end region of the curve is more appropriate for forecasting [30].

The main benefit of technological forecasting, in general, refers to the disclosure of emerging technological alternatives, their characteristics, and their potential impact. Forecasting supports the mapping of relevant knowledge gaps and technological challenges which, in turn, supply R&D programs with additional insights. Similarly, technological forecasting supports firms in predicting the future performance level of a technology – from a competitors' or a customers' view – anticipating alternative technical approaches that are capable of achieving a given performance level, and sensing signals that indicate the end-of-life of a certain technology (e.g., caused by the emergence of substitutes).

Apart from these benefits, technological forecasting is not without limitations. For instance, techniques for technological forecasting might not be suitable for anticipating major scientific discoveries. This is because it is limited when it comes to predicting interactions between different technologies, between technological developments and demand. Furthermore, technological forecasting is somewhat ill-equipped for assessing self-amplifying effects of certain radical technological changes on demand [27]. As a result, technological forecasting is more suitable as a source of incremental innovations.

5.3 Organizing for Innovation

In the first section of this chapter, we highlighted that innovation is not only an outcome but also a process. But what is the foundation for innovation? As stated by Van de Ven and Poole (1989) [6: 32]: "people [...] develop and implement their innovation ideas *within an institutional context*."[7] Managing innovations requires institutionalizing innovation activities, that is, to establish a formal organizational structure that clearly identifies a competent authority. In this section, we address the question of how to structure innovation activities within the firm, and at the boundary between the firm and its environment.

5.3.1 The Innovation System

When considering the organization of internal innovations activities, it is essential to note that there cannot be anything such as a recipe about the ideal structure to manage innovations. Far more, organizing for innovation has less

---

7 See also Section 5.1.2. Italics added by the present authors.

to do with control and regulation than with managing relationships between different entities: people, organizational units, and even machines. In this perspective, organizing means designing the "innovation system" of the firm, namely, taking actions in order to develop a coherent interplay among all stakeholders that are potentially involved in the firm's innovation activities [5]. Compared with the organization of traditional routine tasks (e.g., production, accounting), it is crucial that the innovation system allows a high degree of self-organization. Hence, new relationships between innovation players may emerge and stakeholders organize spontaneously. Such a perspective is more appropriate to the entrepreneurial nature of innovation than the sole process control approach. The different components of the innovation system are shown in Figure 5.4. We have left aside the management of single innovation projects as well as the issue of developing an innovation culture in order to focus on the institution of a specialized innovation function.

As soon as innovations gain in importance, firms start to organize an internal innovation function. Following the logic of the division of labor, they tend to transfer the responsibility of innovation activities to a specialized unit. In this manner, firms aim at capitalizing on the benefits of specialization, such as concentration of competences, or increase in skill level and work efficiency. The risk here is to concentrate all responsibilities on one person, for example, by dedicating a specific innovation position reporting directly to the board of directors or a business-line head, who would then be in charge of steering every innovation project in a certain area. This approach is unlikely to succeed, because one person is unable to handle such a workload on his or her own. Moreover, such an innovation function would never be empowered to overcome the inevitable barriers encountered when innovating [35]. However, these remarks do not diminish the need for a clearly identified organizational

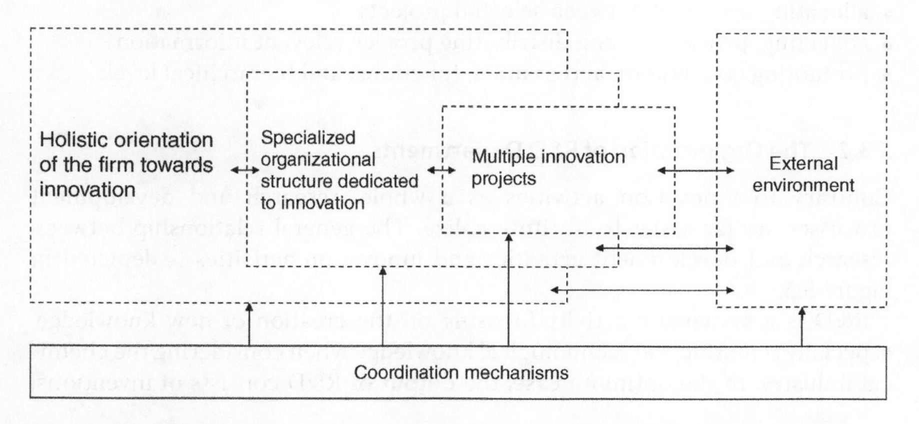

Figure 5.4 Elements of the innovation system [5]. Adapted from: Hauschildt J and Salomo S. 2011. *Innovationsmanagement*. 5th edn. Vahlen: Munich

unit that is in charge of innovation activities. O'Connor (2008) [36] even suggests that such an organizational structure is a key requirement for firms aiming at developing a "major innovation dynamic capability"; that is, a meta-capability allowing them to develop really new and radical innovations.[8]

Besides, this need to specialize in multiple innovation projects does not contradict the idea of a holistic orientation of the firm towards innovation as prescribed in the concept of innovation system (see Figure 5.4). Indeed, the primary role of a dedicated entity responsible for the management of innovation is to align and coordinate innovative efforts towards the achievement of the firm's objectives. While providing coherence and orientation to innovation activities, such an entity should allow innovation to take place outside its boundaries, that is, anywhere else in the firm, and even include external initiatives [5].

A final element to take into consideration is the central role of the project-based organization in the management of single innovation processes [5]. Innovations are particular endeavors characterized by the unique conditions under which they take place: they unfold over a limited timeframe – even if it is not always possible to anticipate its exact duration – and fulfill a clearly defined purpose. Thus, projects are particularly suitable organizational forms to manage the development of single innovations. However, when firms seek to institutionalize innovation activities, they consider innovation as a permanent activity. This implies continuously pursuing new projects, sometimes even several in parallel. Therefore, the main responsibility of a central innovation unit should consist of managing the innovation project-portfolio. According to Hauschildt and Salomo (2011) [5], this multi-project management unit should have responsibility for the following tasks:

- selecting relevant projects and supporting the creation of new projects in line with the firm's strategic goals
- allocating resources between selected projects
- collecting, processing, and distributing project-relevant information
- promoting cooperation across units, functions, and hierarchical levels.

5.3.2 The Organization of R&D Departments

Contrary to innovation activities as a whole, research and development processes are far easier to institutionalize. The general relationship between research and development activities and innovation activities is depicted in Figure 5.5.

R&D is a systematic activity focusing on the creation of new knowledge, especially scientific and technological knowledge when considering the chemical industry. In the optimum case, the output of R&D consists of inventions,

8 See Chapter 3 for an extended definition of dynamic capabilities.

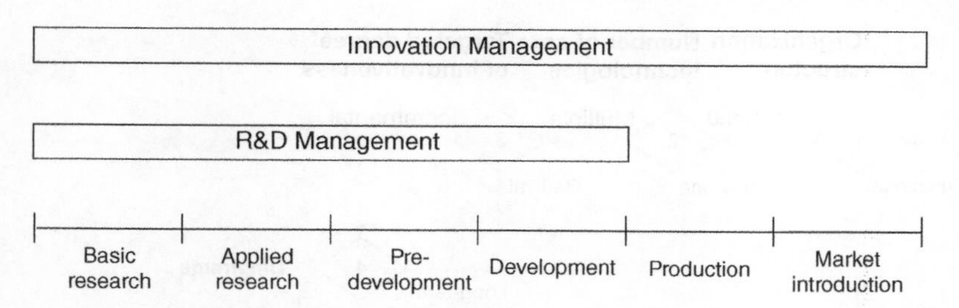

Figure 5.5 Innovation versus research and development [37]. Reproduced with permission of Vahs and Burmester (2005)

whereas the innovation process delivers a marketed product or process. Further, innovations are one-off processes, in contrast to R&D, which involves many repeatable procedures. As a result, the delineation of R&D activities in terms of time and content is much clearer and can be better developed into a routine. It is then possible to reap the full benefit of specialization, which justifies the organization of R&D into specific departments. As for the real problem of organizing R&D activities, whether the organization of R&D results in a centralized or a decentralized structure is a central issue.

The question of centralization, in fact, involves two aspects: first, the extent to which R&D activities should be integrated in one location or distributed across different entities; second, at which organizational level should the responsibility for these activities rest. Thus, a centralized R&D organization is usually related to a unique structure reporting directly to the board. A decentralized R&D organization refers to several entities distributed among lower organizational levels. In between, there exists a multitude of possible hybrid structures. For instance, one option consists in centralizing all fundamental and applied research activities at a corporate level, while development work remains under the responsibility of the business units.

In order to guide decisions in favor of one or the other types of structure, Hauschildt (1997) [35] developed a decision model that is presented in Figure 5.6.

The following factors determine to what extent R&D should be organized as a centralized rather than a decentralized function:

1) **Overall structure of the firm:** The stronger the orientation towards structures relying on stand-alone profit centers (i.e., towards a divisional structure), the higher the tendency towards decentralization. Obviously, firm size plays a significant role. Smaller firms have a more limited product and customer portfolio and tend to have a functional organization[9] with a centralized R&D.

9 Functional organization: the structure of an organization is divided into several departments each taking care of just one task, for example, sales, customer, R&D, administration.

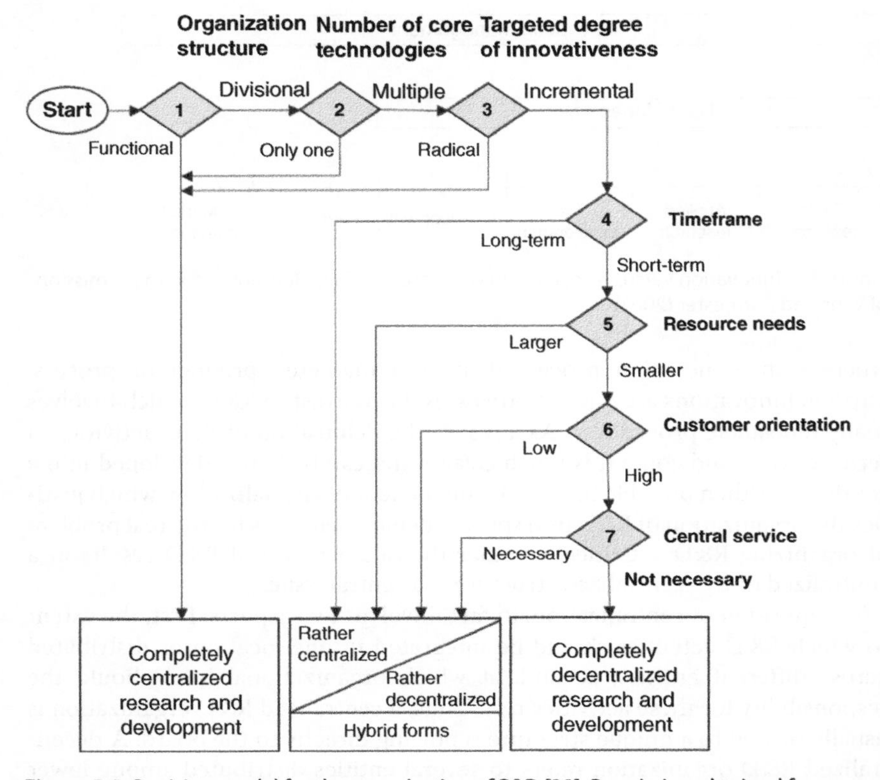

Figure 5.6 Decision model for the organization of R&D [38]. Permission obtained from Salomo S. 2016. *Innovationsmanagement*. 6th edn. Vahlen: Munich, p. 120 (figure 5.2)

2) **Number of core technologies:** Firms pursuing R&D activities in multiple technological fields should favor decentralization. The precedent remark about firm size applies equally here since small firms are also likely to focus on one core technology, and thus favor a centralized R&D organization.
3) **Targeted degree of innovativeness:** When developing more radical innovations, a centralized R&D structure has many advantages. In terms of resources, for example, their concentration in one location allows both the high risk of failure and the important financial commitment associated with radical innovations to be coped with; also in terms of attention and visibility, as radical projects needs more management support to overcome barriers than incremental ones.
4) **Timeframe:** This factor considers the length of the R&D activities up to successful market introduction. The longer the expected timeframe, the stronger the recommendation to transfer R&D activities to a centralized unit.
5) **Resource need:** The larger the necessary "critical mass" with regards to qualified personnel, equipment, and financial resources, the stronger the tendency towards centralization.

6) **Customer orientation:** When research and development occurs in reaction to market demand or in cooperation with customers, decentralization is favored. A diversified customer portfolio with many different needs strengthens this tendency.
7) **Need for a central service unit:** It is more cost-efficient to centralize administrative services, such as information services (patent monitoring, archives, documentation, etc.). This is in fact true for all types of services on which R&D employees systematically rely, but which do not belong to their core competences.

It is important to note that even if current trends in the chemical industry (e.g., shorter innovation cycles, increased cooperation with customers) seem to favor decentralized structures, a complete decentralization of R&D seldom occurs in practice. Hybrid structures consisting of both centralized and decentralized units are quite common, in particular in large chemical companies. In this regard, Bröring and Herzog (2008) [39] investigated the organization of innovation activities at Degussa (now Evonik Industries). They identified four types of a more or less centralized structure:

- **Traditional R&D:** Most R&D is carried out by R&D groups within operational business units. These activities focus on short times-to-market and draw on existing competences.
- **Corporate-funded projects:** For slightly more innovative initiatives which are required to advance the technological knowledge of the business units but still with a short-term commercialization goal; corporate funded projects present an alternative solution. These projects take place within the business units for two years with funds from both the business unit and a central corporate R&D entity.
- **Project houses:** The goal of these hybrid structures is to develop new technology platforms. It combines the competencies of employees from several business units that are working in a separated unit for around three years, outside existing structures. Like corporate-funded projects, project houses are financed in equal proportion by the business units involved and corporate R&D. However, they report essentially to the central R&D unit.
- **Science-to-business centers:** These structures extend the concept of a project house to address emerging markets and technologies with a long-term horizon. They are located at the central R&D facility and rely on corporate as well as public funding.

5.3.3 Closed and Open Innovation

Throughout the twentieth century, the logic underlying the organization of the innovation function in chemical firms, as in most industrial companies, has been almost exclusively internally oriented. Successful firms like Dupont or Edison's General Electric have built their fortune using the same formula: they relied on large centralized R&D centers to generate ideas and develop new products, and then manufactured, sold, distributed, and serviced them, all by themselves. This approach is what Chesbrough (2006) [40] called the "Closed Innovation" approach. It assumes that successful innovation requires control. In this logic, the innovation process takes place strictly within the firm's boundaries, using the firm's own resources and competences to nurture innovations stepwise from idea to market. This model allows firms to capture the sales revenue from breakthrough discoveries and, thus, to perpetuate the cycle of R&D investments, innovations, profits. Consequently, attracting key scientists, protecting IP, and avoiding knowledge spillovers outside the firm are essential for preventing others from benefiting from these critical R&D assets. However, according to Chesbrough (2006) [41], this approach no longer works efficiently. A paradigm shift is at work, which redefines how industrial companies deal with knowledge and engage in innovation activities.

The new paradigm is called "Open Innovation." It draws on the idea that firms should drive their innovation process so as to benefit equally from idea sources and commercial channels located outside the firm. Hence, those following the Open Innovation logic use both internally and externally developed ideas to sustain their flow of innovation activities. Conversely, they rely on internal and external market pathways to commercialize inventions. Open Innovation discards the old view that conceived of the firm's outer limits as a guarded containment wall. Instead, corporate boundaries are now similar to a porous interface through which knowledge can freely flow. At any time during the innovation process, external ideas, technologies, and even ready-to-market concepts can integrate the firm's activities. Similarly, at any stage of development, internally initiated projects may be licensed, divested or spun off to benefit from external commercialization channels and reach additional markets. In summary, Open Innovation is "the use of purposive inflows and outflows of knowledge to accelerate internal innovation, and expand the markets for external use of innovation, respectively" [41]. Figure 5.7 provides an overview of some of the key assumptions underpinning the Closed and Open Innovation paradigms.

Even though the chemical industry had already been using some elements of Open Innovation – partnering with universities to identify or develop new molecules, for example – long before the concept gained importance among management scholars, you may still wonder: What makes the Closed Innovation logic obsolete? Why follow an Open Innovation approach? Beyond the promises of reduced development time and cost, Herzog and Leker (2010) [43] identified six reasons in the literature supporting the shift to an Open Innovation process:

- Research always becomes more resource-intensive. Technology development steadily gains in complexity and single firms are less capable or willing to support the resulting risks alone.
- Outsourcing of R&D tasks is common practice. The market for innovative technology suppliers grows.

Closed Innovation logic	Open Innovation logic
We have the best people in our field working for us.	There are also brilliant people outside our company. We need to work with both in-house and external specialists.
We profit from R&D only if we carry it out all by ourselves.	External and internal R&D are complementary and contribute both to create and capture value.
The first to invent is the first on the market.	We can profit from R&D we did not initiate.
We need to generate the most and the best ideas to win.	We need to rely on external and internal ideas to win.
Our competitors should not benefit from our ideas; IP protection is crucial.	As long as we can profit from it, we should let others use our IP, and buy IP from others.

Figure 5.7 Close versus Open Innovation logic [42]. Adapted from Chesborough, 2003

- Highly-qualified workers are increasingly mobile. Simultaneously, higher education around the world expands and always brings more skilled personnel to the job market.
- The growing presence of innovation intermediaries (e.g., yet2.com, InnoCentive), acting as technology brokers between innovation partners, facilitates inter-organizational exchanges.
- Venture capital is more readily accessible. Therefore, individual inventors are more inclined to establish their own start-up instead of joining traditional R&D organizations.
- Industry convergence blurs the limits between previously unrelated areas. Converging value-propositions, technologies, and markets creates new inter-industry segments (e.g., nutraceuticals or functional foods) and forces firms to seek support from other industries [43].

As mentioned earlier, the Open Innovation concept relies on two fundamental principles. On the one side firms should use external sources to advance their innovation projects. On the other side they should consider external commercialization pathways as alternatives to their own market channels. Thus, the main Open Innovation activities include inbound, outbound, and coupled activities. Figure 5.8 gives a graphical representation of these different activities in the innovation process. Next we will present some examples for each type of activity.

Inbound, or inside-in activities, relate to "the ability to gain and explore knowledge from external partners" [45: 1237]. Potential external sources are manifold and may involve users, suppliers, competitors, start-ups, government agencies, universities, consultancy companies or research institutes. The range

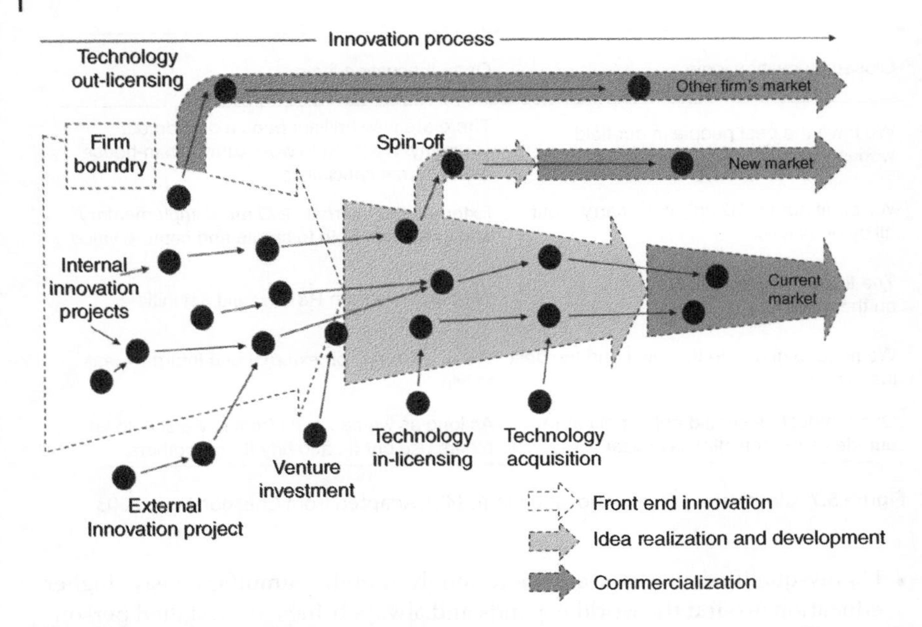

Figure 5.8 Open Innovation model [44]. Reproduced with permission of Herzog, 2011

of activities is also broad, from the involvement of lead-users to acquisitions, through to idea crowdsourcing and technology scouts. We will leave aside the classical R&D cooperation, which is widespread in the chemical industry, and focus on two more representative inbound activities:

- **In-licensing:** A licensing contract is a legal arrangement allowing a firm to exploit another firm's intellectual property for a certain period of time and under specific conditions. Thus, the licensor (seller) grants a license to a licensee (buyer) in exchange for payment of a fee, usually consisting of a one-off upfront payment in addition to yearly royalties. License terms are basically open to negotiation and vary greatly. In the case of technologies, know-how and prototypes may be transferred along with intellectual property. Typical licensing terms specify the applications and markets where the licensed intellectual property may be used, define commercialization milestones that the licensee must achieve to retain its right, and may grant to the licensee an exclusive right to improve the technology. For the chemical industry, licenses are particularly relevant for accelerating technology access [43].
- **Corporate venture capital (CVC):** This approach tries to secure minority equity stakes in innovative start-ups that are not yet publicly traded, but are seeking capital to pursue their growth. To that end, established firms create dedicated venture capital units, which are assigned the task of identifying

and investing in promising technology start-ups within a given budget. A central advantage of CVC investments is their reversibility [43]. When dealing with emerging technologies, it is extremely difficult to anticipate whether it will turn mainstream or not. If the technology does not develop as expected, the minority stake can be easily sold. Hence, CVC allows a firm to learn about emerging technological fields while minimizing risks. Further, in the case where technological developments are successful, the investing firm has the opportunity to increase its investment and gain returns from commercialization.

Outbound, or outside-in activities, refer to "activities involved in external exploitation of internal ideas, for example by licensing out, selling of knowledge, and divestment of parts of the firm, such as spinning off innovation projects into new create innovative firms" [45:1237]. In the following, we address out-licensing and spin-offs in more detail:

- **Out-licensing:** In most technology-intensive industries like the chemical industry, patenting inventive research results is systematic. However, firms often develop and patent technologies that are never commercialized for a variety of reasons. Out-licensing is an option to generate additional revenues from unused intellectual property. Herzog and Leker (2010) [43] mention that chemical firms may earn up to 10% of net operating income from out-licensing activities. However, motivations to license out a technology should not be limited to disinvesting non-core intellectual property or simply reaction to external requests; it may be also part of a broader marketing strategy aiming, for example, at establishing a technology as the industry standard.
- **Spin-offs:** Contrary to the divestment of complete business units or business lines which actually obey general strategic considerations, technology spin-offs are typically motivated by technological reasons [43]. When a technology does not fit into the firm's portfolio because it does not relate to existing operations and has low strategic priority, the team developing that technology may not be able to access sufficient resources to proceed to commercialization. Similarly, expensive and risky emerging technologies may have a hard time finding support within the mainstream business. Thus, creating a legally independent structure to commercialize these types of research outcomes can prove a viable solution. The parent company may retain a minority or a majority share, depending on the strategic role of the venture.

Coupled activities combine inbound and outbound activities. They encompass "collaborative activities between different actors in the innovation," including "co-creation with complementary partners through alliances, cooperation, and joint ventures" [45:1237]. In this way, firms collaborate in order to both develop and commercialize innovations. For instance, it can take the form of an R&D collaboration, resulting later in a joint commercialization of the outcomes. This is exactly what happened between Evonik Industries and KraussMaffei when they developed the CoverForm technology. Combining Evonik's polyacrylate chemistry knowledge and KraussMaffei's know-how in designing injection molding processes, the two firms invented a new one-step, scratch-resistant coating technology. Marketed under a common brand, CoverForm, it consists of a special Plexiglas formulation supplied exclusively by Evonik in association with KraussMaffei's equipment [46].

To conclude this overview of the Open Innovation approach, we wish to echo the view held by management researchers on the need to combine all three types of activities. This is all the more important since Open Innovation activities have been proven to significantly impact the performance of a firm, especially regarding innovativeness and financial performance [45].

5.4 Managing the Innovation Process: Stage-Gate®

In many industries the Stage-Gate® concept of Cooper is well established for new product development. The reason is that the development of new products will be done in a structured way, the requirements are transparent, and the risk can be minimized. Stage-Gate focuses on efficiency and differs from project management: it is a meta-process to ensure that bad projects are killed off and good ones fueled with resources. The process is composed of "stages," corresponding to a specific set of activities, and "gates," where decisions are made on whether to continue or stop the project. Often the "classical" model of the Stage-Gate process will be adopted for different businesses according to the specific needs and requirements for new product development. In our example we are going to describe a new product development process for additives used in various industries, particularly in the paint and plastics industry. The following description will focus on the practical experiences of a Stage-Gate process for the development of a new product to be used as an additive in the chemical industry. The different stages and gates are illustrated in Figure 5.9.

The new product development process comprises five stages and five corresponding gates. In the case presented here, the process starts with the "Ideas Management" phase, followed by the "Feasibility" phase. Gate 3 is the decision

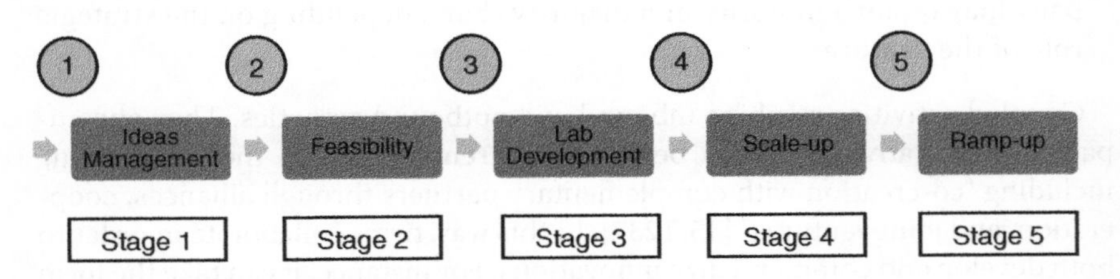

Figure 5.9 Overview of the Stage-Gate process

point to start the lab-work (Stage 3); after having developed a suitable product on a lab scale, the process for manufacturing has to be up-scaled to production and finally the product has to be launched into the market. The commercial development will be monitored within the "Ramp-up" phase. However, depending on the overall risk of the project, a different set of stages and gates could be applied. For example, in case of low-risk projects, a light version with only two stages can be used. Therefore, the design of the Stage-Gate can be implemented in a flexible way, according to the project characteristics.

5.4.1 Stage 1 "Ideas Management"

Within the "Ideas Management" stage, all product-related ideas are collected on a global basis. Employees of the company are invited to bring up new ideas regarding new product development, new technologies, new applications, and new market opportunities. All incoming ideas are stored and handled in a database based on Microsoft SharePoint technology. The owner of the idea could either send the idea description via e-mail to a central Ideas Management Office or dispatch the idea directly into the system. It is very important to note that the use of an extended form sheet has proven to be unsuccessful compared with sending the new idea via e-mail. People do not typically have all the information available at this moment to fill out such form sheets. In practice, they will not dispatch the form sheet and the idea would be lost to the company. Instead of filling out form sheets, sending an e-mail with the new product idea is well accepted and can be processed.

In a first step the incoming new idea is reviewed according to the following questions: Is it a new product idea? Is the idea clear and comprehensible? Are there existing ideas on the same subject already? If the idea is not clear, then the idea owner is asked to submit more detailed information. If similar ideas happen to already be in the market place, the newly submitted idea will be linked to the existing idea(s). Additionally, attributes will be attached to the new idea (e.g., country of origin, application areas, etc.) for further evaluations.

At Gate 1, all incoming ideas are evaluated by marketing as well as by research and development. It was decided to use the feedback from the lab managers rather than middle- or top-management in order to ensure that the experience and practical knowledge from the experts is taken into account. The Ideas Management Platform on SharePoint covers not only the administration of the ideas, but it also includes a task management system. This combination is absolutely essential, to avoid only collecting information on new product ideas.

In addition to the platform the operation of a central Ideas Management Office has proven to be of great value. In order to promote ideas to the next stage, idea owners very often need support from an expert group. Conference calls and meetings are organized by the Ideas Management Office to discuss "face-to-face" the new ideas and to encourage the participants to initiate further actions.

At the end of this process a decision has to be made: to take no further actions or to start a pre-project or a project. This decision is made basically on two criteria: market attractiveness and technology fit. The input comes from the marketing organization and from research and development. At this stage, a qualitative input is sufficient. New projects are not just initiated by single ideas. Similar inputs are often clustered according to their application area and give rise to unexpected new projects. The Ideas Management Office coordinates all activities and keeps records of the final decision in the Ideas Management Platform. To ensure a high acceptance within the innovation community, it is very important that all steps and decisions are transparent and visible, especially to the idea owner.

5.4.2 Stage 2 "Feasibility"

After having decided to progress with the new product idea the "Feasibility" stage is initiated. To this end, a formal request has to be sent to the Innovation Management Office, containing the following items: the target of the feasibility study, the timeframe, and the estimated budget expressed in man-hours. In general, the time frame should not exceed a period of 6 months and the budget should be limited to 300 hours. A request coming from the marketing organization is reviewed by the corresponding Business Manager; in the case of a technology pre-project, the project is reviewed by the Chief Technology Officer (CTO). The process is accompanied by a formal approval including a signature (a requirement for ISO (International Organization for Standardization) certification). After assigning a head for the pre-project, the pre-project head nominates several team members who are crucial for accomplishing the project goals. The pre-project will be set-up in the Project SharePoint Platform. All activities are coordinated by the Innovation Management Office.

Within the "Feasibility" stage, two main topics are central: firstly, lab experiments should exhibit the basic possibility of synthesizing new products with the desired properties; secondly, market opportunities should be detailed by analyzing the given market and contacting potential customers. All information is collected and documented in the Project SharePoint Platform. At this point, the chemistry of the new product is not yet decided; therefore, potential issues regarding regulatory compliance (REACH) are not addressed. The final target of this stage is to fill out the Project Application Form ("PAF"). The PAF consists of several chapters, including marketing, technical, and strategic criteria. In addition, environmental aspects have to be indicated (requirements for classification, emissions, use of renewable raw materials, etc.). Further, PAF covers all key characteristics of the project (budget, timeline, project team, risk attributes).

After accomplishing the PAF, the pre-project head can apply for a presentation to the Steering Committee for New Product Development (Gate 2).

This Committee meets every month and consists of the following members: Board of Management, Business Line Managers, Head of Marketing, Chief Technology Officer, and Head of Innovation Management. Since it is difficult to pre-determine the exact timing and the budget, a target corridor for decision criteria is used by the committee to evaluate the projects.

Once the application for the new project has been presented and discussed, the committee makes a decision according to the following criteria: market attractiveness, technology fit, and strategic fit. The result can be a rejection, resubmission or approval. All types of decisions are explained to the presenter and documented in the Project SharePoint Platform. A formal set of documents have to be signed (requirement for ISO certification).

5.4.3 Stage 3 "Lab Development"

The "Lab Development" stage could be considered as the "heart" of the New Product Development. Within this phase the new substances and/or the new formulations are developed. The "PAF" describes all required product criteria in detail, which indicate the targets for the development of the new product and, thus, represent an important input for the scientist in the lab.

Based on experience, literature, and patents, scientists begin to synthesize the new product on a typical lab scale (approximately 100–200 g). At the beginning, they search for the right chemistry by selecting appropriate substances and formulate them in suitable solvent(s). These products are tested in the application laboratories according to characteristics and how appropriate they are. During all steps, it is crucial that scientists and technicians from the application lab work closely together. Having selected the appropriate chemistry, the project team starts to optimize the chemical structure and the formulation.

Apart from synthesizing the new product, the following aspects have to be considered during the "Lab Development" stage: (i) availability of the raw materials, (ii) production procedure, (iii) storage stability of the product itself as well as in the final application (e.g., paint system), and (iv) the commercial situation. Special attention has to been given to the patent situation immediately after initiating the project. Here, two aspects have to be considered: Can we patent it or do we infringe a patent ("Freedom-to-operate"). It is of strategic interest to file a patent as soon as possible to protect the business opportunity.

The progress of the project is monitored quarterly by means of a short review meeting. The current status of the project is discussed and reviewed according to the original objectives (timeline, budget, product requirements). Depending on the extent to which the current progress deviates from the targets, the status is changed from "Green" to "Yellow" or to "Red." The "Red" status implies that a presentation at the next Steering Committee has to be given in order to discuss the situation and decide on the next steps. These actions could include expansion of the manpower, adjustment of targets, or termination of the project. All decisions and background information are documented in the Project SharePoint Platform.

At the "present stage," scientists have their final opportunity to make any adaptions to the underlying chemistry since at the end of this stage the product concept will be frozen. This is the basis for further evaluations, that is, the selection of appropriate production sites, commercial calculations, and decisions on specification of the core product and raw materials. At this point in time, the patent situation also has to have been clarified.

At the end of this stage the decision is made to pass over the project to phase 4 "Scale-up" within a formal meeting (Gate 3). Gate Keepers are the corresponding Business Line Manager or the Chief Technology Officer for technology projects, respectively. The decision is made on detailed information provided by the project team: cost analysis (raw materials, manufacturing, legal aspects), patent situation (freedom-to-operate, own patent protection), regulatory affairs (required efforts for registrations and notifications), production site (eventual investment for new equipment), and review of business plan (market potential, competitive situation). This decision is the most critical of the entire stage process as it implies significant financial commitment. It is expected that the project would not be discontinued after this point. Only the most attractive and promising projects should pass the gate. This is particularly true for the chemical industry because it is very capital intensive. All activities are monitored by the Innovation Management Office and documented in the Project SharePoint Platform.

5.4.4 Stage 4 "Scale-up"

While the previous stage implied the development of a new product in small quantities (around 500 mL), the major task of the "Scale-up" stage is to scale-up the development/production process to technical quantities of up to several metric tons. Further, all other requirements that are necessary to launch the new product into the market have to be fulfilled, such as regulatory compliance in all relevant markets, materials management issues, and the strategy for market introduction.

To obtain precise process safety data, all new products and technologies have to pass the so-called "Mini Plant." This consists of 2 or 6 L glass reactors that are fully equipped with condensers and filling systems and that allow the reaction to run automatically. All necessary reaction steps are programmed and are executed and monitored by a computer system. These experiments ensure highly reproducible results under strictly controlled conditions. Furthermore, this system allows measurement of the energy that is consumed or released ("exothermic reactions"). The latter value is of greatest interest to evaluate the safety conditions for the process in larger quantities. Only reactions that have a maximum energy for the exothermic reaction far below the cooling capacity of the reactor will be approved for transfer to production sites; otherwise process optimization is initiated. All data are documented in a comprehensive safety report.

Next, scale-up step reactions are executed in a pilot plant, which consists of several reactor systems of up to 120 L. Further process optimization is done without changing the performance properties of the product. Here, close cooperation between pilot plant, synthesis department, and the application testing lab is necessary. Samples are usually sent from the pilot plan batches to selected customers. Feedback from customers is very important at this stage, because any deviations in quality have to be detected during the scale-up process.

Finally, the new product is transferred to production scale. The reactor size is typically between 1 and 10 m^3. Samples from the production batches are carefully analyzed and tested in the application laboratories. It is essential to ensure that the performance of the new product coming from the production batches complies with the characteristics of the original lab product.

Over the last decade, regulatory compliance has also become an important issue during the "Scale-up" stage. Especially when operating on a global basis, the new product must be registered or notified in all relevant countries of commercial interest. In Europe, the substances of the formulation have to comply with REACH (Registration, Evaluation, Authorization of Chemicals); in the United States the product has to be registered according TSCA (Toxic Substance Control Act), and in Japan the product must comply with the MITI regulations (Ministry of International Trade and Industry) (to mention just a few). If the product is supposed to fulfill a specific field of application, additional (specialized) compliance checks are required. For instance, all products that are in any contact with food have to be approved by several legislations (PIM in Europe, FDA in the United States). Testing for registrations or notifications can take several months and can cost several €100 000. For example, the expenditure for REACH testing for a substance greater than 100 tons per year is roughly in the range of €400 000–€500 000 and will take about 18 months.

A successful introduction of the new product into the market requires an adequate marketing plan (target segment, pricing, promotion, distribution). Typically, the new product is presented to the sales forces during global conferences or via Web-Ex meetings. The project is closed after a successful market introduction and the first production batches (Gate 4). The results are documented by the project team and stored in the Project SharePoint system.

5.4.5 Stage 5 "Ramp-up"

After having introduced the new product into the market, the response from the customers and the commercial development is monitored by the Innovation Management Office. The feedback from the customers is collected and evaluated against the original Project Application Form (PAF). As the customer decides whether the new product becomes an innovation or not, it is crucial to include the customers' views during the whole process. Monitoring sales figures represents the basis for evaluating the commercial success of the new product. In so doing, firms can adjust their value offering by means of early modifications to the existing product or by initiating new product developments.

During the ramp-up phase a post-project review session is organized. Here, the overall project performance is discussed and improvements are proposed. The commercial development is monitored over a period of 5 years.

5.5 Summary

- **Innovation starts with an invention, but to cover the full panoply of innovation the invention must be developed, brought to market, and finally successfully accepted by the customer.**
- **The key dimensions of innovation are: (1) temporality, (2) content, (3) subjectivity, (4) intensity, and (5) normativity.** Innovation in a company should be seen as an iterative process, from the invention to the successful commercialization. The most common forms of innovation within the chemical industry comprise product and process innovations, whereby the novelty of innovations needs to be segmented from either a company's view, customer perspective, industry view, or by a firm's executive perspective. Innovations can be clustered into incremental, really new, and radical innovations, but all of these innovations need to be commercialized successfully and should improve a specific situation for the customer.
- **Sources of innovation can be classified into practice and theory by technology-push, demand-pull, or from a functional perspective.** Internal innovation activities remain a key driver for new products and processes, while environmental scanning, causal models, the Delphi method, and extrapolations constitute technological forecasting methods to disclose emerging technological alternatives, their characteristics, and their potential impact, not just on the company.
- **Innovation within a company can be managed.** By institutionalizing innovation activities, innovation becomes a permanent activity pursued by the whole organization. A dedicated entity is responsible for aligning innovation efforts with corporate strategy and needs to decide either to pursue centralized or decentralized R&D attempts. In addition, an open innovation framework might enhance a company's innovation capabilities.
- **A common tool to manage innovation is the Stage-Gate process.** The Stage-Gate process consists of five stages and five gates. In the first stage,

"Idea Management," all ideas are collected and evaluated by marketing and by research and development. The most attractive ideas progress to the second stage, "Feasibility," in which lab experiments are conducted to synthesize the product and market opportunities should be developed. In the next stage, "Lab Development," the new substances are developed and optimized, while the availability of raw materials, the production procedure, storage stability, and the commercial situation also need to be analyzed. Scaling-up the development/production process to technical quantities is part of Stage 4, "Scale-up," before feedback from customers is collected and evaluated in Stage 5, "Ramp-up."

References

1 Hicks J. 2010. The pursuit of sweet: A history of saccharin. *Chemical Heritage Magazine*, **28**(1): 1–3. http://www.chemheritage.org/discover/media/magazine/articles/28-1-the-pursuit-of-sweet.asp (accessed 16 June 2015).

2 Fahlberg C and Remsen I. 1879. Ueber die Oxydation des Orthotoluolsulfamids. *Berichte der deutschen chemischen Gesellschaft*, **12**(1): 469–473.

3 Calantone RJ, Harmancioglu N, and Droge C. 2010. Inconclusive innovation "returns": A meta-analysis of research on innovation in new product development. *Journal of Product Innovation Management*, **27**(7): 1065–1081.

4 Bröring S. 2005. *The Front End of Innovation in Converging Industries*. Springer Verlag: Berlin.

5 Hauschildt J and Salomo S. 2011. *Innovationsmanagement*. 5th edn. Vahlen: Munich.

6 Van de Ven AH and Poole MS. 1989. *Methods for Studying Innovation Processes*. Research on the management of innovation: The Minnesota studies. University Press: Oxford.

7 Song XM and Montoya-Weiss MM. 1998. Critical development activities for really new versus incremental products. *Journal of Product Innovation Management*, **15**(2): 124–135.

8 Tidd J and Bessant J. 2009. *Managing Innovation: Integrating Technological, Market and Organizational Change*. 4th edn. John Wiley and Sons, Ltd: Chicester.

9 Garcia R and Calantone R. 2002. A critical look at technological innovation typology and innovativeness terminology: A literature review. *Journal of Product Innovation Management*, **19**(2): 110–132.

10 Utterback JM and Abernathy WJ. 1975. A dynamic model of process and product innovation. *Omega*, **3**(6): 639–656.

11 Hutcheson P, Pearson AW, and Ball DF. 1995. Innovation in process plant: A case study of ethylene. *Journal of Product Innovation Management*, **12**(5): 415–430.

12 Guile BR and Brooks H. 1987. *Technology and Global Industry: Companies and Nations in the World Economy*. National Academies Press: Washington, DC.
13 Miremadi M, Musso C, and Oxgaard J. 2013. *Chemical Innovation: An Investment for the Ages*. http://www.mckinsey.com/industries/chemicals/our-insights/chemical-innovation-an-investment-for-the-ages (accessed 16 June 2015).
14 McCarthy IP, Tsinopoulos C, Allen P, and Rose-Anderssen C. 2006. New product development as a complex adaptive system of decisions. *Journal of Product Innovation Management*, **23**(5): 437–456.
15 Seidel VP. 2007. Concept shifting and the radical product development process. *Journal of Product Innovation Management*, **24**(6): 522–533.
16 Calantone RJ, Chan K, and Cui AS. 2006. Decomposing product innovativeness and its effects on new product success. *Journal of Product Innovation Management*, **23**(5): 408–421.
17 Kock A, Gemünden HG, Salomo S, and Schultz C. 2011. The mixed blessings of technological innovativeness for the commercial success of new products. *Journal of Product Innovation Management*, **28**(s1): 28–43.
18 Nemet GF. 2009. Demand-pull, technology-push, and government-led incentives for non-incremental technical change. *Research Policy*, **38**(5): 700–709.
19 Chidamber SR and Kon HB. 1994. A research retrospective of innovation inception and success: The technology–push, demand–pull question. *International Journal of Technology Management*, **9**(1): 94–112.
20 Walsh V. 1984. Invention and innovation in the chemical industry: Demand-pull or discovery-push? *Research Policy*, **13**(4): 211–234.
21 Freeman C. 1979. The determinants of innovation: Market demand, technology, and the response to social problems. *Futures*, **11**(3): 206–215.
22 Dronsfield A. *Chemhistory: Mauveine*. http://www.rsc.org/learn-chemistry/resource/res00002119/chemhistory-mauveine (accessed 1 November 2015).
23 Travis T. 2006. A forgotten anniversary? *Education in Chemistry*, **43**(5): 128–130. http://www.rsc.org/education/eic/issues/2006Sept/ForgottenAnniversary.asp (accessed 1 November 2015).
24 Dosi G. 1982. Technological paradigms and technological trajectories: A suggested interpretation of the determinants and directions of technical change. *Research Policy*, **11**(3): 147–162.
25 Christensen C. 2013. *The Innovator's Dilemma: When New Technologies Cause Great Firms to Fail*. Harvard Business Review Press: Cambridge, MA.
26 Von Hippel E. 1988 *The Sources of Innovation*. Oxford University Press: New York.
27 Quinn JB. 1967. Technological forecasting. *Harvard Business Review*, **45**(2): 89–106.
28 Roman DD. 1970. Technological forecasting in the decision process. *Academy of Management Journal*, **13**(2): 127–138.

29 Evonik. 2011. Corporate foresight: A strategic look into the next decade. *Elements*, **37**(4): 24. http://corporate.evonik.de/sites/dc/Downloadcenter/Evonik/Global/en/Magazines/elements/elements-37.pdf (accessed 1 November 2015).

30 Martino JP. 2003. A review of selected recent advances in technological forecasting. *Technological Forecasting and Social Change*, **70**(8): 719–733.

31 Daim TU, Rueda G, Martin H, and Gerdsri P. 2006. Forecasting emerging technologies: Use of bibliometrics and patent analysis. *Technological Forecasting and Social Change*, **73**(8): 981–1012.

32 Wagner R, Preschitschek N, Passerini S, Leker J, and Winter M. 2013. Current research trends and prospects among the various materials and designs used in lithium-based batteries. *Journal of Applied Electrochemistry*, **43**(5): 481–496.

33 Sick N, Golembiewski B, and Leker J. 2013. The influence of raw material prices on renewables diffusion, *Foresight*, **15**(6): p. 477–491.

34 Martino JP. 1993. *Technological Forecasting for Decision Making*. McGraw-Hill, Inc.: New York.

35 Hauschildt J. 1997. *Innovationsmanagement*. 2nd edn. Vahlen: Munich.

36 O'Connor GC. 2008. Major innovation as a dynamic capability: A systems approach. *Journal of Product Innovation Management*, **25**(4): 313–330.

37 Vahs D and Burmester R. 2005. *Innovationsmanagement: Von der Produktidee bis zur erfolgreiche Vermarktung*. 3rd edn. Auflage. Schäffer-Poeschel, p. 50.

38 Salomo S. 2016. *Innovationsmanagement*. 6th edn. Vahlen: Munich, p. 120.

39 Bröring S and Herzog P. 2008. Organising new business development: Open innovation at Degussa. *European Journal of Innovation Management*, **11**(3): 330–348.

40 Chesbrough HW. 2006. *Open Innovation: The New Imperative for Creating and Profiting from Technology*. Harvard Business Press: Cambridge, MA.

41 Chesbrough H. 2006. Open innovation: a new paradigm for understanding industrial innovation. *Open Innovation: Researching a New Paradigm*: 1–12.

42 Chesbrough H. 2003. *Open Innovation: The New Imperative for Creating and Profiting from Technology*. Harvard Business School Press: Boston, MA, p. xxvi.

43 Herzog P and Leker J. 2010. Open and closed innovation – different innovation cultures for different strategies. *International Journal of Technology Management*, **52**(3/4): 322–343.

44 Herzog P. 2011. *Open and Closed Innovation. Different Cultures for Different Strategies*. Gabler Verlag: Wiesbaden, p. 23.

45 Cheng CC and Huizingh EK. 2014. When is open innovation beneficial? The role of strategic orientation. *Journal of Product Innovation Management*, **31**(6): 1235–1253.

46 Evonik. 2010. Press Release: CoverForm® Competence Center opened – experience innovations first-hand. http://www.coverform-info.de/#!de-downloads/ctyf (accessed 4 December 2015).

6

New Business Development – Recognizing and Establishing New Business Opportunities

Daniel Witthaut[1] *and Stephan von Delft*[2]

[1] *Evonik Industries AG, Corporate Innovation Strategy*
[2] *University of Glasgow, Adam Smith Business School*

> *I have missed more than a thousand shots in my career. I have lost almost three hundred games. Twenty-six times I have been trusted to take the game-winning shot and missed. I have failed over and over again in my life. And that is why I succeed.*
>
> Michael Jordan, former basketball player, entrepreneur, and chairman of the Charlotte Hornets

The development of a new business is an exciting entrepreneurial task. This chapter describes a model that has been proven in the chemical industry for how to recognize new business opportunities and how to develop them into growing, profitable businesses. The authors believe that it is not only the money spent on innovation that is critical for the success of one's actions but a well-structured process that gives guidance on the one side but leaves enough flexibility for creativity and entrepreneurship on the other. As we will show later, new business development in the chemical industry is a process that takes several years, often has a high rate of failure and typically requires a substantial amount of resources (e.g. cash, R&D personnel). Larger corporations therefore seize most new business opportunities in the chemical industry. In this chapter we will hence focus on how established chemical companies recognize and establish new businesses. In doing so, we do not intend to understate the role of innovative start-ups in the chemical industry, but because most new products are developed by established players and as start-ups often seek partnerships with these firms, from our point of view a focus on established firms is appropriate in the context of the chemical industry.

Business Chemistry: How to Build and Sustain Thriving Businesses in the Chemical Industry, First Edition. Edited by Jens Leker, Carsten Gelhard, and Stephan von Delft.

In this chapter you will learn about the context in which new business development occurs, about how and where to search for new business opportunities and how to implement them, and you will learn that failure is an intrinsic part of the new business development process. We will demonstrate that new business development matters but it does not happen automatically – new business development is an entrepreneurial endeavour that requires openness, creativity, analytical skills and a network with related industries.

6.1 New Business Development: Management in Unknown Areas

New business development (NBD) is a process – the search, selection, implementation and capture of a new business opportunity – that can be organized and managed, and which occurs within the context of a firm's innovation strategy and its organizational structure and culture. This context is a distinguishing factor between the creation of a new business by an independent entrepreneur and the creation of a new business inside an established organization. In this chapter, we will address how to manage new business opportunities in an established firm and how to align NBD activities with the context of the firm.

The NBD model depicted in Figure 6.1 is not only the basis for the structure of this chapter but has also proven to be a valuable guide for new business development activities in a leading specialty chemicals company. According to the model, a firm's innovation strategy and its organizational structure and culture can be understood as the framework in which NBD happens. NBD projects in two distinct companies that rely on a similar NBD process may

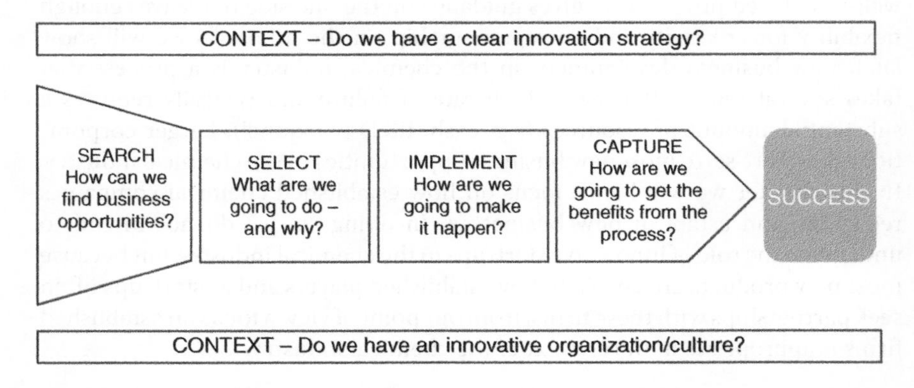

Figure 6.1 A model for new business development (NBD) [1]. *Source:* Adapted from: Tidd J and Bessant J. 2011. Reproduced with permission of John Wiley and Sons

flourish or perish depending on the context in which the projects evolve. Thus, every organization needs both an appropriate context, shaped by the leaders of the organization, and an appropriate process for NBD, executed by its managers. Indeed, developing and implementing the firm's innovation strategy as well as shaping an organizational culture in which innovation and entrepreneurial thinking can thrive are clearly a core leadership task. The first part of this chapter will therefore be about the role of innovation strategy and organizational factors, namely structure and culture, and how they affect the NBD process.

6.2 Innovation Strategy

A firm's innovation strategy is, like its operation or marketing strategy, closely linked to its business strategy,[1] and it shapes and influences the NBD process. Since innovation should be aligned with the business strategy, it needs to be derived from and serve the business strategy. In fact, innovation strategy is vital to successfully implementing business strategy. Integrating the two is therefore important. A chemical company that runs a commodity business where products are sold in large quantities could, for example, have the strategic goal of creating high cash flows. As a consequence, the firm's innovation strategy could focus on ensuring a competitive cost position and process development may be a dominant aspect in its innovation activities. In such a case the budget for NBD activities will be more restricted compared with an organization that focuses on growth. This may sound trivial to the experienced reader, however, it is crucial to clearly define the innovation strategy and make sure that it is communicated and understood throughout the whole organization. Without a clear innovation strategy, the chances of creating the right size of the NBD team, determining the necessary capital investments as well as selecting the right search area and scope of NBD projects are low.

Because the firm's innovation strategy should be linked and aligned with the choices and consequences made at the top of the firm, a close look at the organization's mission and vision and its strategic goals and objectives is a good starting point for developing the innovation strategy. The mission of the Dow Chemical Company, for instance, is "to passionately create innovation for our stakeholders at the intersection of chemistry, biology and physics" and its vision is "to be the most valuable and respected science company in the world" [2]. Dow's mission statement highlights the importance of innovation and it sets a broad framework for innovation and NBD: its innovation and NBD activities should be focused at the intersection of chemistry, biology and physics. Similarly, Dow's vision statement indicates that the company has a strong focus on R&D ("Science Company"), which shows that R&D expertise is a core factor

1 Also known as competitive strategy.

for its innovation process. Taking a closer look at Dow's strategic priorities allows further determination of what related innovation goals the company should pursue. As an example, the company aims to "launch new products in Dow Packaging & Specialty Plastics, Dow AgroSciences, Dow Electronic Materials and Dow Coating Materials, and deliver new margins in these businesses in 2014/2015" [3], which further clarifies the innovation framework, as priority areas are more clearly defined compared with the company's mission statement. The analysis of the organization's strategic priorities may thus start with broad statements like "become the market leader through rapid product introduction," complemented by an analysis of objectives such as "achieving 25% of total sales from sales of products and applications introduced in the last 5 years by 2020". Derived from such business goals and objectives, specific innovation goals and objectives can be developed. The goal of this exercise is to ensure that the innovation strategy answers the question: How do NBD activities fit into the organization's overall plan? In turn, another essential aspect of the innovation strategy is to prioritize possible new business opportunities that organizations can pursue. The better an NBD project fits into the business strategy of the organization, the more likely it is to get approval for investment in the project – something that is particularly important in chemical and pharmaceutical companies where large capital expenditures are typically required to implement new business ideas. As an example, BASF has identified a number of growth areas, such as batteries for mobility, e-power management, organic electronics and functional crop care, from broader global needs (resources; environment and climate; food and nutrition; quality of life) on a corporate level. The projects that BASF's new business subsidiary, BASF New Business, has developed, for example, high-temperature superconductivity (e-power management) and OLEDs for displays and lighting (organic electronics), fit into these growth fields. Similarly, Evonik Industries has identified a number of global megatrends, such as resource efficiency and health and nutrition. Evonik has organized its operating structure around these megatrends in the form of independent companies, such as Evonik Nutrition & Care GmbH or Evonik Resource Efficiency GmbH. Accordingly, the company invests in NBD projects that, for instance, help to utilize resources more efficiently, and its corporate venture capital activities (bundled together in a unit called Evonik Venture Capital) focuses on investments in young companies that align with Evonik's growth areas.

Without such clear growth areas, the search for specific new business opportunities is unfocused, projects would most likely remain unrelated and the firm's innovation and NBD efforts would not be effective. Companies may also further specify growth opportunities of global megatrends. As depicted in Table 6.1, the chemical company Wacker distinguishes between broader economic opportunities and more precise strategic and performance-related opportunities. Statements such as "higher plant productivity" indicate that

Table 6.1 Growth opportunities at WACKER.

Overview of Wacker's business opportunities	
Overall economic opportunities	Growth in Asia and other emerging countries
Sector-specific opportunities	Good product portfolio for megatrends, such as energy, greater prosperity, urbanization and digitization
Strategic opportunities	• Expansion of our production capacities • New high-quality products via innovations
Performance-related opportunities	• Higher plant productivity • Extension of our sales organization and establishment of technical competence centres • Region-specific product development via complete supply chain for dispersions and dispersible polymer powders

Data source: [4]

process innovation is an important aspect of Wacker's innovation efforts, and statements along the lines of "region-specific product development" point at how the company structures its global innovation activities.

In summary, the innovation strategy indicates how NBD (or innovation in general) helps the organization to achieve its goals and objectives. This also serves as an important guiding principle and motivation for everyone involved in NBD activities: knowing where and why to build new businesses is a fundamental principle for the success of an NBD project. Moreover, the role of innovation and NBD in achieving the overall business goal should not only be recognized by NBD team members or R&D staff but also by other members of the organization so that everyone has a common understanding. Therefore, communicating the innovation strategy is a top priority within every organization.

Because innovation goals and objectives cannot stand in isolation, an analysis of the current competitive position of the organization in terms of innovation and NBD activities is another step in developing the innovation strategy [5]. To this end, the general strategic planning tools that you have learned about in the earlier chapters of this book, for example, Porter's Five Forces framework and SWOT analysis, can also be applied for formulating the innovation strategy. By analysing the current competitive situation, companies can identify what their current offering lacks compared with competitors' offerings or future customer requirements and thus determine whether incremental or more radical changes to the product/business portfolio are necessary. Overall, a successful innovation strategy takes into account a full range of stakeholders, including suppliers, competitors, distributors, employees and, particularly, existing and potential customers. Thus, everyone who has an interest in the outcome of the innovation strategy (or in other words who will judge the success or failure) should be considered. For example, a chemical company won't embrace a new additive for aqueous coating systems that significantly improves substrate wetting, but which would demand a new coating technology, unless customers are willing to adapt their coating technology as well. Similarly, people like senior management who affect innovation and NBD activities, contribute resources (time, people, cash, etc.), and eventually gain benefits from a successful innovation strategy and its outcomes should be taken into account.

Finally, developing the innovation strategy should be repeated on a regular basis since the strategic goals of a company and the competitive environment of the firm change over time – in other words, innovation strategy needs to be adaptive to change. Internal and external circumstances (over which a company's management has no control) demand interim decisions that affect the innovation strategy of the company. Moreover, because no chemical or pharmaceutical company can possibly foresee the timing or even the nature of technological change, innovation strategy needs room for adaptation. For example, in 2000, the life science company DSM introduced a strategic initiative called "Vision 2005", which focused on transforming DSM's portfolio from bulk chemicals to specialties, while its "DSM in Motion" initiative, introduced in 2010, focused on maximizing sustainable, profitable growth. Both goals certainly have important but different consequences for DSM's innovation strategy and in turn for its portfolio of NBD projects.

6.3 Organizational Structure and Culture

NBD teams in the chemical industry typically bring together people with different characteristics, for example, age, functional background and personality, who work together to identify and develop new business opportunities. In fact, numerous innovation studies have shown that team heterogeneity is an important driver of innovation [6–9]. Some companies such as the Swiss health-care company Roche have begun to recognize differences in team characteristics and structure their NBD activities in accordance with the team members' personalities, that is, such companies create fit between the content of an NBD activity (e.g. searching) and certain personalities. In doing so, these companies recognize that some people are exceptionally good in searching and selecting new business ideas, for example due to their job experience or their ambition to try something new, while others are exceptionally good at implementing these ideas, for example because they are good negotiators and know how to communicate with stakeholders. As an example of such a practice, BASF has organized activities in its new business unit into two sub-units: "Scouting & Incubation" and "Business Build Up". In recognizing such differences and separating the activities of search and implementation into different organizational sub-units, companies can support the NBD process. The task of the NBD team leader would then be to ensure that team members of different sub-units interact.

When combining different sub-units into one NBD unit, most chemical companies structurally separate their NBD units from the operating business. The previously mentioned BASF New Business unit is, for instance, organized as a subsidiary of BASF, that is, it is an NBD unit on a corporate level. In addition to such corporate NBD units, companies may also have NBD units at a business unit level. Whether a business unit has its own NBD unit depends on the characteristics of the respective business unit. Business units that are characterized by high growth prospects and a high market share are more likely to have an NBD unit than business units which have a high market but low growth prospects. In the case of a diversified company, exploiting (and possibly creating) synergies between the NBD units of different business units would be an objective for the NBD unit at the corporate level.

The reason for separating NBD activities from the every-day operation of a company is differences in short- versus long-term orientation, management principles and culture. For instance, operating units typically face a strong pressure to fulfil short-term performance goals, have to make sure that they generate today's income and may have to focus on saving costs. NBD units in contrast have a long-term orientation, have to make sure that they generate the future income of the firm and NBD projects are initially not profitable. However, even though such a structural separation is advisable, the NBD unit cannot stand in isolation. Frequent collaboration and interaction with other units such as marketing, sales and production, as well as communication with the general management are important – NBD is in its very essence a coordination function. NBD teams naturally have to connect to internal and external parties: internally to functions such as R&D, marketing, sales, controlling, the firm's legal department and corporate functions, such as corporate strategy; and externally, to parties such as other companies, universities, research institutions, suppliers and industry associations.

As explained, running an existing business and developing a new business are very different from each other. Whereas in an existing business, customers, competitors and products are well known, new businesses often start from scratch, move into uncharted territory and naturally face a high level of uncertainty. Because of these unknowns, new business development is characterized by a higher degree of failure. A recent study among 108 business units in the chemical industry with more than US$2 billion in sales, by the consulting company McKinsey, shows that the average success rate for a typical NBD project is only 15% [10]. In other words, 85% of NBD projects fail (see Table 6.2). Moreover, the time to commercialization is substantially higher for NBD projects compared with extensions of existing businesses (e.g. a product-line extension into an existing market), and may take 10 or (particularly in pharmaceuticals) even 20 years.

Table 6.2 Success rates and time to commercialization for different innovation projects (McKinsey study). Adapted from [10].

	Product-line extensions into existing markets	Product-line extensions into new markets	New-product launches in existing markets	New-product launches in new markets
Familiarity with market	High	Low	High	Low
Familiarity with technology	High	High	Low	Low
Time to commercialization	2–5 years (average 4)	2–7 years (average 5)	6–15 years (average 11)	8–19 years (average 14)
Success rate	40–50%	30–40%	30–40%	15–20%
Margin	0–5%	0–10%	0–10%	0–60%

Given the high failure rates, NBD team members need to be able to accept failures and have a high tolerance to frustration. Furthermore, it is not only the NBD team that needs to be aware of high failure rates but also the senior management of a company, responsible for both existing and future business of the company (regardless of whether they have experience in building new businesses or not). Of course, high failure rates should not be a general excuse for failed NBD projects but as long as failure results in learning ("never make the same mistake twice"), management needs to accept a higher failure rate in NBD while also ensuring successful projects. Overall, the associated uncertainty and low success rates of NBD projects demand an organizational culture that is not only characterized by low risk aversion and tolerance to failures but accepts failures as a normal aspect of seizing a new business opportunity.

Organizational culture refers to a "complex set of values, beliefs, assumptions and symbols that define the way in which a firm conducts its business" [11:657] and describes the extent to which values are shared by members of the organization [12]. The culture of a company has important implications for NBD activities. Evonik Industries, for instance, states that the "culture of a company determines whether—and how fast—employees are able to drive forward good ideas and convert them into a profitable business" [13]. Hence, an organizational culture that is characterized by high control, high stability and a high need for predictability, will likely impede individual creativity, hamper motivation of R&D and NBD team members and ultimately prevent the identification of new business opportunities, because employees will keep ideas to themselves to avoid risks. Many chemical and pharmaceutical companies recognize that such an organizational culture will corrupt attempts to develop new businesses and new products. Many companies therefore promote an

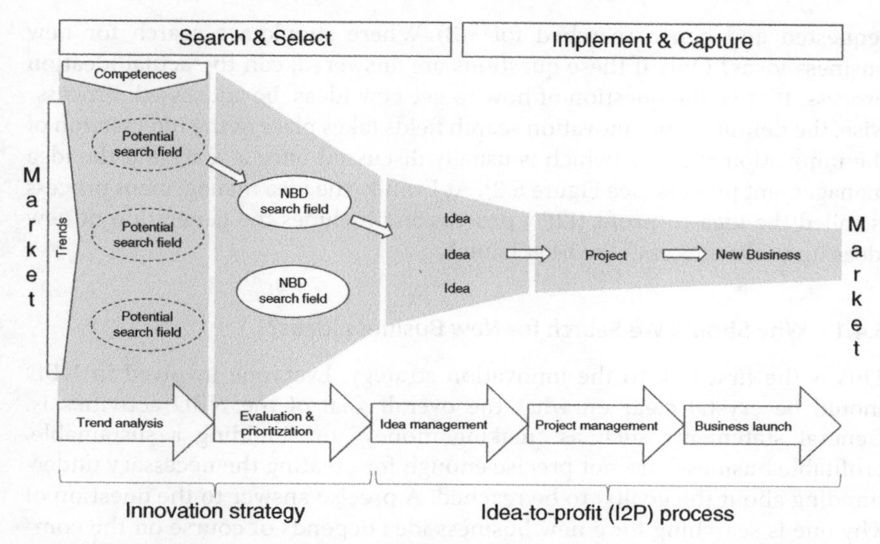

Figure 6.2 NBD process. *Source:* Evonik Industries [10]. Adapted from: Miremadi M, Musso C, and Oxgaard J. 2013. *McKinsey on Chemicals – Chemical Innovation: An Investment for the Ages*

organizational culture that welcomes creativity and innovation in all its facets. One example of this is Evonik's recent innovation initiative. According to Evonik's Chief Innovation Officer (CIO), "we're addressing ever shorter innovation cycles, more complex problems, and more demanding conditions. This is why we launched our Leading Innovation initiative last fall. With our global Leading Innovation initiative we are promoting a culture of innovation, with courage to break new ground, in which our employees are encouraged to take risks and which is based on trust, close cooperation and openness and which also rewards courage to innovate. Everyone at Evonik has to embrace innovation" [14].

In summary, the first part of this chapter demonstrates that structure should be managed simultaneously with innovation strategy and organizational culture to facilitate new business development. In the next part of the chapter, we will now take a look at the NBD process as depicted in Figure 6.2.

6.4 Searching for New Business Opportunities

The search for new business opportunities is more than just an application of different idea generating tools. It rather starts with connecting to the innovation strategy of the company and answering three basic questions: (1) Why should we search for new business ideas? (2) What kinds of business ideas are requested and hence searched for? (3) Where should we search for new business ideas? Only if these questions are answered, can the actual ideation process, that is, the question of how to get new ideas, be addressed. Process-wise, the definition of innovation search fields takes place at the intersection of the innovation strategy (which is usually discussed once a year) and the idea management process (see Figure 6.2). At Evonik, the idea management process is called the idea-to-profit (I2P®) process and includes the generation of new ideas up to the successful market launch.

6.4.1 Why Should We Search for New Business Ideas?

This is the first link to the innovation strategy. Everyone involved in NBD should be crystal clear on what the overall goal of the NBD activities is. General statements such as "making money" or "building a sustainable, profitable business" are not precise enough for creating the necessary understanding about the goal(s) to be reached. A precise answer to the question of why one is searching for a new business idea depends of course on the company itself. Consider a company that has ambitious growth targets. Depending on the time horizon of these targets, the company might consider growth through a number of acquisitions, a combination of acquisitions and NBD, or primarily through NBD.[2]

6.4.2 What Kinds of Business Ideas Are Requested and Hence Searched for?

This is the second link to the innovation strategy. The answer to this question depends on the general expectations a new business needs to fulfil. These conditions build the framework for searching for new ideas and include, but are not limited to, expectations regarding collaboration with external partners, demands about time-to-market or time-to-breakeven, budget restrictions, or requirements with respect to the profitability of the business, for example, certain EBITDA margins that need to be achieved.

6.4.3 Where Do You Search for New Business Ideas?

This question can also be derived from the innovation strategy with respect to the market segment or field of application where a new business should be built. As depicted in Figure 6.2, the search process starts with an analysis of market trends (external analysis) and analysis of existing competencies (internal analysis). There are two general ways to assess the internal competencies in relation to market trends: traditionally, one would compare the existing competencies (or means the

2 For a state-of-the-art overview on mergers and acquisitions (M&A) see [15].

company has) to those required to seize a given market opportunity. If competencies are missing, the company would have to acquire or build new competencies. This approach can be classified as an **outside-in** approach because the starting point is the external analysis – the environment determines if the existing competencies are sufficient. Another way to start the trend analysis is to define the unique competencies that already exist and search for potential fields that can be seized by means of these competencies. While both approaches might look very similar, they are very different in their nature for identifying potential NBD search fields. Consider the simple picture of cooking a meal: the **outside-in** approach is to have the recipe (the market trends derived from an external analysis) and then acquire the necessary ingredients (the competencies) to cook the desired meal. In contrast, the **inside-out** approach starts with the existing ingredients (the competencies) that can be combined to create various types of new meals; an approach that one can term "creative cooking". The reality will be between the outside-in and inside-out approaches because it is not only the environment of a firm that determines its path (with the firm having no influence at all) nor is it the firm that can always shape or influence its environment. Hence, defining new search fields is a combination of market trends and competencies with an emphasis on the fact that such a combination can be realized with existing and non-existing market access and with existing or non-existing competencies. But, a company should have either access to a market or possess a competence – one of the two should be present. In other words, without any means in-house and without any access to a market, internal development through NBD is questionable and an acquisition might be more promising.

To illustrate the application of the Why-, What- and Where-question, consider an example from Evonik Industries, former[3] business unit Advanced Intermediates (e.g. active oxygens, specialty catalysts and polymer additives). In order to enable additional growth, the unit considered the idea of growing its product portfolio of "advanced intermediates" by focusing on methods to replace petroleum-based chemicals. The industrial branch devoted to the use of living cells and enzymes to synthesize products that can replace petroleum-based chemicals is known as white biotechnology. Because white biotechnology has prospered in recent years and because Evonik has strong competencies in this area, especially in its animal nutrition business, the idea was to search for new business opportunities in the area of bio-based advanced intermediates. However, as all of these opportunities would have required investments in the three digit million Euro range, the idea did not fit with the overall business strategy of Evonik and was, therefore, not developed further. As an alternative, Evonik considered building on its existing technology and product know-how in the field of amine chemistry and applying its competencies in a different application field. The idea was to offer customers a chemical absorbent in the

3 Today part of Evonik's Performance Materials business unit.

area of acid gas removal that would not require process modifications. This idea, later commercialized as CAPLUS®, required much lower investments and, thus, had a better fit with the firm's innovation strategy. Hence, while the "What"-question did not fit with Evonik's strategy in the case of bio-based chemicals, its newly developed specialty amine formulation CAPLUS did fit. In order to be successful with an initial idea, it has to demonstrate fit in all three dimensions of the Why-, Where- and What- questions. If an idea does not fit for any (significant) reason, like the mismatch of the required and available investment for bio-based chemicals, the idea should not be further developed and the resources allocated to other ideas that better fit with a company's strategy.

6.4.4 Looking Outside the Boundaries of the Firm

In addition to collaborating with internal partners,[4] new business development involves connecting the firm to external partners such as suppliers, industry associations and universities. Hence, new business development entails the creation of a network beyond the boundaries of the firm. To build such a network, the new business development manager needs to talk to (existing and potential) customers, attend industry conferences and science fairs, consult with experts inside the chemical industry but also with those outside the chemical industry, for example, in the automotive or construction industries. It also entails reading scientific articles, industry magazines and management literature and studying patent landscapes.

Networking with a diverse group of external partners enhances business intelligence (e.g. by obtaining customer know-how), provides access to a wide range of ideas and complementary competencies and increases flexibility in the business development process. In fact, inter-firm networks and strategic alliances are increasingly a source of competitive advantage [16–21]. This includes concepts such as "Open Innovation" – a term promoted by Henry Chesbrough from the Haas School of Business – that assume that "firms can and should use external ideas as well as internal ideas and internal and external paths to market" [22: 1]. Open Innovation is a holistic approach to innovation management that not only means opening a firm's innovation process to external parties (with ideas flowing inside and outside the firm) but goes beyond simply using external sources of ideas and technologies [23]. Open Innovation can be used to change the way companies manage intellectual property and guide the integration of exploration-oriented firm-level capabilities and resources (e.g. internal R&D capabilities) with exploitation-oriented capabilities and resources (e.g. supply chain management). The concept, therefore, has a significant strategic element as it concerns the strategic capabilities of the firm, such as a firm's ability to manage alliances.

4 Within business units and across business units (e.g. NBD teams of other business units).

To better utilize external sources of innovation, chemical and pharmaceutical companies increasingly employ a variety of techniques to connect to external parties. The chemical company Evonik, for example, regularly holds so-called "Evonik Meets Science" events in Germany, China, Japan and the United States to strengthen its networks with leading international researchers. In these events managers from Evonik exchange knowledge with scientists from different disciplines and institutions [24]. The consumer goods company Henkel organizes a competitive fair, called the "Henkel Innovation Challenge", where students from all over the world come together in teams of two and imagine being a new business development manager. The students "develop their own vision of life in 2050 and develop a new Henkel product or technology for the future. With their concepts, which also take into account Henkel's sustainability strategy, they compete with international students around the world: Who has the best idea and who will succeed in the national and international final?" [25]. The pharmaceutical company GlaxoSmith Kline (GSK) operates a website where it publishes so-called "Wants", that is, technologies or innovations the company is looking for. If external parties (e.g. individual researchers or a team of researchers) have an innovative idea that matches GSK's wants, they can submit the idea and potentially partner with GSK. According to the company, "[t]he ultimate goal for GSK [...] is to in-licence and commercialize innovative technologies matching our strategic wants. Through appropriate legal agreements, we hope to link external innovative technology to our global brands—a process that can result in a win for GSK and a win for innovators" [26]. On average, the early stage review for new ideas submitted to GSK takes 3 months, and12–18 months to plan a development programme and execute legal agreements in the field of consumer healthcare. Box 6.1 discusses another approach to finding external ideas – ideation jams.

6.5 Selecting New Business Opportunities

After collecting several new business ideas, the most promising ideas need to be selected (because of resource constraints a chemical company cannot pursue every idea). The challenge, of course, is to determine which business opportunities or projects are worth pursuing and which are not. In this section we will present four approaches to selecting new business opportunities that have proven to be successful in business practice. While you have already learned about certain criteria to select new product ideas in Chapter 5 (see e.g. Cooper's Stage-Gate® model[5]), we will now focus on managing the selection of new business opportunities in areas where uncertainty is higher and where the firm has limited experience in terms of a technology and/or market. In

5 Stage-Gate® is a trademark of Product Development Institute and Innovation Management U3.

Box 6.1 Using ideation jams to search for new ideas

A method to find external business ideas that has gained prominence in the chemical industry over recent years is external ideation jams. In contrast to internal ideation jams and ideation jams with customers or suppliers, external ideation jams target people a company does not collaborate with at present (sometimes called "the crowd", hence, external ideation jams are also known as crowdsourcing events). In 2013, Evonik started an external ideation jam called "Evonik Call for Research Proposals" (ECRP) and invited about 100 well-known scientists in Germany to submit proposals for collaborative projects on a given chemical question. The goal was to find solutions for the synthesis of DL-methionine or L-methionine that do not require hydrocyanic acid (HCN). The best ideas were presented in front of an expert panel and awarded with cash prizes. By combining external sources of innovation with internal R&D capabilities Evonik is better able to cope with shortening product life cycles and the growing complexity of innovation in chemicals. The company estimates that the integration of external partners into its innovation activities, for example, in the form of external ideation jams, will increase in the future. The ECRP initiative is also a means of increasing Evonik's visibility among universities and positioning itself as an attractive employer.

contrast to methods such as discounted cash flow analysis, which are biased against late returns and uncertainty of NBD projects, the methods presented here are particularly suitable in the context of new business development. The selection process is primarily focused on the first part of the I2P process (see Figure 6.2), where the number of ideas from the fuzzy front end is reduced to a manageable number of projects.

6.5.1 The R-W-W Screen

While the growth needs of large chemical companies demand investments into radical product innovation and new business development, studies show that the majority of projects in firms' development portfolios focus on continuous improvements of existing product lines and established businesses [27]. Hence, despite the need to invest in a few but risky NBD projects, firms often rather tend to approve a large number of safe projects that result in incremental innovations or opportunities that improve businesses that already exist.

Although the tendency of established firms to bet on projects with certain outcomes is not surprising, focusing on incremental developments during the selection process has two key disadvantages. Firstly, though necessary for continuous improvement, safe projects don't contribute much to the growth needs of the firm and its profitability targets. Secondly, selecting a large number of safe projects congests the firm's project pipeline, resulting in "traffic jams" that delay NBD projects. As a result, firms fail to achieve their revenue goals and may even risk future competitive advantage. Accordingly, what companies need is a systematic and disciplined selection method, resulting in a balanced portfolio of safe projects that sustain the present businesses of the firm (e.g. through incremental product innovations) and more risky projects that build future competitive advantage and contribute significantly to the growth of the corporation.

Such a tool is the so called R-W-W ("real, win, worth it") screen – a method developed by George Day from the Wharton School – which lays out a series of questions about every NBD project, its potential market and a firm's capabilities – "not for making go/no-go decisions but, rather, [...] to expose faulty assumptions, gaps in knowledge, and potential sources of risk" [27: 114]. The R-W-W method is used by companies such as Evonik, Honeywell, Novartis and 3M to assess the potential of business ideas and select individual new business development projects [27]. In the following we will summarize Day's method, apply it to the context of the chemical industry and highlight key aspects.

Based on the R-W-W method, the potential of an NBD project is evaluated in three categories: product-market relation (*Is it real?*), competition (*Can we win?*) and value (*Is it worth pursuing?*). Following these three first-column questions (see Figure 6.3), the screen guides the NBD team to ask six fundamental questions (second column of Figure 6.3): *Is the market real? Is the product/service real? Can the product/service be competitive? Can our firm be competitive in this business? Will the product/service be profitable at an acceptable risk? Does launching the NBD project make strategic sense?*

Usually, not all of the second-column questions (see Figure 6.3) can be answered at an early stage of developing a business opportunity but these questions are a helpful guide to navigating through the unknown jungle of unproven business concepts and should be repeatedly asked throughout the different stages of the NBD process. Specifically, the R-W-W screen points at the type of questions to be asked and where an NBD team should focus its resources in order to determine the viability of a business idea. As an example, when the initial answer to the question "Is the market real?" is "maybe", the NBD team has to explore ways to turn the answer into a yes. If, after a thorough investigation, the definite answer to any second-column question is no (i.e. the team cannot identify any way to change it into a yes), then the project will typically[6] be terminated. In such a case, one should document the reasons why it has been stopped (in order to come back to the project when market conditions have changed or, in case a project has been terminated but later successfully exploited by a competitor in order to explore why the project was dismissed).

---

6 When a project has passed the first five questions and only the last question ("Does launching make sense?") is answered with a no, the firm might still select the idea and pursue the project.

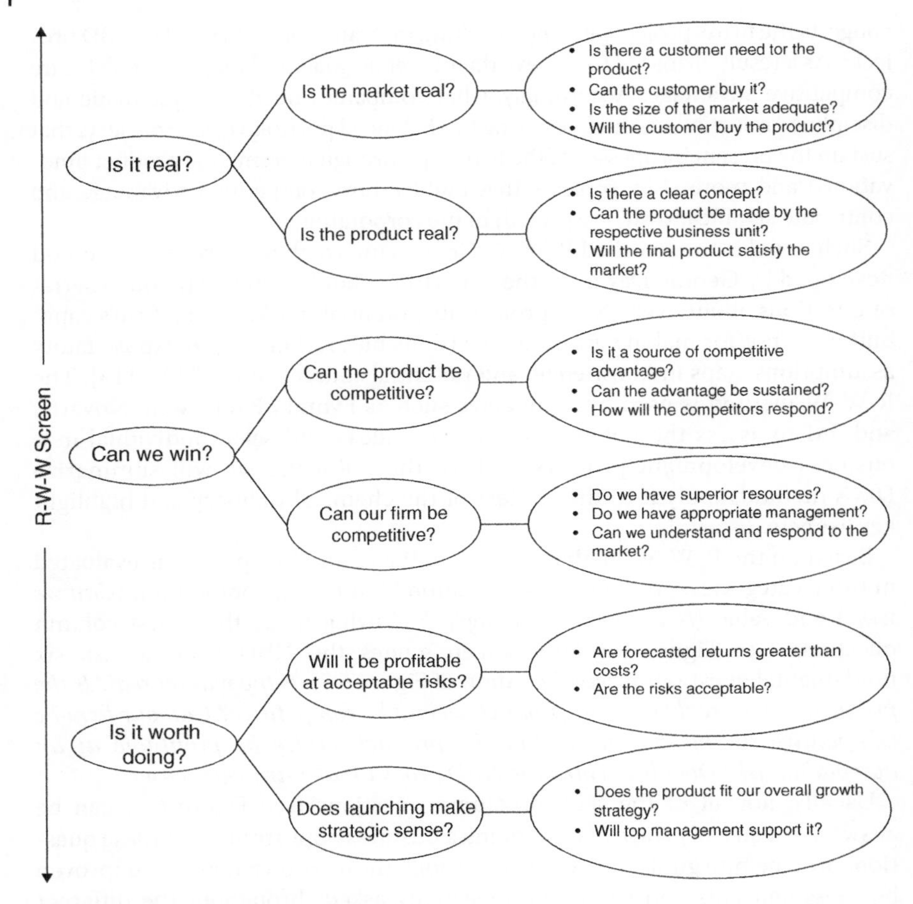

Figure 6.3 The R-W-W-screen. *Source:* [27]

In supporting the NBD team answering the six core questions, the team can deal with a number of sub-questions (third column in Figure 6.3)[7]:

1) **Is the market real?** To determine whether a market exists is the first step towards making the decision to pursue an NBD project or not. In general, a market exists when four conditions are fulfilled:
   - *Is there a customer need for the product?* The proposed product/service has to solve a customer problem better than the available alternatives (best case, there are no alternatives available in the market). Market research can be used to effectively determine whether there is a desire in the market, what motivates customers to demand a solution and how customers behave.

7 See [27].

- *Can the customer buy it?* The customer must be able to buy the proposed product/service. The goal in answering this question is to explore purchasing barriers such as regulations, existing long-term contracts that prevent customers from switching, or distribution problems (e.g. providing health care solutions in developing countries is often difficult due to a lack of infrastructure).
- *Is the size of the market adequate?* A market opportunity is not real if the customers' demand for the product/service is not big enough. Particularly in the chemical industry it is often not the number of customers but rather the quantity they order that determines the size of the market.
- *Will the customer buy the product?* Having a real customer need and ability to purchase the product/service is insufficient if customers are not willing to buy the offering. Substitutes or other aspects such as perceived risks, brand loyalty, or customers' assessment of the price–performance ratio influence the purchasing decision. In the end, it is the customer who compares the benefits (as viewed by the customer, not the NBD team) in relation to associated costs and compares the offering to alternatives in the market before making the purchase decision.

2) ***Is the product/service real?*** After the NBD team has established the reality of the market, it needs to evaluate the feasibility of the concept. The following questions foster the analysis:
   - *Is there a clear concept?* At the early stages of NBD, technology and performance requirements need to be assessed in order to determine what exactly the project is about. Besides technical dimensions (e.g. reaction speed, required energy), social and environmental dimensions (e.g. waste, emissions, fresh water consumption) of the concept should also be evaluated.
   - *Can the product/service be made by the respective business unit?* The product/service can only be created when the business unit has the legal rights (e.g. in the form of patents) to produce it, can source the required raw materials at appropriate costs, can establish a value chain (i.e. the primary activities and their linkage to create value – from inbound logistics to outbound logistics, marketing and sales, and services – as well as secondary activities such as training) and has the financial resources to do so. Moreover, answering this question also requires determining whether external partners such as biotechnology firms are needed to produce the product/service.
   - *Will the final product/service satisfy the market, therefore the customers' needs?* During the development of a new business several attributes of the offering (e.g. modification of product features) and the surrounding business conditions (e.g. availability of staff) naturally change. It is therefore important to monitor these changes during a new business development.

3) **Can the product be competitive?** In a next step the NBD team needs to assess the company's ability to gain and hold market share with the initiative. This can be done by answering three questions:
   - *Is it a source of competitive advantage?* In general, this advantage can be achieved in one of two ways: by supplying a product/service at lower costs, in which case the product/service provides the basis for a cost advantage, or by supplying a differentiated product/service in a way that the customer is willing to pay a price premium, in which case it provides the basis for a differentiation advantage. Hence, the NBD team must investigate whether a competitor can offer the same benefit to customers at lower operational costs (often at a lower price) or in a more differentiated way (e.g. through better customer service or a well-known brand).
   - *Can the advantage be sustained?* Having an advantage for only a short period of time is (typically) not desirable. The NBD team must therefore explore what means exist to defend the new business against imitators. This concerns questions of IP and knowledge protection and continuous R&D efforts, but also questions about how to ensure not only continuous but maybe even exclusive supply of raw materials and how to attract and keep the best people to work on the project.
   - *How will the competitors respond?* A common mistake in NBD is to assume that competitors stand still. Considering the reactions of rivals is therefore an important aspect in analysing NBD projects. A medium-sized chemical company should, for example, carefully consider retaliation from large chemical companies such as BASF or Dow Chemical (market leaders could, for instance, escalate marketing activities or start price wars to fend-off market entrants). Even among equals, historic sentiments (e.g. visible in statements such as "this business has always been our flagship") might result in severe reactions from competitors. As a result, industry prices that have been assumed in the project's pro forma calculations might no longer hold.
4) **Can our firm be competitive?** Having a competitive product is insufficient to succeed in NBD if the company is not able to successfully compete in the marketplace. In order to create sustainable competitive advantage, the firm must – relative to its competitors – possess superior resources and capabilities, superior market insight and an appropriate management. The following questions help to evaluate the competitive position of the firm:
   - *Do we have superior resources?* Superior resources, such as know-how in chemical engineering, logistics, technical sales professionals, or information networks (e.g. customer relationship management) are crucial to be competitive. This also concerns the strategic capabilities of the respective business unit. Does the business unit have capabilities that are valued by the customer and provide potential competitive advantage? Do current or potential competitors possess the same or similar capabilities? Are the capabilities difficult to imitate? Is there a risk of substitution?

- *Do we have appropriate management?* The NBD team must also raise important questions about having the right people in place to commercialize and scale the project. Do managers of the company have experience in the market and can they handle a complex project? Overall, does the NBD project fit with the culture of the company?
- *Can we understand and respond to the market?* NBD requires a good deal of marketing know-how, particularly an understanding of customers' desires, needs and concerns. Managers involved in NBD need to be open to customer response and able to share customer intelligence among team members. Customer feedback is important to refine the concept.

5) **Will it be profitable at acceptable risks?** At this stage of the screening, the NBD team investigates the financial value of the project. Two questions can guide this rigorous analysis:
   - *Are forecasted returns greater than costs?* A firm's top management typically will not approve an NBD project if its forecasted returns are smaller than the cost. The project's capital outlays, expenses, expected margins, time-to-breakeven and other aspects need to be clarified.
   - *Are the risks acceptable?* In order to assess the associated risks of an NBD project, a sensitivity test can be carried out. At the core, this test should reveal how small changes in, for example, market share, price, or timing of market introduction affect cash flows and time-to-breakeven.
6) **Does launching make strategic sense?** At the final stage of the screening, the NBD team needs to consider the strategic value of the project by asking two more questions:
   - *Does the product fit our overall growth strategy?* Here, the NBD team should analyse whether the project fits with the business strategy of the respective business unit (e.g. is the project meeting expected growth rates and profits?).
   - *Will top management support it?* It is not only the initial but the ongoing support and commitment of a firm's top management that plays an important role in developing a new business. Having a respected supporter at the top of a firm is naturally a big advantage for an NBD project. The rigorous assessment of an NBD project by means of the R-W-W method will provide a solid base to gain such support. Moreover, the NBD team should be aware of the top management team's characteristics, such as functional backgrounds when presenting project proposals. Building a common ground with those who make the final decision, may eventually be the final ingredient for project approval.

At Evonik Industries, the R-W-W screen is applied especially in the first three stages of the idea-to-profit (I2P) process (see Figure 6.2). Therefore, project leaders would need to answer a few questions in Stage 1, revisit these questions in Stage 2 and add some more questions and repeat the process in Stage 3, to be able to answer all of the previous questions that are often based on reasonable assumptions.

6.5.2 Understanding and Mapping the Whole Value Chain

In order to create and deliver a product or a service, a specific set of activities needs to be performed. This set of activities is, in line with the process-view of organizations, known as the value chain [28]. The value chain describes what activities a firm needs to undertake to create value.[8] But, the value chain is not only about choosing activities, it is also about how to link (e.g. sequence) activities. When engaging in NBD, it is important to understand this chain of activities and their linkages because it allows evaluation of the feasibility of an NBD opportunity and, thus, selection of promising projects.

Compared with building a new business in a business-to-consumer (B2C) context, such as online retailing, chemical companies (primarily) build businesses in a business-to-business (B2B) context that is characterized by smaller numbers of potential customers (e.g. 1–50) and by smaller numbers of suppliers from which one can source raw materials (e.g. 1–30). The starting point in constructing the value chain is the customers and their needs (i.e. what does the customer actually want?). Discussions with potential customers at very early stages of NBD allow creation of insights on product requirements, potential quantities and product prices that, in turn, enable firms to analyse the consequences on the value chain and might also lead to joint development agreements with customers. Thus, based on an analysis of customers' needs, one can start to think about the components required to meet those needs (e.g. production facilities with a certain capacity) and the design of the processes for creating, communicating and delivering the offering (e.g. given the requirements, do we have the marketing and sales capacity to serve our customers in a reproducible fashion?).

Furthermore, the value chain should not only be analysed in the direction of the outputs, which is forward towards the customer, but also in the direction of the inputs, which is backward towards the suppliers. In NBD, one needs to understand where one can source the required raw materials, at what price and in what quantities. Supplier integration during the early stages of NBD may not only allow a better understanding of inbound-logistics in one's value chain but also reduce development time and improve quality of new products [29, 30]. Integrating both customers and suppliers is, hence, an important and advisable aspect of NBD to understand and fine-tune a business model to the needs and restrictions of the market one aims to enter (for more insights into customer and supplier integration see Chapter 8).

To illustrate the role of understanding and mapping the value chain for NBD, consider the following examples from Evonik Industries:

- Together with a start-up, Evonik has developed a catalyst that can transform a specific class of linear olefines into polyolefines. The unique selling point

---

8 Related but not identical to a firm's value chain is the supply chain. The supply chain links individual value chains and can, hence, be considered as an industry value chain.

of this catalyst is that it can transfer bio-based linear olefines – which are in this specific case not that pure compared with fossil-based linear olefines – into polyolefines. In contrast, the catalyst that is traditionally used in this application field cannot transfer bio-based linear olefines into (bio-based) polyolefines. Producers of the given type of polyolefines were very keen on sourcing bio-based polyolefines in order to fill the supply–demand gap of polyolefines. While Evonik had all the resources to produce the catalyst, the missing piece in the value chain was on the supply side: only a very limited number of companies are able to produce the required type of bio-based linear olefines and plans for building further production capacities for bio-based linear olefines were dragging on. Despite a well-performing catalyst, the project was put on hold, as it was unclear if the customer of Evonik (the polyolefines producer) would be able to source bio-based linear olefines in sufficient amounts.

- As a leading company in the production of hydrogen peroxide, Evonik must not only understand the hydrogen peroxide market and the technology that enables its customers to produce propylene oxide from propene and hydrogen peroxide, but it must also understand the propylene oxide business (e.g. volumes and prices) and the market for propene (e.g. access to the required volumes of propene and the prices). It is not sufficient to understand the hydrogen peroxide market; Evonik must also understand the forward part of the value chain.

Both examples illustrate that it is not sufficient to understand the technology or the product alone, but companies also need to understand the whole value chain, especially with respect to the sourcing of raw materials and the markets their customers are selling to. Hence, NBD projects should be assessed in terms of how well the value chain is understood towards the supply side, that is, suppliers, and how well the value chain is understood towards the demand side, that is, the customers.

6.5.3 Discovery-driven Planning

Another way to approach the selection of NBD projects is discovery-driven planning. This method – developed by Rita McGrath from the Columbia Business School and Ian MacMillan from the Wharton School – focuses on testing and re-evaluating assumptions and is, hence, particularly suitable for the unpredictable setting of NBD projects [31]. In the following we summarize McGrath and MacMillan's method, apply it to the context of the chemical industry and highlight key aspects.

Discovery-driven planning is a tool that takes the fundamental differences between planning for an established business and planning for an NBD project into account [32]. Conventional planning tools such as discounted cash flow or net present value analysis are based on the premise that future returns can be extrapolated from past experience. These methods are therefore appropriate in the context of established businesses, where technology is well understood, forecasts are reliable and market conditions are relatively safe. By contrast, the context of an NBD project, particularly during the early stages of development, is by its very nature uncertain, difficult to predict and not obvious. Instead of relying on existing knowledge that doesn't change during the development of a project, NBD projects rely on assumptions that often turn out to be wrong. Conventional planning tools that require reliable data are hence in general not suitable to select NBD projects. What is needed is a tool that allows discovering the potential of a project as new data are uncovered. Discovery-driven planning is such a tool that "systematically converts assumptions into knowledge" [31:44]. As depicted in Figure 6.4, discovery-driven planning is based on a five-step process to develop and select NBD projects.

The first step is to create a reverse income statement. Instead of estimating the revenues and then calculating the profits for each project, a reverse income

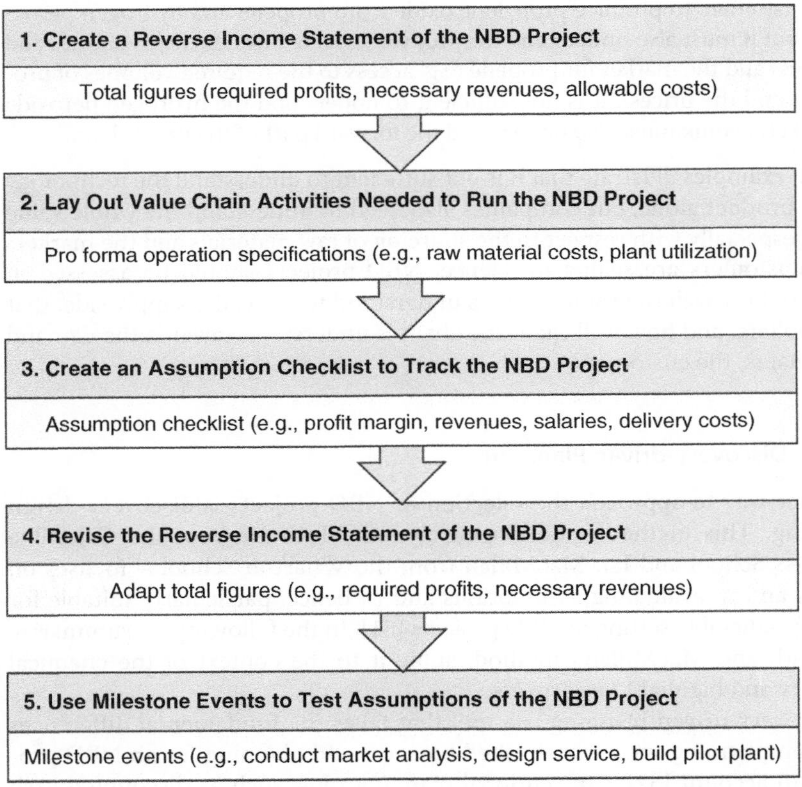

1. Create a Reverse Income Statement of the NBD Project

Total figures (required profits, necessary revenues, allowable costs)

2. Lay Out Value Chain Activities Needed to Run the NBD Project

Pro forma operation specifications (e.g., raw material costs, plant utilization)

3. Create an Assumption Checklist to Track the NBD Project

Assumption checklist (e.g., profit margin, revenues, salaries, delivery costs)

4. Revise the Reverse Income Statement of the NBD Project

Adapt total figures (e.g., required profits, necessary revenues)

5. Use Milestone Events to Test Assumptions of the NBD Project

Milestone events (e.g., conduct market analysis, design service, build pilot plant)

Figure 6.4 McGrath's discovery-driven planning – an effective tool for NBD. *Source:* own figure

statement starts with the required profits. Following this approach, the NBD team will begin with the required profits and then work backwards to the revenue necessary to deliver the required profits and the allowable costs (required profits = necessary revenue – allowable costs). Accordingly, the team has to define what the profit margin and the absolute profit should be at a minimum. To calculate the required profits, we must understand that the project should: (i) significantly contribute to the total profits of the firm and (ii) compensate for the investment risk (alternatively, the chemical company could invest in an established business). If, for example, the existing business has an EBITDA[9] margin of 20% then the new business should have a margin of about 25% (i.e. in this example an NBD project must deliver a premium of 5% to compensate for investment risks). Here, the absolute profit should be in balance with the amount of resources and the expected risk. If the upfront costs are, for instance, several million Euros over several years, the targeted absolute profit, such as EBITDA, should be a two digit Euro million figure. Based on these two figures one can then calculate the necessary revenues and derive from this the necessary volumes and prices. Already at this stage one can double check if the necessary volumes lead to a reasonable and achievable market share. Going further one can now also calculate the allowable cost for production, general and administrative expenses, and marketing and sales expenses. For example, in the case of the previously mentioned CAPLUS project (see also Box 6.2), analysing how fast Evonik can ramp-up production volumes if customers are willing to replace their existing amine scrubbing agent, enables the NBD team to estimate what resources (e.g. sales forces) are necessary to fulfil the demand. Such an analysis not only creates a more detailed but also a more realistic picture of the business opportunity compared with deriving volumes from aspired market shares.

In a second step, the value chain is laid out, describing key activities necessary to run the NBD project. Activities that create value are, for example, inbound logistics (e.g. receiving raw materials), production, marketing and sales and services (e.g. training). The operations specifications also allow identifying activities that are most significant in terms of operating costs and test how the existing value chain of the business unit is affected by the NBD project. Looking at related chemical businesses allows identifying typical operation standards such as plant utilization, asset to sales ratios, or other key performance indicators that can be used to determine whether a project holds together and to test and adjust underlying assumptions. The goal of this exercise is not to create the most accurate but a reasonable model of the logistical and economic reality of the new business.

The third step of discovery-driven planning is to create an assumption checklist in order to track assumptions. This is an important step, as the model will be based on a number of assumptions due to the natural uncertainty that

9 EBITDA = earnings before interest, taxes, depreciation and amortization.

comes with an NBD project. Keeping this checklist is an important discipline to ensure that each assumption is flagged and tested as NBD projects unfold. This is a key advantage of the discovery-driven planning, not only because it allows tracking of the assumptions but also because the tool facilitates the discussion with senior management. It lays out clearly not only the promised additional sales and profits (which every manager will happily accept) but also the necessary resources with respect to cash and employees that need to be committed in order to capture the benefits. Furthermore when, for example, assumed sales need to be adjusted, discussing the reason for this adjustment allows making changes in assumptions transparent and, therefore, sets the basis for constructive discussions.

Next, the company uses the new data to loop back the entire process into a revised income statement. At this step, the NBD team needs to ensure that all the assumptions are in line with the figures in the income (or profit and loss (P&L)) statement. Having transparent assumptions and figures, one can see if the entire business proposition hangs together or not. If, after several tests, the performance requirements cannot be met, the NBD project should be terminated and another project selected.

The fifth and final step in discovery-driven planning is to use milestone events to test assumptions. This allows NBD managers to formally plan while ensuring learning.[10] In contrast to conventional planning tools, the idea of milestone planning is not to focus an NBD manager on meeting a plan but rather to monitor the progress of the new business opportunity. Essentials are to further update the assumption list (if necessary) and to postpone major resource commitments until evidence from a previous milestone indicates that assumptions are still valid. If an NBD project does not pass a milestone test and the company does not see any way to redirect the project, the project would normally be terminated.

6.5.4 Portfolio Management

Another view on the selection process is not looking at a single idea but rather at the overall portfolio of ideas and projects. Portfolio concepts such as the BCG-matrix (Boston Consulting Group) in the case of corporate strategy are frequently used in management practice and can also be applied to the selection of new business opportunities. In the context of NBD, the following concepts have been used by a leading specialty chemical company:

- **Ansoff matrix:** A tool that is frequently used to illustrate the basic directions of firm growth is the so-called Ansoff product/market growth matrix [34]. Like the BCG-matrix, the Ansoff matrix is normally used as a corporate strategy framework but can be adapted to the context of NBD. In the context

---

10 The basic idea behind using milestone events is described by Zenas Block and Ian MacMillian in the book *Corporate Venturing*, Harvard Business School Press, 1993 [33].

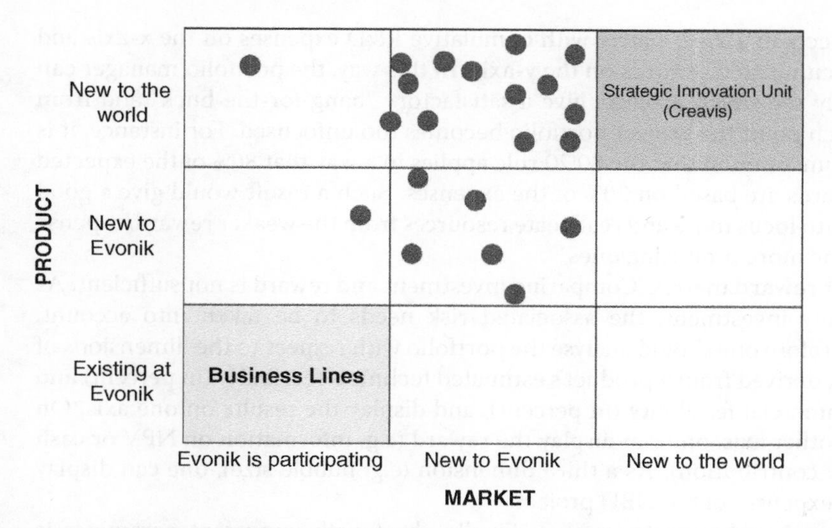

Figure 6.5 An NBD portfolio at a specialty chemical company. *Source:* Evonik Industries

of NBD, the matrix has nine zones based on different technology–market combinations (see Figure 6.5). Businesses in the lower left part target an existing market with an existing technology. This is the zone where the established business lines of a company are active. Future businesses that bring together a technology that is new to the world and a market that does not yet exist belong to the upper right part of the matrix. In the case of Evonik Industries, this is the zone where the strategic innovation unit of the firm (called Creavis) is active. Figure 6.5 also shows an NBD portfolio from Evonik. The majority of the presented NBD projects combine a market that is new to Evonik with a technology that is either new to the world or to Evonik. No NBD projects are in a zone where a market is new to the world and, hence, not existing yet because it is often difficult to anticipate new markets in the chemical industry. As a result, NBD projects often target existing or potential customers in existing markets. If a business idea brings together a technology and a market that are both new to the world, the idea would not be selected for NBD but could be recommended to be screened by the strategic innovation unit of Evonik.

- **Comparison of reward to expense**: Another way of looking at the project portfolio is to sort the projects by the size of the rewards,[11] for example, net present value (NPV) or expected cash flow in 10 years, and display the

---

11 As mentioned when we described the advantages of discovery-driven planning, financial tools are not always suitable to evaluate individual NBD projects (particularly at early stages of development). We present portfolio concepts here because they are commonly used but the reader should note that portfolio concepts should – due to their limitations – not be used exclusively. Rather, a combination of different tools will help to determine the feasibility of NBD projects.

projects in a 2 × 2 matrix with cumulative R&D expenses on the x-axis and the cumulated rewards on the y-axis. In this way, the portfolio manager can easily see which projects give a satisfactory "bang-for-the-buck" and from which point the project portfolio becomes too unfocused. For instance, it is not uncommon that the 80:20 rule applies in a way that 80% of the expected rewards are based on 20% of the expenses. Such a result would give a good hint to focus more and reallocate resources from the weaker reward/expense to the more promising ones.

- **Risk reward matrix**: Comparing investment and reward is not sufficient. As in any investment, the associated risk needs to be taken into account. Therefore one should analyse the portfolio with respect to the dimensions of risk, derived from a product's estimated technical feasibility (in percent) and commercial feasibility (in percent), and display the results on one axis. On the other axis, one can display the reward (e.g. information on NPV or cash flow contribution). As a third dimension (e.g. bubble size), one can display the expenses of the NBD project.
- **Reward and time-to-market**: Finally, the fourth important parameter is when to expect rewards out of the overall portfolio and in what amounts. Therefore, a good way to display this is to use time-to-market on the x-axis and a reward variable (e.g. NPV, CF) on the y-axis. Again, the third dimension can be used to display the size of the NBD project with respect to its expenses.

6.6 Implementing the New Business Concept

Having identified and selected a promising business concept one could assume that from this point the progress is linear and foreseeable. One could imagine that managing an NBD project is straightforward after core questions about product–market relations, competition and value are addressed, and milestone planning is used to test initial assumptions. But this is simply not the case. The uncertainty around the project is still very high and one should expect many loops, delays and detours until the new business becomes real and a scalable business model unfolds.

Implementing a new business demands experimenting rather than elaborative planning, using customer feedback early rather than relying on intuition (or even a dream) for too long, and it requires spending less time and energy on preparing the perfect product/service and more on piloting a new way of doing business. A methodology that takes these differences into account is the "lean start-up" concept from Eric Ries.[12] The lean start-up method recognizes that

12 For latest insights, see: http://theleanstartup.com/.

managing new businesses is different from managing established businesses. Despite its name, it is not only start-ups that can benefit from this method. In fact, scholars like Steve Blank emphasize that "some of its biggest payoffs may be gained by the large companies that embrace it" [35:66]. In the following we will summarize Ries's method, apply it to the context of the chemical industry and highlight key aspects.

To illustrate the idea behind the lean start-up method, consider a chemical company that has selected an NBD project on asymmetric catalysis for implementation. The company knows that the demand for enantiopure compounds – particularly in the life sciences – is growing and has established a reverse income statement for this project following discovery-driven planning.[13] Also, together with colleagues in the firm's R&D department, the NBD team has confirmed that the proposed synthesis is in principle highly selective and should be robust for a number of natural products, the team knows that the size of the market is adequate and it has also established an agreement with a small biotechnology firm that holds an important patent and has certain process know-how believed to be relevant for the final production of the new chiral catalyst used in the reaction. Together with its partner, the firm is now ready to proceed with the project. Traditionally, the company would now ask to sequence project activities, such as further developing and testing the synthesis in the firm's R&D labs, creating 5- to 10-year forecasts for profits and cash flows, to identify potential bottlenecks and the critical path of the project, to monitor and control adherence to the project plan and, finally, to launch the product after all technical difficulties have been solved. Thus, the company uses a project or business plan to execute the NBD project. The downside of this approach is that only after launching the new catalyst, that is, when the company's sales forces are asked to sell it, does the chemical company get feedback from customers. Hence at that point, our chemical company learns the hard way that customers have no need for a new catalyst, cannot or are not willing to adapt their production processes despite the technical superiority of the new catalyst, or that it had targeted the wrong customer segment.

Of course, our example is simplified, but in the context of NBD such a situation occurs more often than one might expect. Far too often companies work month after month or even for years on implementing an NBD project in accordance with the project plan, before the brutal reality of the market destroys all their aspirations (or we might even call it "dreams"). This happens for two reasons. First, because companies wait too long before they involve the customer in NBD. Instead of spending months behind an office desk,

---

13 It should be noted that the lean start-up method and discovery-driven planning have a lot in common. This is in fact why we present both concepts in this chapter.

writing the perfect project plan and then executing this plan for years, companies need to ensure that they get early feedback from customers. The problem in NBD is that it is sometimes not that easy to understand what the customer actually wants (e.g. needs cannot be clearly expressed or the customer does not know what to demand in five years' time) or, more importantly in the uncertain context of NBD, it is often simply not clear if the initially targeted customer is going to be the final customer. Who should be asked for feedback if the chemical company does not know who the customer is in the first place? The second reason is that companies often focus too much on executing the project plan after they have selected a promising NBD project. This problem occurs because established companies are usually very efficient in executing a project plan and therefore have a tendency to utilize this strength. However, in the context of NBD it is important to understand that focus should be on effectiveness rather than efficiency or, put simply, you need to make sure that you implement the *right it* before implementing *it right*. Executing a detailed project plan that includes a 5- to 10-year forecast for profits and cash flows, a definition of the problem to be solved and a description of the offering that solves the problem, is often no more than a static document reflecting the assumptions of its creators. Because no one in the chemical or pharmaceutical industry can possibly foresee the technological or economic context of a new business that ventures into the unknown, it is crucial to first discover what the right business is before efficiently executing it. A method that takes both these problems into consideration is the lean start-up method.

Lean[14] start-up is a scientific approach to developing and implementing new businesses that focuses on testing two key hypotheses: the value hypothesis and the growth hypothesis. Using the term hypothesis reminds us that every NBD project is based on a series of assumptions and we need to conduct experiments to find out if the project is on the path to becoming a sustainable business [36]. The value hypothesis is about discovering whether customers find the offering valuable or not. Thus, the NBD team needs evidence – not a plausible premise or a good story – that customers are willing to buy the new product. The growth hypothesis is about questioning how to scale the business, given that the value hypothesis has been confirmed. Thus, to implement an NBD project: (1) evidence must exist that customers find the business concept convincing and valuable and (2) the business must be repeatable and scalable. Testing the value and growth hypothesis early allows

14 The word lean comes from *lean manufacturing*, a production philosophy pioneered by Taiichi Ohno and best known to be applied in the Toyota Production System in Japan. The lean start-up takes ideas from lean manufacturing, such as reducing cycle times and eliminating waste and applies it to the new business development process.

starting validated learning before too many assumptions about the project have accumulated.

To test the value and the growth hypothesis, the lean start-up approach suggests engaging in **customer development** and, closely linked to customer development, **agile development**. Customer development focuses on going out of the company and asking potential customers and partners for feedback about the company's initial assumption of how it creates and captures value with the new businesses, including product characteristics such as functionality and price, the revenue model, brand identity and potential strategic alliances and supplier relationships. The focus is on rapid adjustment of the initial hypothesis: NBD teams that follow the lean start-up method create a minimum viable product and get customer feedback to revise assumptions. Sometimes the minimum viable product is a product that has the same or very similar characteristic as the existing product that is currently in use ("Would you buy a catalyst with the following features at the following price?") or an offer to buy a product on a website. A chemical company producing hydrophobing agents could, for example, offer a new agent on its website to see if customers are willing to klick on the purchase button. If so, the company has learned that there is a general interest in the product. There is no doubt that setting up a website is not sufficient to get detailed customer feedback, but if not a single customer klicks on the purchase button, then a cheap experiment can reveal the same information as the chemical company from our previous example obtained after trying to launch its NBD project and spending hundreds of thousands of Euros. Lean start-ups use customer feedback early on to adapt their offering and start the cycle all over again. Thus, at its heart, customer development is about testing a hypothesis and making small adjustments (iterations) or more substantial changes (pivots) to NBD projects.

Closely linked to customer development is agile development, which focuses – in contrast to traditional linear new product development – on building products in short, repeated cycles. Rather than splitting an NBD project into big pieces (e.g. R&D, design, improve, deploy), this method emphasizes dividing the project into smaller activities that enable continuous improvement, in order to develop functionality or performance aspects of a new product fast (e.g. a chemical company producing dispersing additives would focus on one specific functionality of a new additive or a specific part of the formula instead of developing the entire additive before receiving customer feedback for the first time). Companies like Evonik that successfully implemented this lean approach focus on continuously improving the NBD flow and optimizing the whole project into small, iterative steps rather than building production capacity, scaling-up production and then launching the new offering after years of development (see Box 6.2 for an example).

Box 6.2 The lean start-up methodology at Evonik Industries

Many industrial gas streams, such as natural gas, biogas, synthesis gas and flue gas have to be scrubbed prior to downstream processing to remove acid gases, such as carbon dioxide and hydrogen sulphide. In most cases, amines are used as absorbents. Conventional amines cause a variety of technical problems during gas scrubbing. First, because of low chemical and thermal stability, conventional amines have to be regularly topped up, and, in some cases, changed completely. Also, they generate numerous secondary reactions and the decomposition products can lead to foaming and corrosion. In addition, amine regeneration consumes high amounts of energy and conventional amines can cause (dramatic) fluctuations in the acid gas content, which can reduce the plant capacity. An NBD team at Evonik recognized that a new amine developed by the company had the potential to outperform conventional products and create higher value for customers than market alternatives.

Instead of preparing the production of the new amine, scaling-up the production and then offering the amine on the market (ultimately named CAPLUS®), the team engaged in customer discovery and agile development. The team members left their office desks behind and met face-to-face with several potential customers around the globe to explore applications and better understand the customers' points of view. In this way the team discovered that despite the technical advantages of the new amine, customers in the oil and gas industry are generally quite conservative and, to a certain extent, resistant to changes. Customers, for instance, were afraid of plant shutdowns caused by an unproven new amine that could cause significant financial losses and damage to a company's image. Listening to these concerns, the team at Evonik adjusted the business model behind the new amine in an agile fashion and implemented customer feedback early on. Using customer feedback, the team revised their assumptions and ultimately positioned Evonik as a complete technology supplier with an intense focus on the customer's plant and a comprehensive package of services. In doing so, Evonik was able to address obstacles and customers' concerns and is now selling CAPLUS to natural-gas producers in Southeast Asia, the Middle East/South Africa and South America. From the first idea to the first reference plants and, therefore, first sales, it took Evonik six years to build the business. Although this may seem long compared with other industries, considering development cycles in the chemical industry, the lean start-up approach helped Evonik to significantly speed up the process.

6.7 Learning: Capturing the Value from Lessons Learned

As you have learned, developing new business opportunities in the chemical industry takes several years and the failure rate is quite high. It is therefore necessary, first, to ensure that the organization learns from its failures and, second, to use appropriate performance measures to evaluate an NBD unit's progress.

6.7.1 Learning from Failures: Post-completion Audits

Given the high failure rate of NBD projects, it is important for companies to conduct post-project reviews to analyse the causes of failure or positive developments that should be applied again. A retrospective analysis of terminated projects enables firms to examine what has been learned from failure and to apply the lessons learned in future projects. However, despite the relevance of learning from failures, studies show that many companies do not review projects at all (e.g. due to time and management constraints) and the small number of projects that are reviewed, are reviewed without established guidelines [37]. In the following, we present the essentials of a project learning or debriefing process that works well, which is based on Evonik's recommendations for post-completion audits (PCA) that can be used as a guideline for post-project reviews in NBD. Table 6.3 summarizes different types of post-project reviews.

The PCA should take place from a certain project stage onwards, for example, in Cooper's Stage-Gate model after Stage 3 (i.e. "Development"). A PCA should be conducted when an NBD project is terminated, when an NBD project is launched (Stage 5: "Launch") and approximately two to five years after the launch. The PCA is usually prepared by the project team and presented to the steering committee of the project.

The PCA should contain a business case review and a process review:

1) **Business case review**
   In the business case review, the last business case and previous versions of the business case are analysed with respect to deviations in market entry date, sales, project cost, manufacturing cost, investment (capital expenditures), profits (contribution margin, EBIT(DA) or cash flow) and NPV. The analysis is also based on additional information such as the market and competitive environment, competitive strategy, competencies and customers' feedback (i.e. the "voice of the customer").
2) **Process review**
   In the process review, the project is analysed in terms of its development. This review includes, for example, the experiences made, the interactions of the project team members, critical success factors and the main challenges of the project. These factors are analysed with respect to what can be

Table 6.3 Overview of post-completion audits.

Parameter	Project review/ audit	Post-control	Post-project appraisal	After action review
Time of execution	After project completion or during each project phase	Exclusively at the end of the project	Approximately 2 years after project completion	During work process
Carried out by	Review: moderator Audit: external	Project manager	External post-project appraisal unit	Facilitator
Participants	Project team and third parties that were involved	Project manager	Project team and third parties that were involved	Project team
Purpose	Status classification, early recognition of possible problems	Addition to a more formal project end	Learning from mistakes, knowledge transfer	Learning from mistakes, knowledge transfer
Benefits	Improvement of team discipline, decreasing vulnerability and validation of strategies	Result is a formal document with project aims, goals, milestones, budget (estimated versus actual)	Best practice generation for large-scale projects	Immediate reflection of the team's past actions to improve future actions
Interaction mode	Face-to-face meeting	Non-cooperative form: conducted by project manager	Document analysis, face-to-face meetings	Cooperative team meeting
Codification	Partly in reports	Partly in reports	Booklets as well as in person	Flip charts

Source: own table

recommended for future projects and what can be improved (e.g. "What went well?" or "Options for improvement"). In order to have a meaningful review, it is essential to have a culture in place that allows an open discussion about mistakes without fear of personal disadvantages and a top management that supports continuous learning.

6.7.2 KPIs for Measuring the Success of an NBD Unit

Employees in an NBD unit can be doing a great job; however, as the nature of the development process takes several years to reach the first sales, the normal metrics to measure the success of an established business will not work in this

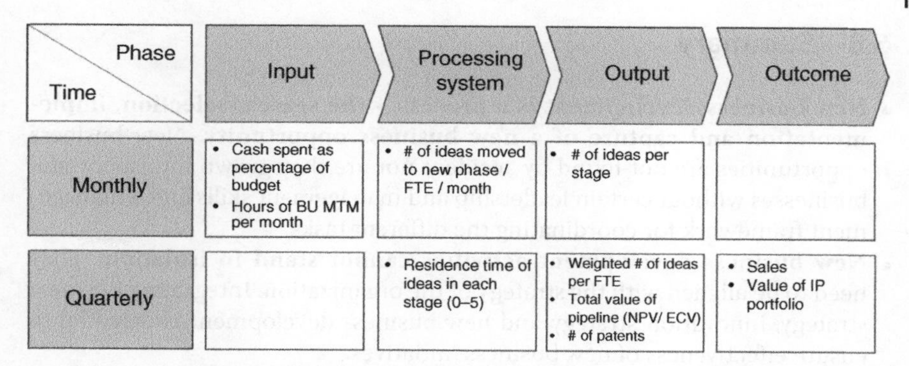

Figure 6.6 Key performance indicators in NBD (BU = business unit, ECV = expected commercial value, FTE = full-time employees, IP = intellectual property, NPV = net present value, MTM = management). *Source:* Evonik Industries

case. In order to measure the success of the NBD unit's efforts, different metrics have been developed by one of the present authors, to give an overview of the performance of a respective NBD unit without considering the details of several highly complex projects the unit is pursuing. This task is fulfilled by defining and measuring a number of key performance indicators (KPIs). As depicted in Figure 6.6, depending on the development phase, different KPIs are used. As the term KPI indicates, these indicators focus on "key performance" metrics and it is, thus, important to keep in mind that they do not create a comprehensive picture. KPIs are "only" indicators and reality might be more complex. For example, the KPI "Number of ideas moved to new phase per FTE and per month" indicates the progress of the innovation pipeline in relation to the number of employees in the unit. Whereas a static view on the innovation pipeline only shows how many projects are at which stage, this KPI shows how many projects are really "alive" and being worked on. Furthermore, the KPI is not only an incentive to push projects forward to higher stages but it also takes terminated projects into consideration. Therefore, it incorporates the reality of a high failure rate, as discussed earlier, and still shows that the unit produces productive outcomes (a terminated project could save the company significant amounts of money, especially when done at an early stage).

The KPI "Weighted number of ideas per FTE" shows the workload of the group and its development over time. Furthermore, it could be benchmarked between similar groups in and outside of a chemical company. Whereas an idea in an early stage needs less time and resources, projects in higher stages require more time and resources and are usually worked on by several employees. Therefore, ideas and projects in different stages need to be weighted in order to have meaningful numbers that can be compared.

6.8 Summary

- **New business development is a process – the search, selection, implementation and capture of a new business opportunity.** New business opportunities are not found by accident nor are they grown into successful businesses without certain leadership and management skills and a management framework for coordinating the different tasks.
- **New business development activities cannot stand in isolation.** They need to be aligned with the strategy of the organization. Integrating business strategy, innovation strategy and new business development is essential to ensure effectiveness of new business initiatives.
- **Methods for screening new business opportunities are discovery-driven, expose faulty assumptions, and potential sources of risk.** In addition to the overall framework for new business development, tools presented in this chapter such as the R-W-W screen, discovery-driven planning and lean start-up have been proven successful in the chemical industry for the development of new businesses.
- **Running an established business is different from developing a new business.** In contrast to established businesses, the context of a new business development project is by its very nature uncertain, difficult to predict and not obvious. As initial assumptions often turn out to be wrong, many projects that looked promising at an early stage of development will not be successful.
- **Failure is an intrinsic part of new business development.** To be successful in new business development, companies need a culture that enables risk-taking behaviour and accepts failures. To capture the lessons learned from failed projects, companies need to organize learning processes and ensure that such learning is transferred to future projects.

References

1 Tidd J and Bessant J. 2011. *Innovation and Entrepreneurship*. John Wiley & Sons Ltd: Chichester, p. 29.

2 Dow Chemical. 2015. *Mission and Vision*. http://www.dow.com/company/aboutdow/vision.htm (accessed 26 October 2015).

3 Fitterling J. 2014. *Credit Suisse Basic Materials Conference – The Dow Chemical Company*. http://www.dow.com/~/media/DowCom/Corporate/PDF/investor-relations/Presentations/CS_Chemical_Conference.ashx?la = en-US (accessed 26 October 2015), p. 3.

4 Wacker. 2015. *Market Opportunities*. http://www.wacker.com/cms/en/investor-relations/profile/prospects/market-opportunities/opportunities.jsp (accessed 26 October 2015).

5 Adner R. 2006. Match your innovation strategy to your innovation ecosystem. *Harvard Business Review*, **84**(4): 98.
6 Talke K, Salomo S, and Kock A. 2011. Top management team diversity and strategic innovation orientation: The relationship and consequences for innovativeness and performance. *Journal of Product Innovation Management*, **28**(6): 819–832.
7 Alexiev AS, Jansen JJ, Van den Bosch FA, and Volberda HW. 2010. Top management team advice seeking and exploratory innovation: The moderating role of TMT heterogeneity. *Journal of Management Studies*, **47**(7): 1343–1364.
8 Hambrick DC. 2007. Upper echelons theory: An update. *Academy of Management Review*, **32**(2): 334–343.
9 Hambrick DC, Cho TS, and Ming-Jer C. 1996. The influence of top management team heterogeneity on firms' competitive moves. *Administrative Science Quarterly*, **41**(4): 659–684.
10 Miremadi M, Musso C, and Oxgaard J. 2013. *McKinsey on Chemicals – Chemical Innovation: An Investment for the Ages*. http://www.mckinsey.com/~/media/mckinsey/dotcom/client_service/chemicals/pdfs/chemical_innovation_an_investment_for_the_ages.ashx (accessed 26 October 2015).
11 Barney JB. 1986. Organizational culture: Can it be a source of sustained competitive advantage? *Academy of Management Review*, **11**(3): 656–665.
12 Saffold GS III. 1988. Culture traits, strength and organizational performance: Moving beyond "strong" culture. *Academy of Management Review*, **13**(4): 546–558.
13 Evonik Industries. 2015. *A Breeding Ground for Creativity*. http://corporate.evonik.de/en/company/research-development/innovation-culture/pages/default.aspx (accessed 26 October 2015).
14 Evonik Industries. 2014. *Evonik Starts Innovation Campaign*. http://corporate.evonik.com/en/media/focus/innovation-campaign/Pages/innovation-campaign.aspx (accessed 26 October 2015).
15 Faulkner D, Teerikangas S, and Joseph R.J. 2012. *Handbook of Mergers and Acquisitions*. Oxford University Press: Oxford.
16 Hamel G, Doz YL, and Prahalad CK. 1989. Collaborate with your competitors and win. *Harvard Business Review*, **67**(1): 133–139.
17 Bamford J, Ernst D, and Fubini DG. 2004. Launching a world-class joint venture. *Harvard Business Review*, **82**(2): 90–100.
18 Lei D and Slocum Jr JW. 1992. Global strategy, competence-building and strategic alliances. *California Management Review*, **35**(1): 81–97.
19 Kortmann S and Piller FT. 2016. Open business models and closed-loop value chains: Redefining the firm-consumer relationship. *California Management Review*, **58**(3): 88–108.
20 Deeds DL and Hill CW. 1996. Strategic alliances and the rate of new product development: An empirical study of entrepreneurial biotechnology firms. *Journal of Business Venturing*, **11**(1): 41–55.

21 von Delft S. 2013. Inter-industry innovations in terms of electric mobility: Should firms take a look outside their industry? *Journal of Business Chemistry*, **10**(2): 67–84.
22 Chesbrough HW. 2006. *Open Innovation: The New Imperative for Creating and Profiting from Technology*. Harvard Business Press: Cambridge, MA, p. xxiv.
23 Chesbrough H. 2004. Managing open innovation. *Research-Technology Management*, **47**(1): 23–26.
24 Evonik Industries. 2015. *Open to New Ideas*. http://corporate.evonik.com/en/company/research-development/open-innovation/pages/default.aspx (accessed 26 October 2015).
25 Henkel. 2015. *Henkel Innovation Challenge*. http://www.henkel.com/careers/students/henkel-innovation-challenge (accessed 1 July 2015).
26 GSK. 2015. *Partner with GSK: Our Open Innovation Pathway*. http://innovation.gsk.com/new-product-development-process.aspx (accessed 8 November 2015).
27 Day GS. 2007. Is it real? Can we win? Is it worth doing? *Harvard Business Review*, **85**(12): 110–120.
28 Porter ME. 1985. *Competitive Advantage*. The Free Press: New York.
29 Petersen KJ, Handfield RB, and Ragatz GL. 2005. Supplier integration into new product development: Coordinating product, process and supply chain design. *Journal of Operations Management*, **23**(3): 371–388.
30 Petersen KJ, Handfield RB, and Ragatz GL. 2003. A model of supplier integration into new product development. *Journal of Product Innovation Management*, **20**(4): 284–299.
31 McGrath RG and MacMillan IC. 1995. Discovery-driven planning. *Harvard Business Review*, **73**(4): 44–54.
32 Rice MP, O'Connor GC, Pierantozzi R. 2008. Counter project uncertainty. *MIT Sloan Management Review*, **49**(2): 54–62.
33 Block Z and MacMillian I. 1993. *Corporate Venturing*. Harvard Business School Press: Boston, MA.
34 Ansoff HI. 1988. *Corporate Strategy*. Penguin Books: London, chap. 6.
35 Blank S. 2013. Why the lean start-up changes everything. *Harvard Business Review*, **91**(5): 63–72.
36 Euchner J. 2013. What large companies can learn from start-ups: An interview with Eric Ries. *Research-Technology Management*, **56**(4), 12.
37 Von Zedtwitz M. 2002. Organizational learning through post-project reviews in R&D. *R&D Management*, **32**(3): 255–268.

7

Designing and Transforming Business Models

Stephan von Delft[1]

[1] *University of Glasgow, Adam Smith Business School*

> *Over the course of my career, I've come to see business model innovation not as a static process but as a systemic and reliable capability, one that leaders need to build, strengthen, and eventually turn into a sustainable competitive advantage.*
>
> A.G. Lafley, Chairman and CEO of Procter & Gamble

Since its spinoff from Bayer in 2004, the German chemical company Lanxess has experienced several years of upward momentum: rising sales, increasing earnings and the admission to the DAX 30 blue-chip index in 2012 are only a few examples that indicate its success. In March 2013, Axel Heitmann, then CEO of Lanxess, explained that the company had just experienced "the best in our growth story so far," and further explicated "our *business model* proved itself once again" [1]. Only one year later, the situation had dramatically changed. Lanxess's customers were destocking, prices declining, its balance sheet for 2013 showed a net income loss of €159 million, and Heitmann was replaced. Lanxess's new CEO, Matthias Zachert, concluded after a corporate-wide analysis in August 2014, that the "role-out of a *new business model*" is necessary, meanwhile asserting that Lanxess needs to become "significantly more competitive" and customer oriented [2].

As the case of Lanxess illustrates, determining the competitiveness of a business model, sustaining a successful business model, and transforming business models that are threatened to become obsolete is an important yet difficult managerial task for executives. In this chapter you will learn what a business model is, why business models matter, what the difference between business models and strategy is, how firms can successfully engage in business model innovation and why chemical and pharmaceutical companies should engage in business model thinking.

Business Chemistry: How to Build and Sustain Thriving Businesses in the Chemical Industry, First Edition. Edited by Jens Leker, Carsten Gelhard, and Stephan von Delft.

7.1 Business Model Design: Essential Management Decisions

In the chemical and particularly in the pharmaceutical industry one can observe increasing attention towards business models in general and business model innovation in particular. However, while a firm's top management often has (and should have) a thorough understanding of how the firm's business model works and why the capability to innovate a business model is important to maintain competitive advantage, managers three to four steps down the hierarchy, namely those who actually operate the business model, can often not explain the firm's business model or articulate how a business model could potentially be changed, such as in response to shifts in competition. Moreover, many managers do confuse business models with strategy or other management concepts. For example, an experienced marketing manager at a major chemical company once explained that the business model of BASF is *verbund* and the business model of Procter & Gamble is scaling. While BASF's *verbund* is part of the answer why BASF's business model – or more precisely its business models – is successful, *verbund* is not a business model. Neither is scaling (though being able to scale a business model is without any doubt a capability worth building on). Arguably, a lot of confusion about what a business model is and what it is not exists. So what actually is a business model? In essence, a business model is *the story* of how a company operates [3]. In so doing, a business model answers questions in two core areas that are essential for any business (see Figure 7.1): (1) Who is our customer, and what does the customer value? and (2) How do we make money in our business? The latter explains the economic logic of businesses, that is, how to make a profit while creating value for the firm's customer(s) at appropriate costs. Hence, business models enable managers and entrepreneurs alike to think about basic choices firms have to

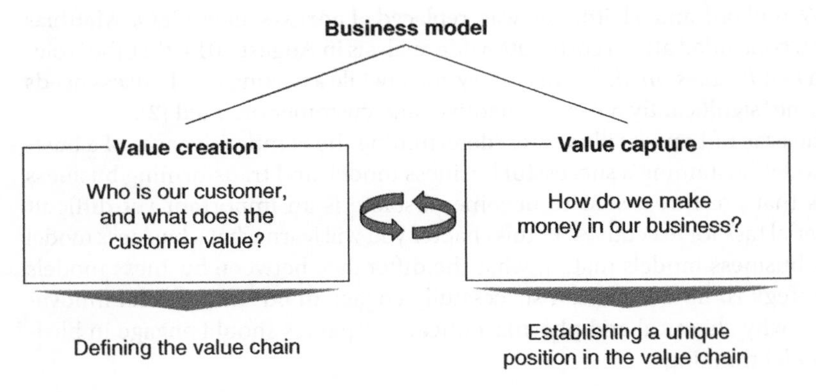

Figure 7.1 The two functions of a business model. *Source:* own figure

make when it comes to their targeted customer segment, their revenue model, their cost structure and make-or-buy decisions [4]. A business model is thus about the managerial choices of *how* to operate an organization. These choices comprise, but are not limited to, compensation practices, location of facilities, partnerships and so on.

In short, a business model describes how a company creates and captures value. Value creating activities in companies may comprise defining an opportunity or formulating a solution to a given problem, and value capturing activities may involve the collection of fees for rendered services or the collection of rent from the ownership of property rights. In simple terms, the petroleum exploration business, for instance, is to acquire exploration licences, identify the best area for petroleum extraction and then sell the right to exploit a particular area to a client. Thus, the firm creates value in the form of an opportunity for petroleum exploitation and appropriates value by selling the right to exploit the field. Similarly, a biopharmaceutical company might focus on the identification of promising molecular compounds, define an opportunity by testing its safety and efficacy and then sell or licence the property right(s) on that compound to a large pharmaceutical company that develops the drug further.

Value creation and value capture are closely interlinked; this is basically because a business model that doesn't create value for a firm's customers doesn't create value for the firm. Although both value creation and value capture are equally important functions of a business model, we observe that managers often overemphasize the value creation aspect and tend to underemphasize the value capture function. To better grasp both functions of a business model equally and to foster the development of a common language about business models inside companies, scholars have developed frameworks to describe core aspects of a firm's business model. These frameworks often refer to components or so-called design elements that constitute a firm's business model. While these conceptualizations have certain limitations [5], we believe that they are valuable because they allow companies to improve the execution of their business model, enhance resilience by enabling management to think about alternative ways to design their business model, allow executives to communicate management priorities in their business model and, overall allow sense to be made of firms in action. A particularly useful business model conceptualization is offered by Mark Johnson, Clay Christensen and Henning Kagermann [6]. According to their concept, a business model consists of four interlocking elements, namely the customer value proposition, the profit formula, the key resources and the key processes, which taken together explain how companies create and capture value (see Table 7.1).

The customer value proposition, an established concept in the marketing literature, refers to "the combination of end-result benefits and price to a prospective customer from purchasing a particular product" [7: 533]. Accordingly, a specifically selected bundle of products and/or services that caters to the

Table 7.1 Johnson, Christensen and Kagermann's business model definition.

Customer Value Proposition	Profit Formula
• What important problem do we solve for a customer, how do we help our customer in getting an important job done? • What satisfies the problem or job?	• What is our revenue model; how much money can we make (quantity × price)? • What is our cost structure? • What unit margin do we target; what is the contribution needed per transaction to achieve the desired profit level? • What will the resource velocity be, how fast do we need to turnover resources to achieve target volume?
Key Processes	**Key Resources**
• What processes (e.g. design, product development, manufacturing, sales, marketing, etc.) do we need to deliver the value proposition that is scalable and repeatable? • What are our business rules and metrics for success (e.g. margin requirements)? • What is our culture, what are our norms?	• What resources (e.g. brand, equipment, IP rights, distribution channels, partnerships and alliances, staff, etc.) do we need to deliver the value proposition profitably?

requirements of the customer is part of a firm's value proposition. In turn, understanding customers' preferences and needs is reflected in the customer value proposition. In this context it is important to note that designing a compelling customer value proposition requires that companies "must stop trying to figure out what kinds of products people are trying to buy and instead work out what [job] they are trying to get done" [8:26]. Hence, customers' needs should not be defined in relation to a product. Instead, companies need to figure out what type of job their customers want to get done and then construct a "blueprint" of how to fulfil that job profitably. By job we mean a problem the customer wants to solve, such as more efficient stereoselective synthesis of an alkene. Therefore, the outcome or the benefit, for example, cost or time savings, a customer gains from your offering is distinct from the job the customer wants to get done. Companies need to recognize both the job and the outcome. For this reason, companies must entirely understand the customer experience life cycle, namely the discovery, purchase, first use, ongoing use, management (e.g. maintenance, repair and acquisition of add-on products) and disposal of an offering, and not only the result or the outcome of consumption. Studies, however, show that companies regularly forget about the pre- and post-use phase of products and thus neglect the overall context in which customers use a given product or service [9].

Benefits that contribute to a customer value proposition can be quantitative (e.g. product performance) and/or qualitative (e.g. product design) in nature, depending on the customer's requirements. Qualitative benefits such as emotional or social aspects should not be underestimated. For example, when a company like Beiersdorf develops a skin-moisturizing active substance for one of its skincare products, the social job their customer wants to achieve is to look good or younger. Although emotional or social aspects play a more obvious role for companies such as Beiersdorf, Henkel, P&G or Unilever that offer personal care products, cosmetics or consumer goods, chemical manufacturing companies should consider these aspects as well, for example, with respect to the jobs the customers of their customer want to achieve. The better a company understands the customers' preferences and needs, and the better a company recognizes customers' benefits, the better it can create a compelling customer value proposition. We can conclude that a compelling customer value proposition consists of one or two points of difference an offering has relative to the next best available alternative on the market (the most profitable situation occurs when no alternative is available). To create a compelling customer value proposition firms have to: (1) understand what elements of the offering are most relevant for the customer, (2) demonstrate the superior performance of the offering compared with its alternatives and (3) communicate the benefits in accordance with the customer's priorities. The last implies that product characteristics as the firm's R&D department, for example, defines them can, and most often do, differ with from the customers' perception of the product or service characteristics and the customers' priorities. It is crucial to understand that objective product or service characteristics are often inconsistent with subjective perceptions of the customer. Again, it is therefore often advisable not to define customers' needs in relation to a product or service.

To better understand specific customer needs and subsequently create a more compelling customer value proposition, firms can group customers into different customer segments. For example, some business models are tailored to serve one single customer segment in a niche market with a specific value proposition, while other business models target two or more interdependent customer segments (e.g. the business model of credit card companies, which depends on credit card holders and merchants that accept the credit cards). The market for specialty chemicals for example consists of hundreds of customer segments, such as advanced ceramics materials, cosmetic chemicals and water-soluble polymers. Some chemical companies serve a customer segment that consists of hundreds or more customers, while others may only have two or three customers in a respective segment. Defining customer segments helps managers and entrepreneurs alike in developing a clear target within a market. A targeted customer (segment) should therefore be an element of every business model. Targeting involves analysing the size and the growth of the identified customer segments (e.g. segment sales, growth rates and expected profitability) and the segments' structural attractiveness (e.g. using the tools you have learned about in the first four chapters of the book, such as Porter's Five Forces), evaluated against the company's objectives and resources. The latter concerns questions like: "Do we have the skills and resources to succeed in this segment?"; "Are we allowed to operate in this segment with a given technology? Do we have the necessary intellectual property right(s)?"; or "If we operate in this business, do we have the capability to offer superior value compared with our competitors?" Based upon this evaluation, a firm selects one or more target customer segments. The selection itself is, however, not part of the business model. The business model is tailored towards the target customer segment but the business model does not answer the question of which segment to target. We will come back to that point later when we discuss the difference between strategy and business model. For now we can conclude that designing a business model begins with the identification of an opportunity to satisfy a real customer need and selecting a target customer segment.

The second element in the presented business model definition – the profit formula – is a "blueprint" of how the firm captures a part of the value that it creates for its customers, namely how the company can fulfil the customers' needs profitably. In Johnson, Christensen and Kagermann's view [6], the profit formula consists of the revenue model, the cost structure, the margin model and the resource velocity. The revenue model explains how much revenue the firm generates with its offering (i.e. volume × price). Usage fees for analytical equipment, leasing of manufacturing facilities, or licencing of intellectual property rights are examples of revenue streams in the revenue model of many chemical companies. Accordingly, different revenue streams require different pricing mechanisms, for example, fixed or dynamic pricing. The way to think about "volume" largely depends on the way companies generate their revenue. If a producer of batches for example sells these batches, the company certainly defines "volume" in its revenue model differently from a producer of batches that offers a leasing model to its customers. While the revenue model explains how much revenue the firm generates with its offering, the cost structure reflects the allocation of costs, for example, direct costs or economies of scale. The cost structure hence answers what it costs to operate a business model. One way to develop a successful business model is to base cost targets on strategic prices and consequently force the company to "question virtually every assumption about materials, design, and manufacturing" [10: 135]. Companies can therefore set a total profit target and work back from there to determine what the profit formula should look like. Another important element of a firm's profit formula is the margin model. The margin model refers to the contribution needed from each transaction to achieve the desired profits. Far too often companies do not see the difference between their margin model and their profit formula. Hence, although the margin model is technically reflected in the cost structure, it makes good sense to analyse it as an additional building block of the firm's profit formula. Moreover, analysing revenue model, cost structure and margin model step by step allows to see why low margin opportunities might be worth pursuing. For example, if a pharmaceutical company only focused on high margin opportunities, it would most likely not make any sense to capture the pharma market in emerging economies or business opportunities at the bottom-of-the-pyramid (i.e. the largest, but poorest group of people). However, we observe that markets in emerging economies can be highly profitable for companies willing to design new business models around lower margins instead of simply importing their domestic business model. The last part of the profit formula, resource velocity (sometimes also termed asset productivity), refers to the utilization of resources, namely how fast a company needs to turn over inventory and assets given the expected volume and profits. In other words, resource velocity tells you how many components or products you can develop, manufacture, store, deliver, sell and service for a given amount of time and money. Mark Johnson hence describes resource velocity as the "measure of not how much money flows through your company but how quickly it flows through it" [11].

In summary, the profit formula answers questions about costs, pricing mechanisms, expected margins, overheads, throughput, lead times and so on. In terms of the profit formula, some business models are more cost-driven than others. The US chemical company Dow Corning for instance serves customer segments with very different value propositions, and as a result it operates two business models simultaneously: a cost-driven business model under the Xiameter brand and a value-driven business model under the Dow Corning brand. With its traditional value-driven business model the company sells high-end silicones to customers who expect customized, high-quality products and sophisticated technical support. Xiameter, on the other hand, serves customers that are looking for competitive prices and "no-frills" standardized products sold over the internet. Accordingly, the profit formulas of both business models differ substantially from each other. While the value-driven profit formula is characterized by high-margin, high-overhead retail prices and revenue streams from value-added services, the Xiameter profit formula is based on lower margins, spot-market pricing and high throughput. These differences are also reflected in the other two business model elements, namely the key resources and key processes (see Table 7.2 for a comparison of both business models).

Key resources refer to a firm's assets, such as technologies, distribution channels, equipment, production facilities, the brand and people, required to deliver the customer value proposition to the customer. These examples show that key resources can be physical, financial, intangible (e.g. intellectual property) or human resources. Of course not all resources are equally important among firms. Therefore, the emphasis here is on those resources that in combination are required to create and capture value, that is, resources that create

Table 7.2 Value- and cost-driven business models – the case of Dow Corning.

	Dow Corning (value-driven business model)	Xiameter (cost-driven business model)
Customer value proposition	Customized solutions, negotiated contracts, flexible ship dates	No frills, bulk prices, sold through the internet
Profit formula	High-margin, high-overhead negotiated prices, paying for value-adding services	Spot-market pricing, lower overheads, lower margins, higher throughput, low transportation costs
Key resources	Sales force, R&D staff, large number of products, technology	Web-based platform, dedicated traders, commonly used silicones
Key processes	R&D, material delivery, sophisticated technical service	Auto-order-handling, optimization of plant utilization, operating web-platform

Source: own table

competitive differentiation. Generic resources that do not create competitive differentiation are thus not an element of a business model. It is important to understand that key resources can also be provided by external partners. Partnerships are indeed a core aspect of many business models. For example, consider a distribution channel partner. Distribution channel partners very often play an important role in creating the ultimate customer experience [9]. In such a case it is crucial not only to consider the customer's role during value creation and value appropriation but also the distribution partner's role. Companies should for instance ask themselves what incentives a distribution partner has to market, or appropriately support, the delivery of the product. Firms must hence align channel actions in their business model by understanding and emphasizing their channel partners' experience cycle. To develop fit with the distribution partner, firms can for example analyse how their offering affects profitability and operations at each link in the supply chain. In so doing, they can help distribution partners to integrate the offering.

Leveraging partner capabilities and resources can also result in an open business model, a term developed by Henry Chesbrough to explain how partners can help firms to create and capture value more effectively. On the one side, external partners allow firms to create value more effectively because firms can not only define a series of activities that will result in new products or services based on internal capabilities, but also leverage external capabilities. Companies thus have a broader set of options available. On the other side, they allow firms to capture value more effectively because firms can utilize not only their own resources, assets or position but also those of their partners. Open business models typically have an outside-in and an inside-out pattern [7]. Typical outside-in patterns are resources and capabilities that are required to build gateways to external organizations (key resources), activities that connect a firm's internal business processes and R&D groups to the network of external partners (key processes) and increased R&D productivity, as well as reduced time-to-market, which result in lower costs (cost structure). Typical inside-out patterns are established distribution channels, originally designed to deliver outcomes from internal sources of innovation but now used to deliver outcomes from external sources of innovation to customers (key resources), R&D outputs that cannot be commercialized in the focal industry but are valuable in other industries (customer value proposition) and additional revenue streams from the exploitation of internal ideas outside the firm (profit formula).

Accordingly, firms may generate new revenue streams and achieve cost and time savings with an open business model. The concept of open business models is related to Chesbrough's concept of Open Innovation (see Chapter 5 on Open Innovation). Chesbrough argues that firms aiming to exploit the full potential of Open Innovation should also consider opening-up their business models. He reasons that "companies must open their business models by actively searching for and exploiting outside ideas and by allowing unused internal technologies to flow to the outside, where other firms can unlock their latent economic potential" [12: 22]. The rationale to open the business model is to address the economic pressure on innovation, that is, the rising costs of innovation and shorter product life in the market. The first refers to increasing internal development costs due to increasing scientific complexity, such as multi-disciplinary and cross-functional R&D efforts, limitations to breakthrough discoveries and the rising costs of capital. Rising costs of technology development would imply that only large companies would be able to undertake technology development and commercialization, and as a result become even larger. However, the second force at play – shortening product life cycles – "makes these economics challenging even for the largest firms" [13: 11]. For example, the shipping life of a new drug (while it enjoys patent protection) is today shorter due to longer approval processes by regulation agencies such as the Food and Drug Administration (FDA) agency in the United States. Accordingly, the time to earn a profit with a new drug in the market is shortened. At the same time, successful products are imitated quickly and/or compete with rival products after a short time. As a result of this effect, revenues in the closed model decrease. Together with the rising costs of innovation, "R&D investments under the closed model of innovation [are] increasingly difficult to sustain" [13: 11]. Chesbrough concludes that the economic pressure on innovation reduces the ability of firms to earn a satisfactory return on their investment in innovation. Companies that open their business model can address the described challenges. On the cost side, open business models enable companies to achieve cost and time savings by leveraging external resources, and on the revenue side they allow companies to generate new revenues, for example from licencing fees. Since this notion is similar to the one of Open Innovation, one may come to the conclusion that the concept of open business models is "Open Innovation wrapped into the business model concept."

However, though similarities between Open Innovation and open business models exist, the concept of open business models is much broader and strategically distinct from the concept of Open Innovation. Firms with open business models co-create value with their external partners, which may result in product or service innovations (e.g. open business model in combination with Open Innovation), but the concept is not limited to the innovation process. Companies that make use of open business models can co-create value with their upstream (e.g. supplier) and downstream partners (e.g. distribution channel partner) as a key resource of their business model or they can co-create value with their customers or other stakeholders, such as NGOs.

The final element of a business model is the key processes. Key processes are operational and managerial processes, for example, product development, production and distribution, which enable a firm to deliver value in a way that can successfully be repeated and increased in scale. In the case of Dow Corning, key processes of the value-driven business model are R&D, sales and services, whereas key processes of the cost-driven business model are low-cost processes characterized by a high degree of automation, such as online ordering of silicones. Key processes also comprise a firm's rules, metrics and norms. Examples include financial measures such as gross margins, unit margins, net present value calculations and time to breakeven, operational measures such as quality of supply, lead times, or throughput, and other measures such as product development life cycles, brand parameters or incentive systems. These rules, metrics and norms are often the last aspect that emerges during the development of a business model. Moreover, they protect the status quo and are hence a typical obstacle against business model innovation. For example, if a pharmaceutical company demands that the "opportunity size to invest into a new business must be €100 million or more," this rule might prevent the discovery of new business opportunities that initially do not meet this criterion.

In summary, the customer value proposition and the profit formula define value creation for both the customer and the company, and key processes and key resources describe how that value is delivered. Johnson, Christensen and Kagermann conclude that "as simple as this framework may seem, its power lies in the complex interdependencies of its parts" [6:53]. Thus, the four elements cannot be viewed as being independent of each other. Scholars for example argue that "an appropriate business model involves choosing the right mix of alternatives" [14:66] and observe that when an interlocking element of a business model is missing or if one or more elements do not fit with the product offering, market failure is likely to occur. How well different elements of a business model complement each other, that is, the internal consistency of choices, therefore determines the effectiveness of a business model.

Now that we know what a business model is, we can take a look at the way companies in the chemical industry define and manage their business models. To illustrate how companies manage their business models we will take a look at the plastics business of BASF and the consumer goods business of Procter & Gamble (P&G).

7.1.1 Business Models at BASF

In 2001, BASF announced a new business strategy for one of its core businesses, its plastics segment (e.g. elastomers and epoxy resins), in response to a difficult economic climate characterized by weak margins, plant closures and divestitures – a situation similar to the current economic environment and therefore particularly suitable for an analysis. Back then, the company declared three goals to achieve profitable growth in its plastics segment: sharing success with customers, optimization of regional portfolios and development of new business models (see Figure 7.2). To share success with customers, on the one hand, BASF expanded in areas in which it could offer customers "clear advantages" while growing its own business profitably. On the other hand, BASF closed or divested businesses that were (due to capacity, technology or location) no longer competitive, that is, they could as a consequence no longer provide customers with any advantages. The optimization of regional portfolios was estimated to be achieved by improving processes and cost structures (and hence efficiency) in Europe and North America, while continuing expansion in Asia. As a third pillar, BASF planned to re-align standard polymers as well as specialties businesses due to changing market conditions by building new business models. In the following paragraph several examples will be used

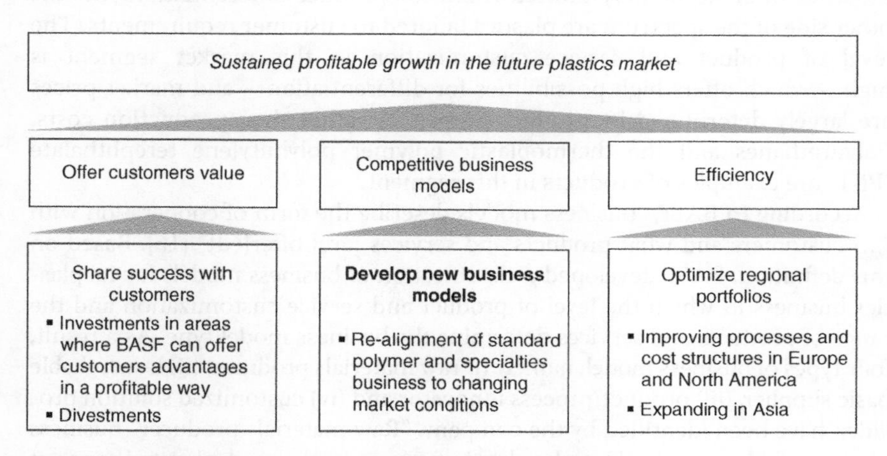

Figure 7.2 BASF's strategy for its plastic segment in 2001 [16]. Adapted from: Feldmann (2004)

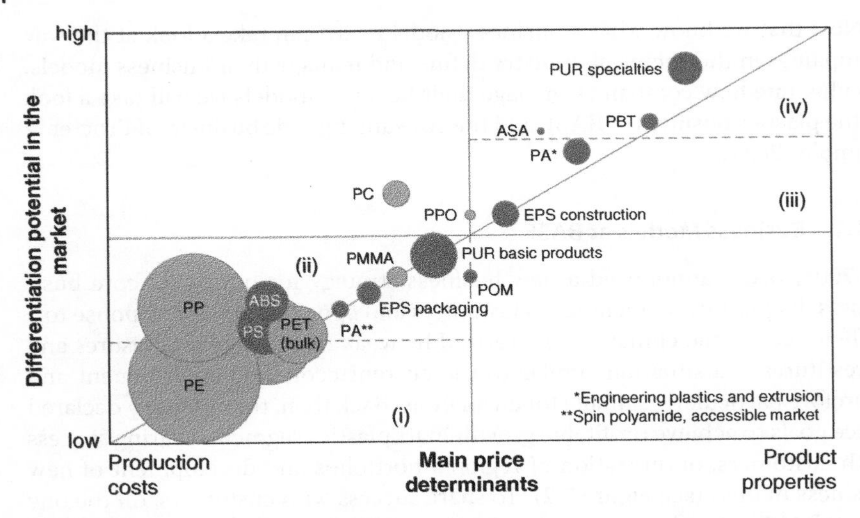

Figure 7.3 Market segments for plastics (bubble size = consumption in 2006). Estimation based on data from: Feldmann (2006) [15]

to demonstrate the interdependence between these goals and the development of competitive business models in particular.

Plastics can be characterized by means of main price determinants and differentiation potential in the market (see Figure 7.3). The market price of high-volume mass-produced plastics such as polyethylene (PE), polypropylene (PP) and poly(vinyl chloride) (PVC) is largely determined by production costs (which are primarily determined by costs of the raw materials), and these commodities offer no or only limited room for product differentiation. On the other side of the spectrum are plastics tailored to customer requirements. The level of product and service customization in this market segment is high – which offers high possibilities for differentiation – and market prices are largely determined by product properties rather than production costs. Polyurethanes and the thermoplastic polymer polybutylene terephthalate (PBT) are examples of products in this segment.

According to BASF, "business models describe the form of cooperation with [...] customers and what products and services [are] offer[ed]" [16]. Based on this definition, BASF developed a classification of business models for its plastics business in which the level of product and service customization and the range of products and services determine the business model type. As a result, four types of business model, namely (i) raw materials producer, (ii) lean/reliable basic supplier, (iii) product/process innovator and (iv) customized solution provider, have been identified by the company. "Raw materials producer" business models are characterized by a low level of customization and a limited range of products (see quadrant (i) in Figure 7.3). Such business models focus on low production costs and are typically employed for commodities such as PE and PP. BASF left the PE and PP markets in the early 2000s and thus does not operate a "raw materials producer" business model any longer. "Lean/reliable basic supplier" business models are designed to enable a higher level of customization compared with standard products and product price is determined more by product properties than in the case of typical commodities (see quadrant (ii) in Figure 7.3). Standard products with an appropriate availability and quality constancy are typically delivered with this business model. Furthermore, this business model type is characterized by reliability of supply with lowest costs, while short-term pricing takes regional and global supply demand into account. BASF employs this type of business model for spin polyamides, polyamide intermediates, styrene, polystyrene (PS), acrylonitrile butadiene styrene (ABS), styrene foam and polyurethane (PUR) basic products.

ABS is an example of how BASF re-aligned one of its business models in the early 2000s in accordance with its strategic goal to create competitive business models. ABS is a resistant, common thermoplastic widely used in products such as toys, for example LEGO® bricks are made of ABS, and consumer goods. BASF traditionally produced ABS exclusively at its location in Ludwigshafen and offered customers more than 1500 ABS products, ranging from full standard products to specialties. This model worked for several years. However, in the early 2000s BASF's ABS business model was under pressure. Firstly, production solely in Ludwigshafen resulted in complex global logistics and packaging requirements. Secondly, more and more applications were no longer dependent on specialties since the quality of standard products, produced at lower costs, had increased over time. In other words, standard ABS products became "good enough" to meet customers' quality requirements. According to John Feldmann, then member of BASF's Board of Executive Directors, "[t]his development was a direct assault on our business model, whose costs were [...] rising continually [...] It was obvious to us that a new business model was needed" [16]. BASF re-aligned its traditional business model in the following years to a "lean/reliable basic supplier" business model by dramatically decreasing the number of ABS products offered through its business model, enabling online ordering, opening new world-wide production facilities to enable supply from local production plants, developing a new production technology for the low-cost production of ABS and introducing a new colouring concept together with supply chain partners. Accordingly, several supply chain practices enabled the implementation of the new business model. On the one hand, the new business model was less complex due to the reduced number of product offerings, convenient online ordering and opening of new production plants. Focused production and efficient logistics thus became key aspects of BASF's business model. On the other hand, BASF began to offer its customers a colouring concept that allowed customers to colour their resin at their own facilities by using batches offered by selected partners. In this model, BASF supplies uncoloured ABS and partner companies, such as ALBIS Plastic and Clariant, supply the colour batch. Integration of partners hence became another key aspect of BASF's ABS business model.

At the next level of customization and product/service range are "product/process innovators" business models (see quadrant (iii) in Figure 7.3). These business models are characterized by joint development/collaboration with customers or clients, such as original equipment manufacturers (OEMs) in the automotive and electronics sectors. BASF for instance integrates its customers into its business model in the case of engineering plastics. At the highest level of customization are "customized solution provider" business models (see quadrant (iv) in Figure 7.3). These models are designed to deliver products that have a high potential for differentiation, and require an ever-closer interaction with customers. In this context, John Feldmann notes that "[f]or many business models, geographic proximity to the customer is [...] of great importance" [16]. "Product/process innovator" and "customized solution provider" business models cannot always be exactly distinguished from each other since they share certain characteristics. For example, both business models are characterized by a strong orientation towards customers, enabling for example the joint development of solutions with customers, the need for sales personnel with technical expertise and a high product/application range. BASF employs these types of business models for polyurethane systems and specialties, engineering plastics and styrenic specialties.

BASF's polyurethane (PUR) specialties business is an example of a "customized solution provider" business model. This model works in a fundamentally different way compared with the business model employed for ABS. A key aspect of its PUR business model is the so-called system houses, which are responsible for formulating tailor-made PUR systems for their customers. BASF offers about 8000 tailored systems to customers as well as services that correspond with the product life cycles of these PUR systems. As John Feldmann explains, "[t]he business model draws on two decisive BASF strengths: firstly, our integrated production facilities with all its economies of scale, and secondly, our network of system houses and development centres that enable us to respond flexibly to develop products that meet the exact requirements of the customer" [16].

In 2006, the majority (~70%) of BASF's plastics business portfolio was based on "lean/reliable basic supplier" business models, followed by "product/process innovator" and "customized solution provider" business models. However, BASF's target for 2010 was to significantly increase the share of "product/process innovator" and "customized solution provider" business models in its portfolio to 40% (whether this target was met cannot be determined from publicly available data since BASF's plastic business was regrouped and today it is an integrated part of its Performance Materials division). Accordingly, the company aimed to strategically shift its portfolio towards specialties.

To further illustrate the relevance of business model management, consider the case of BASF Coatings. BASF Coatings is a division of BASF that operates a "customized solution provider" business model, in which BASF Coatings manages customers' chemical processes. Instead of employing a business model that is entirely based on selling paint to an automotive OEM, BASF Coatings is today engaged to run the OEM's paint line. As the company explains, "combining modern paint processes with special-effect pigments and technologies" allows BASF Coatings to offer a "broad array of colour solutions and development capabilities that enable [customers] to enhance productivity and environmental performance" [17]. In such a business model it is in the interest of BASF Coatings to use and waste (and hence to produce) as little paint as possible – a fundamental reconceptualization of the firm's traditional business logic. Instead of selling as much paint as possible, the firm reversed its logic of value creation. Today BASF Coatings is no longer paid per supplied litre but per painted car, or as Alexander Haunschild, Senior Vice President at BASF Automotive OEM Coatings, explains "we don't simply supply a kilo of paint [...] especially when it comes to automotive OEM coating, processes, paint application and services for all aspects of the finish are indispensable" [18]. This new element of the firm's profit formula not only shifts the emphasis to fixed costs but also turns out to be quite valuable in changing competitive environments. For example, BASF Coatings recently reported that the US car industry has begun to restructure due to changing demand in terms of vehicle size "as consumer purchases were shifting to smaller, more fuel-efficient models" [19]. Logically, smaller cars also have smaller surfaces to be painted. Now while this might have been a problem for BASF's Coatings traditional business model, its new model is much more resilient to this shift in demand. Furthermore, the company expects that North America will strive for more colourful coatings and colour individuality. As Mark Gutjahr, Head of Design Europe at BASF Coatings, notes "[i]t's about time for us to have more colour" [19]. The new business model allows the unit to create and capture more value from this shift in customer demand due to closer customer proximity and customer familiarity.

BASF's coating business is also an example of how the different business models of BASF deliver value to the same customer. The electric car manufacturer Tesla is a customer of various BASF business divisions that employ different business models. Tesla approached BASF in 2010 and was at that time only interested in BASF's coatings capability, but BASF managed to seize "the opportunity and presented not only the benefits offered by a partnership with BASF for coatings but also the opportunities of a cross-business unit approach" [19], including engineering plastics, heat management and other electric vehicle-related offerings. Hence business models operated in different business units simultaneously target the same customer, and potentially reinforce each other.

7.1.2 Business Models at P&G

Every company needs to judge whether it is worth pursuing an internal R&D project or not. Studies suggest that firms very often pursue those R&D projects that fit well with their business model, and discourage those that do not [6, 20]. The consumer brand company Procter & Gamble (P&G) experienced this effect in the late 1990s. At that time the company was primarily pursuing internal R&D projects, while business opportunities that emerged outside P&G's core business were discouraged. By 2000, P&G's innovation success rate (i.e. the percentage of new products that meet financial objectives) was stagnating at 35%, the company's growth objectives could not be met, it duplicated services across regions and its shares lost more than half of their value [21]. A.G. Lafely, CEO of P&G, realized that P&G's innovation culture and its business model were constraining P&G's ability to benefit from new technologies and new application fields that emerged outside of its organizational boundaries. Thus the innovation culture and business model were impeding building a competitive advantage and ensuring the future survival of the firm. Therefore, Lafely made it P&G's goal to generate 50% of its innovations externally in a time where that number was close to 15%. But instead of filling P&G's internal innovation pipeline with acquisitions, selective innovation outsourcing or licencing, Lafely decided that a more radical change inside P&G was necessary to achieve this goal: P&G's innovation culture and its business model needed to change. In a first step, P&G initiated a programme called "Connect and Develop" to build an Open Innovation culture.

In the old organization, P&G had a centralized organizational structure and was focused internally. With respect to its innovation culture, R&D teams were primarily pursuing internally developed ideas and discouraging external ideas. In some cases this attitude was so strong that R&D teams even discouraged internal ideas that originated from other P&G business units [21]. P&G experienced the not-invented-here (NIH) syndrome. The NIH syndrome is a widely recognized innovation barrier that can be defined as "a negative attitude to knowledge that originates from a source outside the own institution" [22: 368]. NIH leads to an overestimation of the firm's innovation performance, scepticism towards outside knowledge and the belief in possessing a knowledge monopoly in a respective field. The central idea behind the Connect and Develop programme was to move P&G's attitude from "not-invented-here" to "proudly-found-elsewhere" by blurring the firm's boundaries [21]. To transform its innovation culture, P&G adjusted incentive systems, corporate structure and encouraged open collaboration (internally between units as well as externally). The company calculated that for every P&G researcher there were 200 external researchers outside P&G with the same level of expertise in a given science or engineering field – in addition to P&G's 7500 internal researchers, 1.5 million external researchers could hence be potential sources of innovation.

Instead of outsourcing innovation to these 1.5 million individuals, it was P&G's goal to find "good ideas and bringing them in to enhance and capitalize on internal capabilities" [21], that is, to combine internal and external competencies to create value. One aspect of this Open Innovation strategy is to collaborate globally with individuals and organizations and to search systematically for new technologies that can be commercialized inside P&G or at a partner company. P&G's open network involves partners in academia, suppliers, retailers, trade partners, universities and public research institutions, private labs, entrepreneurial firms and venture capital (VC) funds, as well as customers and in some cases even competitors. As an example, as part of its Connect and Develop programme, P&G created an IT platform to share technological problems with its suppliers. Chemical suppliers can submit ideas on how to solve a challenge and P&G may then jointly develop the solution with the supplier. Another aspect is the definition of search areas and search criteria (i.e. enablers of a systematic search). P&G decided to search for ideas that build on working technologies or technology prototypes with evidence of consumer interest, and that can further be developed by utilizing P&G resources, such as distribution channels, and capabilities. For example, together with external partners P&G discovered and developed new whitening strips and flosses that were branded under P&G's toothpaste brand Crest. Accordingly, an existing resource – the brand – was used to capitalize on products in new application fields. In another case, P&G partnered with an Italian bakery that had invented an ink-jet method to print edible images on cookies and cakes to produce printed Pringles potato crisps [21].

P&G's Connect and Develop programme is an example of how incumbents aim to counter competition, rejuvenate their existing business and create new growth paths. Specifically, the programme enabled P&G to deliver "its bargaining position in distribution channels" [23]. Today, roughly 50% of P&G's new products result from its Connect and Develop partnerships (see Table 7.3 for product examples that emerged from the programme) [24]. This programme enabled P&G to create an open business model that allows the company to benefit from external technologies and ideas, discover new application fields with partners and move beyond its core business. Based on the success of the Connect and Develop programme, P&G built bridges into its business model to further strengthen its Open Innovation strategy. Specifically, P&G redesigned its closed business model into an open business model. These "bridges" are designed to connect key resources and processes, for example, internal R&D, with key resources from partners such as other companies' IP to leverage internal R&D.

One of these "bridges" are so called technology entrepreneurs. Technology entrepreneurs are senior scientists from P&G's business units who are responsible for developing and maintaining relationships with academic partners, start-ups and other innovation partners. Furthermore, these technology entrepreneurs

Table 7.3 Examples from P&G's Connect & Develop programme.

Brand name	Product category	Description
Tide Pods®	Laundry detergent	P&G developed a patented film technology that wraps cleaning fluids in a clear casing together with the chemical manufacturer MonoSol
Olay Regenerist®	Skin care	The company Sederma developed a new peptide to repair wounds and burns that was, together with P&G, developed into an anti-wrinkling technology
Crest Whitestrips®	Dental care	Together with Corium, P&G developed a new teeth-whitening product
Pantene® Nature Fusion	Hair care (packaging)	P&G collaborated with Braskem to produce renewable, sugarcane-derived plastics for use in P&G packaging

Source: pgconnectdevelop.com (2014)

act as technology scouts who search outside P&G for solutions to internal R&D challenges. If these scouts find a promising technology they fill out an online questionnaire (e.g. How does the technology meet our business needs? Are its patents available? etc.), which is reviewed by general managers, brand managers, R&D teams and others. Meanwhile scouts can also promote the technology directly to management teams in a business line. If the technology receives positive feedback, it will be tested in consumer labs, and P&G's business development group will negotiate the terms of development with the inventor/manufacturer. Finally, the technology enters P&G's internal innovation pipeline for further development and commercialization. P&G estimates that from 100 external ideas, one ends up on the market [21]. Another bridge are internet platforms such as pgconnectdevelop.com or innocentive.com. While the latter acts as a technology broker, P&G's own homepage offers potential partners an overview of P&G's areas of interest, collaboration examples and an online form to submit collaboration proposals. Besides these open platforms, P&G also utilizes networks that connect companies with contract partners. An example of such a bridge in its business model is NineSigma. NineSigma connects companies that have technology-related challenges with contracted partners, for example, universities or consultants, who can submit possible solutions. If the solution is promising, NineSigma connects both partners and they can start a collaboration. As of 2006, P&G had submitted challenges to more than 700 000 individuals in NineSigma's network, resulting in 100 completed projects [21].

In 2003, P&G launched a business called YourEncore, which connects retired scientists with client businesses. Today the firm is independent and connects 8000 high-performing engineers and scientists with 70 of the largest food, consumer product and life sciences companies. This platform offers several advantages for companies like P&G. Most fundamentally, companies get access to experienced experts in a specific field – a valuable cross-disciplinary source of expertise. But firms can also experiment with problem solving approaches outside their traditional knowledge bases, for example by engaging retired aircraft engineers with experience in virtual aircraft design to develop virtual product prototyping, which has low risk and low costs. Retirees, on the other hand, are compensated based on their pre-retirement salaries, adjusted for inflation. Open Innovation bridges such as YourEncore are today key elements of P&G's business model but all started originally as a part of P&G's Open Innovation strategy. Engaging in Open Innovation can thus lead to business model transformation.

To further strengthen its open business model, to support P&G's business units and, in particular, to continuously transform P&G's way of doing business, P&G has launched a new unit called P&G Global Business Services (GBS). This unit operates a business model inside the P&G organization that allows P&G to offer its business unit's employee and business services such as IT, finance, facilities, purchasing as well as business building solutions. GBS is hence a shared service organization. Its 7000 GBS employees assist P&G business units in strengthening the operational efficiency and effectiveness of P&G's business model by delivering cost savings, driving scale, increasing innovativeness and agility and supporting digitalization of the business. Founded in 1999, GBS designed its business model between 2003 and 2005 to support the P&G organization in developing the structural ability to take advantage of economies of scale, provide services from one source instead of duplicating services across regions and to build a basis for making P&G's operations more efficient. Today, GBS offers, for example, virtual solution tools, that is, replacing physical product mock-ups with virtual reality applications, accelerates internal collaboration, such as by operating virtual collaboration studios and digital business spheres that allow P&G to make accurate and timely decisions. This separate business model allows P&G to utilize "talent and expert partners to provide best-in-class business support services at the lowest possible costs to leverage P&G's scale for a winning advantage" [25]. Overall, P&G's Open Innovation strategy enabled the transition to an open business model, whose efficiency is strengthened by means of service businesses inside P&G such as GBS. P&G's transition from a closed to an open business model was successful. The company has significantly increased its R&D productivity to 85% without disproportionally increasing its R&D expenditures [7].

7.2 Strategy, Business Model and Tactics

The relationship between business model and strategy, and subsequently between business model and tactics, has drawn a lot of attention from managers, but has also led to a great deal of confusion. To start with the most important aspect, strategy, business model and tactics are related but are different from each other. Whereas the business model, as a system of interdependent design elements, describes how the single parts of a business fit together to create and capture value in a competitive marketplace, the business model does not explain how a firm plans to compete in such a marketplace nor does it explain which customer segment to target. Although a business model facilitates the analysis, testing and validation of strategic choices, it is not itself a strategy.

Strategy is, by its definition, "the plan to create a unique and valuable position involving a distinctive set of activities" [5: 107] (see also the first four chapters of this book). Accordingly, based on their strategy, firms have made a choice about how to compete in the marketplace. This choice has consequences for a firm's way of value creation and value capture. The business model, as a system of choices and consequences, is therefore a reflection of a firm's strategy but it is not the strategy. Rather, strategy is about which business model a company should choose. This definition implies that competitive strategy contains actions for different contingencies, such as moves from competitors or market change, whether they become a reality or not [5]. Executives can thus either translate strategic choices directly into a single business model or they can consider several business models simultaneously based upon different strategic choices. Arguably, not every firm has such a plan of action for contingencies, but every firm operates according to its system of choices and consequences. Hence, every firm has a business model, whether expressed and understood by its management or not, but not every firm has a (competitive) strategy. This situation has been dramatically in evidence during the early 2000s when the dot-com-bubble burst. Many internet start-ups at that time had basically the same business model but no competitive strategy [3]. One can hence conclude that a well-designed business model is not enough for a company to survive in the long run. This observation allows us to realize that business models are not a silver bullet. True, they are a valuable way to understand how firms act and operate but executives should never forget about the competitive strategy of their firm.

While strategy influences the business model, the business model in turn influences the tactics available to the firm. Tactics refer to the "residual choices open to a company by virtue of the business model that it employs" [5: 107]. Therefore, the business models determine the tactics that are available to a firm. For example, a newspaper company with an ad-sponsored business model (sometimes also termed a freemium business model) cannot use price as a tactic, because the business model dictates that the newspaper must be for free. Accordingly, the business model "connects" strategy and tactics. Ideally, companies start by developing a (competitive) strategy, choose a business model determined by the strategic choice(s) and then have a set of tactics available by virtue of the business model. Figure 7.4 illustrates the relationship between

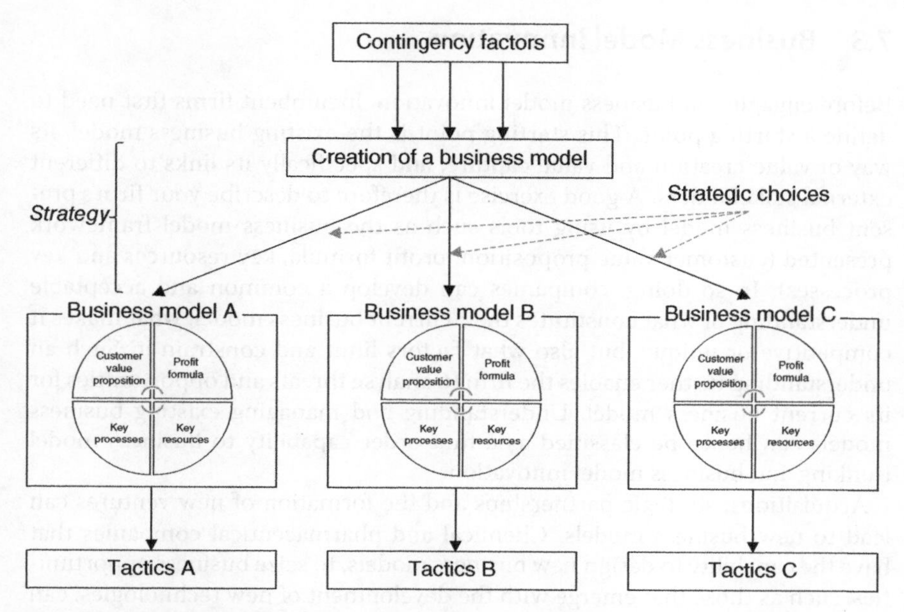

Figure 7.4 The relationship between strategy, business model and tactics.
Source: own figure

strategy, business model and tactics in a simplified way. While strategic choices made by firm's management during the creation of a business model cannot easily be changed, tactical choices, such as prices or minor product modifications, can be changed relatively easily. Strategic choices are accordingly more rigid than tactical choices. As an example, the German chemical company Altana describes its strategy as a "value-added specialty strategy" that is characterized by high spending for R&D. High R&D spending is hence a strategic choice with consequences for Altana's business model that cannot easily be reversed. However, this R&D policy does not prescribe a specific R&D spending, just that R&D spending should be high. Hence, the precise R&D spending can be changed and is thus not a strategic but a tactical choice. What constrains the set of tactics available to Altana is its business model. Elements of the business model set a boundary on how high Altana's R&D spending can go.

As straightforward as this relationship seems, we acknowledge that the business model does not necessarily emerge as a linear process out of the chosen strategy. Some companies, especially entrepreneurial ventures, may start with the development of a business model and later on develop their competitive strategy by virtue of their business model. Hence the business model can also be a blueprint for a strategy. In any case, executives need to understand the relationship between business model and strategy to sustain a successful business.

7.3 Business Model Innovation

Before engaging in business model innovation, incumbent firms first need to define a starting point. This starting point is the existing business model, its way of value creation and value capture, and specifically its links to different external stakeholders. A good exercise is therefore to describe your firm's present business model by using tools such as the business model framework presented (customer value proposition, profit formula, key resources and key processes). In so doing, companies can develop a common and acceptable understanding of what constitutes their current business model, what makes it competitive or unique, but also what factors limit and constrain it. Such an understanding further enables the firm to analyse threats and opportunities for its current business model. Understanding and managing existing business models can hence be classified as a first-order capability to business model thinking and business model innovation.

Acquisitions, strategic partnerships and the formation of new ventures can lead to new business models. Chemical and pharmaceutical companies that have the capability to design new business models, to seize business opportunities, such as those that emerge with the development of new technologies, can increase the total value creation potential by identifying new ways to create and capture value. This can, for example, be achieved by analysing business models in other industries and adapting these models to the industry in which the focal firm operates. Developing and designing new business models can therefore be classified as a second-order capability to business model thinking.

Finally, companies may face a situation in which their established business model is under threat of becoming obsolete. Chemical and pharmaceutical companies that have the capability to anticipate the need to transform their established business model, re-think their traditional business model, namely question the current way of value creation and value appropriation, and the capability to manage the transformation process, will be able to survive in dynamic and competitive environments. Re-thinking and transforming existing business models can thus be classified as a third-order capability to business model thinking.

To illustrate the role of these three capabilities we have associated each of them with a quote from a CEO in Figure 7.5. Later you will find a more detailed description of how to develop these capabilities and you will be provided with some basic questions to test whether your firm is ready for business model innovation. But beforehand you need to understand why business model innovation is important and why business model innovation is often challenging for incumbents.

Companies often invest substantially in the development of product and process innovations to achieve future revenue growth, profits and market advantages. However, creating, developing and then commercializing those innovations is more and more complex and expensive. Firstly, this is because technologies and business contexts change more rapidly. Secondly, existing

1st order capability: Understanding and managing existing business models
"In the new organization, the **bundling of product groups with the same business model** will help management to better focus on the success **factors necessary to be a market leader** both in meeting customer's needs and in operational excellence."
Kurt Bock, Chairman of BASF's Board of Executive Directors

2nd order capability: Developing and designing new business models
"Over the course of my career, I've come to see **business model innovation** not as a static process but as a systemic and reliable **capability**, one that leaders need to build, strengthen, and eventually turn into a **sustainable competitive advantage**."
A.G. Lafley, Chairman and CEO of Procter & Gamble

3rd order capability: Re-thinking and transforming existing business models
"For developed and developing countries alike, the new demands **cannot be met if health care continues to operate in the same way**. What is required are **new business models** that spread risks, take a broader view of health, and address the needs of the world's poorest people."
Joseph Jimenez, CEO of Novartis

Figure 7.5 Core capabilities for chemical and pharmaceutical companies.
Source: own figure

business rules (e.g. assumptions about scale or profit mechanisms) that are built to increase efficiency and revenue are more often in conflict with innovations. Thirdly, innovation takes place more frequently in new, uncharted areas, that is, areas where no precedent products and services exist, overlooked market segments emerge, or the nature of competition is changed by new operational capabilities. In turn, this also implies that companies today may face competition from players that were traditionally not considered a threat at all – not only in terms of competition from the low-end of the market, but also in terms of competition from other industries. As an example of unexpected competition from a different industry, consider the case of Google's self-driving car from the viewpoint of, for instance, Volkswagen. Over recent years, Google has acquired several robotic companies, such as Boston Dynamics and Schaft, and is actively developing capabilities in the field of robotics and in particular in the field of self-driving cars. These activities are not only a hotbed for future technological development, they are Google's way of strategically positioning itself in the markets of the future. In the case of cars, Google is adding robotics and auto-driving to its capabilities and competencies in maps and route planning. Taken together, Google is thus building core strengths for the production of autonomous cars. While established automotive companies such as Volkswagen certainly did not view Google as a potential competitor ten years ago, they should do so today (or at least seriously observe Google's robotic initiatives that, moon-shots aside, are not cute science projects but rather intended to sell products sooner rather than later).

While future returns from investments in innovation are always uncertain, it is today even more difficult (1) to evaluate the potential of new ideas due to increasing complexity and (2) to justify investments in innovation when the expected net revenue is smaller due to rising costs of innovation and shorter product life cycles. Meanwhile, this situation is far from news to companies. Many firms accept that the environment for innovation is today more complex and expensive than it used to be. Though they accept this, it is surprising that companies still struggle to manage innovation (e.g. Blackberry) or even survive in this environment (e.g. Kodak). We find that firms struggle because they view innovation entirely through a product-, service- or process-lens (or worse through a market-capital-lens). Instead, successful firms realize that value creation in this environment goes beyond traditional types of innovation and encompasses new ways of achieving competitive advantage. Research shows that many firms innovate in operations and/or products and services in response to fundamental changes in their industries, but financial outperformers put twice as much emphasis on other forms of value creation and on new ways of achieving competitive advantage as underperformers do [26]. One can conclude that successful firms recognize that:

I) innovation is more than R&D and technology
II) innovation can be a fundamental reconceptualization of what the business is all about.

Together, successful firms engage in business model innovation in response to a fundamental change in their environment.

Many companies realize that shifts in the economy open up opportunities to create breakthrough growth and allocate a lot of energy and resources to the generation of business ideas in such an environment. Meanwhile, the same companies often struggle to convert breakthrough ideas into sustainable businesses. By breakthrough ideas we mean the highest risk, highest return type of innovation. The challenge for incumbents with breakthrough ideas is to realize that generating breakthrough ideas is not enough because the capabilities of the company that "surround" this category of ideas and new technologies "will make or break them" [27:66]. Accordingly, breakthrough ideas can often not be converted into breakthrough growth because they go beyond idea generation and leadership excellence. In a study by Vijay Govindarajan and Chris Trimble, one executive described this phenomenon as follows: "I came to the conclusion [...] that limits to innovation have less to do with technology or creativity than organizational agility" [27:58]. Thus, the problem with breakthrough ideas is that they very often cannot coexist with the established business. As a consequence of this unnatural coexistence, breakthrough ideas often fail, when firms do not have the organizational agility to "cultivate" them. In this context, organizational agility means that firms have to build and successfully launch a new business model around the breakthrough idea when the existing business model is not suitable for commercializing the idea. Business model innovation therefore refers to the introduction of a new business model in an existing market. In summary, truly transformative companies never focus exclusively on R&D and the commercialization of new technologies. Truly transformative companies recognize that innovation includes business models, rather than just products or processes.

A changing environment is not only an opportunity to create and commercialize breakthrough ideas to achieve long-term growth with new businesses, but it can also be a serious threat to the established business of a company. Instead of having an existing business in strategic health and developing a new business (model) to commercialize a breakthrough idea or technology in addition to the established business model, the situation here is that the strategic health of the existing business model is in danger. When the existing business model of the firm is threatened with becoming obsolete, a fundamental reconceptualization of what the business is all about may become necessary. Accordingly, rather than introducing technological innovations (in combination with a new business model), the affected firm has to find a way to "break the rules of the game"[1] in its industry. Business model innovation therefore not only refers to the introduction of a new business model in an existing market but also to a fundamental change in at least one design element of an *established* business model (since the design elements of a business model are interdependent, a fundamental change in one design element, e.g. the profit formula, as a consequence affects the other elements as well). When firms recognize that innovation requires a fundamental reconceptualization of what the business is all about, they engage in business model innovation (sometimes also termed strategic innovation), which "in turn, leads to a dramatically different way of playing the game in an existing business" [28: 32]. As an example, in the 1960s Xerox dominated the photocopier market with a business model designed to serve customers that demand fast-working copying machines that can handle high volumes, that is, large corporations. Elements of Xerox's business model, for example, a professional sales forces (key resources) and leasing (profit formula), were tailored to serve these corporate customers as well as possible. In fact, Xerox's business model was so successful that some entrants, like IBM and Kodak, tried to adopt basically the same model. When Canon entered the market for photocopiers it decided, however, to compete with a fundamentally different business model. Instead of tailoring its business model to serving large corporations, Canon designed a business model to serve small and medium-sized enterprises that were overlooked by Xerox, because they did not demand high-speed copy machines. To make photocopiers affordable for these customers, Canon consequently differentiated itself based on price instead of speed. This different customer value proposition had consequences

1 It should be noted that the question of whether to break the rules or whether it is better to play them is by itself already a difficult question that cannot be analysed extensively in this chapter.

for other elements of its business model. For example, Canon leveraged partner capabilities by selling its machines through a dealer network instead of through its own distribution channel by means of a direct sales force. While entrants like IBM and Kodak were not very successful in imitating Xerox's business model, Canon was able to become the market leader in terms of unit sales because it entered the market with an innovative business model [28].

Accordingly, business model innovations by entrants and business model innovations by incumbents can be two sides of the same coin. On the one side, business model innovations by entrants threaten the business of established firms, and on the other side, established firms can respond to exactly that kind of threat with business model innovation. Consequently, understanding business model innovation and thus realizing innovation can be a fundamental reconceptualization of what the business is all about is relevant for established firms and newcomers alike.

While it may sound obvious to change the established business model before it becomes obsolete, several barriers exist that prevent an early adoption of an incumbent's business model. For example, a prerequisite for business model innovation from established companies is to fundamentally question the firm's present way of doing business. However, as Constantinos Markides notes very well in one of his studies, "advising companies to question their way of playing the game and think of alternative ways, especially when they are successful, is fruitless [...] they simply do not do it, even though they know and agree with the principle" [28: 34]. To understand the reason for this disturbing observation, let us consider a company that has experienced a history of increasing profits, then, after profits have reached a maximum, a profit decline, and finally a financial crisis (net income loss, etc.). More often than one might expect, companies do not begin to question their business model in times of declining profits but only when they are already in a financial crisis. It is not difficult to understand that innovating a business model in the middle of a financial crisis is the worst time to do so. Most obviously, the ideal scenario would be to question a firm's current way of playing the game not when the numbers are red but when profits have reached a maximum. The problem with this intuitive scenario is that financial metrics that are typically applied by companies to monitor the health of a business model will at the point of maximum profits not indicate an upcoming crisis [28]. Accordingly, when firms are successful, many managers will argue against a change, simply because metrics indicate good financial health. This inertia of success is often the reason why firms recognize a necessary change in their business model too late.

Before exploring how to overcome the inertia of success, let us address another common pitfall in the context of business model innovation: business model innovation is "not about what the business will be doing in 20 years; [it is] about the preparations it must make today" [27: 110]. This statement should be framed and placed somewhere in a company's executive meeting room because we frequently observe the opposite understanding of business model innovation in the chemical and pharmaceutical industry. Business model innovation is not about the choice of business model design elements in 20 years, it is about the changes a firm has to make today to still exist in 20 years. Too often we notice that managers confuse business model innovation as a kind of forecasting tool; business model innovation has more to do with the present than most managers realize. This in turn is one reason why business model innovation often fails – at the point when an (unprepared) company recognizes the need to change, it is already too late to alter the course, because the time to make the preparations for a fundamental change in the business model is over. True, examples exist where companies had their backs against the wall and managed to rejuvenate themselves out of a financial crisis by means of business model innovation, for example, Dow Corning with Xiameter, but for one company that manages this transition there are hundreds of firms that did not. The business graveyard is littered with failed business model innovators. This is not only a disaster for the focal firm but for the whole economy. For instance, at its peak Kodak[2] had 160 000 employees and a market capitalization of $28 billion, and Blockbuster (a company that offered video rental services in the United States) had 30 000 employees and a market capitalization of $5 billion at its peak. Both companies missed adapting their business to the digital revolution and went bankrupt (when Blockbuster had to close its last stores in 2013, CNN titled the story "digital killed the video store").

So how can your company become a successful business model innovator? To answer this question, we will break down the three core capabilities (understanding and managing existing business models, developing and designing new business models and re-thinking and transforming existing business models), which we introduced earlier, into six steps:

1) **Analyse your current business model.** The fundamental starting point to business model thinking and business model innovation is to understand your present business model. Particularly in the chemical industry one can observe a lack of communication from top management about the elements of the firm's business model. Business models need to be explained. It is the job of every executive to communicate the business model internally. The better you communicate what your business model is, the better it can be executed, and the more likely it is that people in your company can identify a need to change the business model.

   These questions will help you to develop a thorough understanding of your present business model: What are the key elements of your business model? What value proposition(s) do you offer? How do you deliver value to

2 Kodak invented digital photography in 1975 but was unable to create a breakthrough business around that technology.

your company? What key resources and key processes do you need to deliver your value proposition? What makes your business model unique? What are the strengths and weaknesses of your current business model?

2) **Develop the capability to manage your business model.** Companies need to understand how to manage their present business model before engaging in business model innovation. A firm's business model connects the strategy level of the company with its operational level, for example, the firm's supply chain practices. Successful companies align these interdependent levels by creating strategic fit.

   These questions will help you to identify core management tasks: Do you have product groups with similar business models in your portfolio? How do the business models in your portfolio benefit from each other? How do strategic orientation and the design of your present business model reinforce each other? What contingency factors, such as shifts in competition, changing regulation, disruptive innovation, are a threat to your current business model? Would any of these contingencies require a fundamental change in your business model? If your chemical or pharmaceutical company was Apple (Amazon, Dell, GE, Google, McDonald's, Skype, UPS, Wal-Mart, or any company that operates differently to other companies in your industry), what would your business model look like and how is this different from your current way of doing business?

3) **Develop the capability to sense the need to adapt your business model.** In order to recognize a necessary change, firms need to develop a strategic sensing capability. A first step to developing this capability is to "forget" about the existing business model of the firm. This means that a firm has to "leave behind notions about what skills and competencies are valuable" [27] and subsequently find a new answer to the fundamental questions that define a business. "Throwing assumptions overboard" is something that is easy to say but difficult to achieve because companies have strong sources of organizational memory. Sources of organizational memory may cause companies to assume that what has successfully worked for the established business model will also work in the future. Because firms' existing rules, metrics and norms can interfere substantially with those needed in the future, it is important that firms develop a questioning attitude and systematically begin to sense alternative ways of operating. In this context, a questioning attitude refers to mentally experimenting with a few "what ifs" and "whys." Developing this kind of attitude is especially difficult when your company has well-established performance standards, a history of promoting from the inside and a strong culture of holding people accountable to plans. Moreover, as you have already learned, financial metrics cannot tell you early enough that something is wrong with your business model (don't get it wrong: you absolutely need financial measures such as ROI or gross profits but they do not monitor the

strategic health of a business). Accordingly, you need measures for strategic health and not just for financial health. These strategic measures are company-specific but could be things like customer dissatisfaction, structural changes in the industry, deregulation, distributor and supplier feedback, or even employee morale. In a second step, successful business model innovators challenge the established planning process of the firm. Under stable conditions, the typical annual planning cycle of an established business works, but if business realities begin to change you need a more frequent planning cycle, such as quarterly or even monthly. Moreover, in a typical planning meeting companies spend more time on questions about efficiency, that is, *how* to improve operations, than on asking *who* their customers are and *what* kind of job they really want to get done. Sensing requires focusing more on the who- and what-type of questions. Therefore, a forward-looking executive not only focuses on performance excellence and continuous improvement of the existing business, that is, the execution of the existing business model, but promotes the long-term potential by questioning the status quo [27]. Overall, it requires strong leadership to "see a different future and having the courage to abandon the status quo for something uncertain" [28].

These questions will help you to develop this capability: Who is our customer? What if our customers begin to value different attributes of our offering and what if their priorities among these attributes change? Why do our customers pay the way they do? What if our customers can no longer be reached through our traditional distribution channels? What if we have to operate our business abroad, why would our domestic business model still work in a foreign country? Do we have measures for customer/employee satisfaction/dissatisfaction? In our planning meeting, how much time do we spend on how-type questions and how much time on who- and what-type questions?

4) **Integrate customers and suppliers during business model development.** Business models need to be explained and communicated – not only internally but also externally. Discussing your business model design with your most important customers and key suppliers will help you to transform your present business model. From our point of view, most chemical and pharmaceutical companies already integrate their customers into their new product development efforts on a regular basis. This integration, however, needs to be shifted from a product to a business model level. In the case of supplier integration, there is still a lot of untapped potential in the chemical and pharmaceutical industry. In contrast to, for example, the automotive industry, suppliers are more seldom integrated into innovation efforts in the chemical and pharmaceutical industry. In terms of business model innovation, companies should, however, consider customer and supplier integration equally.

These questions will help you to integrate customers and suppliers into the business model development process: What is the business model of your customer (supplier)? On a scale from 1 to 7, how different is your current business model from those of your key customers (suppliers)? Which three key suppliers are needed to operate your present business model?

5) **Use strategic and operational flexibility to successfully implement a new business model.** Because organizational memory is an obstacle to forgetting how your business currently operates, one might come to the conclusion that the easiest way to ensure that the established organization does not interfere with the implementation of a new business model would be to strictly separate the new from the established business model. Though a separation strategy has advantages, a complete separation does not come without disadvantages and is in fact often impractical. The biggest disadvantage of a complete separation is that it prevents exploitation of synergies between the two business models [29]. Accordingly, firms face a trade-off. On the one side, a new business model has a lot to borrow from the established business, for example, supplier network or people, but on the other side if the new business borrows too much, breaking free of organizational memory becomes more difficult. Thus, companies have to keep a balance between separation and integration of the two business models. This requires strategic and operational flexibility. A rule of thumb is that the new business should only borrow those resources that are crucial to gain competitive advantage. Therefore, a small and carefully selected number of links between the established and the new business has to be established and managed. This requires significant attention from senior management because points of interaction often create tension between the two businesses and their business models. Those tensions might even disintegrate the collaboration between the two businesses and consequently destroy the new business. Firms should hence promote collaboration through individual incentives that reward managers' willingness to collaborate, compensate the established business when it releases resources to the new business and reinforce common values to build trust between the two businesses. Overall, a firm's top management needs to convince unit heads, or those managers in the established business who have to share resources with the new business, of the relevance and usefulness of resource sharing. Because these conflicts tend to be serious, it is advisable to let a senior executive with a proven track record supervise this process. As a general rule, the more intense the anticipated conflicts are, the more separated should the new business model be. We can conclude that companies that have the flexibility to re-allocate resources and reconfigure processes will more likely be able to succeed in business model innovation.

These questions will help you to identify core management tasks for the development of strategic and operational flexibility: How painless is resource sharing, for example, between business units, in your company? Are the new business model and the established business model strategically related (same market, same profit formula, etc.)? What key processes are an inherent part of the new business model and which can better be performed in the established business? What level of autonomy does the new business model require with respect to rules, norms and metrics? Is a new culture required to successfully operate the new business model?

6) **Learn from strategic experiments with business model design elements.** Executing business model innovation does not just require strategic and operational flexibility. Companies that successfully transform their business model engage in strategic learning, that is, they conduct experiments among the design elements of a business model, accept failure but fail fast and learn from failure with speed and discipline. During the implementation of a new business model (or the transformation of your established business model), you will naturally face a lot of uncertainties. It is important to solve these critical uncertainties with speed and discipline because unknowns will make or break the new business model. Strategic learning is a process, starting with "wild guesses" and ending with "reliable forecasts." Because predictions are at the outset imprecise and uninformed, and consequently tend to be wrong, it is quite tempting to forget them or put little effort into their analysis. For example, top managers at the new business may assume that the new business has little time to review discrepancies between planned outcomes and realities because they want to focus on execution (instead of planning) in order to quickly achieve first-mover advantages. However, "predictions are important not because of their accuracy but because of the learning opportunities they present" [27:66]. Companies should therefore carefully analyse disparities (including positive disparities) between predicted and actual outcomes, such as sales or profits. This analysis is a crucial learning step during the ongoing development of a new business model because decision-making based on discarded or overly aggressive predictions can lead to fundamental misjudgements of business conditions, preventing necessary adjustments in the design of the new business model. To improve learning, companies can for example create architectural variety among the design elements of their model by testing different designs, such as at different locations. In so doing, companies can learn which business model designs work and which do not, and act on that knowledge. Joan Magretta once described this process as follows: "Business modelling is, in this sense, the managerial equivalent to the scientific method: You start with a hypothesis, which you then test in action, and revise when necessary" [3:90]. Inspiration for experiments with the design elements of your business model can come from almost everywhere – but in particular from other industries and markets. These experiments do not have to be complicated. Consider an unusual example for an experiment with a firm's

profit formula: imagine a German zoo in December. It's cold, it's grey and it's raining a lot. The zoo in the city of Münster has on average 11 200 visitors in December – the lowest number throughout the whole year. In December 2012 Münster zoo decided to do a small experiment with its profit formula and offered customers the opportunity to pay whatever they liked to get into the zoo. Instead of selling tickets at a regular price (typically around €12 in winter), this *pay-what-you-want*-model enabled customers to pay more or less than the regular price or even nothing. What is easy to guess is that the number of visitors significantly increased in 2012. Indeed, the number of visitors increased by a factor of five. Surprisingly, revenue increased by a factor of 2.5 at the same time. Although the zoo's basic business model remained the same, this experiment with the profit formula was a huge success. Münster zoo tried the same again in 2013 and it was even more successful. In fact, the zoo had so many visitors that it had to stop the pay-what-you-want-model in order to protect the animals from too much stress. One can draw two conclusions from this simple example. (1) Experiments with the business model design do not have to be complicated. (2) Take a look outside your own bubble (company/market/industry) to get inspiration for business model innovation. From our experience, developing a learning attitude is the capability most difficult to build.

These questions will help you to identify core areas in developing a learning capability: On a scale from 1 to 7, how good is your firm in identifying strategies that have not worked? How easy is it to recognize alternative approaches to achieving your firm's objectives? How accepted is it in your company to fail? How fast can you adjust your business model design?

In practice, these six steps will help you not only to develop a business model that is new to your company but one that is also appropriate in the market. Overall you need both novelty and appropriateness (Figure 7.6).

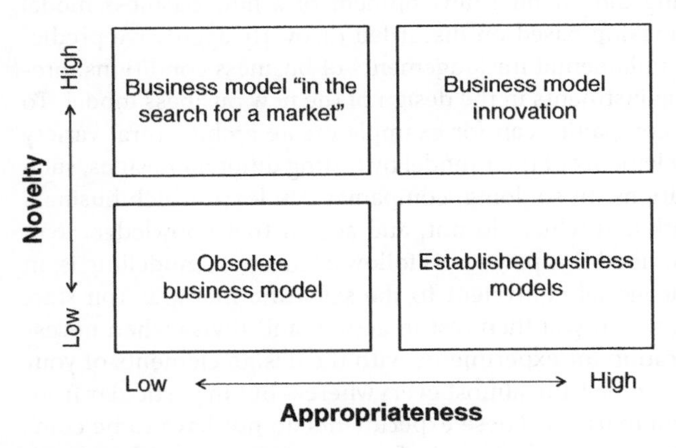

Figure 7.6 Business model framework. *Source:* own figure

7.4 The Role of Business Models in the Chemical and Pharmaceutical Industry

We will close this chapter with a look at why and how chemical and pharmaceutical companies engage in business model thinking and business model innovation. Both the chemical and pharmaceutical industries are facing dynamic change on a global scale. Global megatrends such as growing and aging populations, energy and climate change and inter-industry innovations (e.g. electric mobility) provide growth opportunities for chemical and pharmaceutical companies but also represent challenges.

On the one side chemical and pharmaceutical products that increase sustainability, quality of life and health represent opportunities for downstream expansion. Companies that can leverage their core competencies to seize these opportunities can create value by increasing supply chain sustainability, integrating core businesses and/or by moving beyond current areas of operation. Accordingly, the changing shape of the chemical industry's value chain that has forced companies to diversify and reduce value chain coverage, from the 1990s onwards, is estimated to continue. Some companies will continue to integrate forward, for example, to performance chemicals, while others, particularly chemical companies in emerging economies, will focus upstream on feedstock foundations. Future growth in the chemical and pharmaceutical industry will also be enabled by new functional materials, such as in the area of battery materials or functional crop care and the creation of "system solutions." Companies that have the capability to identify new ways to deliver future technologies to customers and to expand into new application fields, for example, personalized medicine, are estimated to significantly benefit from this development. Furthermore, shifts from developed to emerging economies such as Argentina, Brazil, Malaysia and Thailand, particularly in markets for plastics, agrochemicals, pharmaceuticals and automotive materials, continue to create growth opportunities. Capturing growth in emerging economies requires that firms develop the capability to manage complex global supply chains and operations, balance global and regional strategy execution and create ways to seize business opportunities at the bottom of the pyramid.

On the other hand, the changing industry landscape threatens traditional ways of operation and value capture. Shifts in competition, such as from low-end disruptors, commoditization, "genericization" and new industry entrants that target emerging customer segments, challenge the traditional ways of how incumbents do business. Established chemical and pharmaceutical companies face the challenge of transforming their businesses by increasing agility and flexibility, reducing complexity and improving resilience to shifts in competition. Companies that seize downstream business opportunities for example become more customer-focused but meanwhile increase the complexity of their operations. Closing or divesting upstream businesses in a firm's business portfolio

Table 7.4 The role of business models in a changing industry landscape.

Trend	Aspects	Role of the business model
Value growth in- and out-side the core	Re-organizing corporate structures, M&A, partnerships	Managing business model portfolios
New technologies – new applications	Future growth areas, functional materials, "system solutions"	Developing new business models
Shifts in competition	Low-end disruptors, emerging customer segments	Shorter business model life cycles
New ways of value creation	"Job-to-get-done" emphasis, bottom-of-the-pyramid	Innovating established business models

Source: own table

and meanwhile focusing on downstream opportunities in future growth markets, for example, medicines for rare diseases, OLEDs (organic light-emitting diodes) or lithium-air batteries, may hence not be the appropriate decision. In fact, responding to the defection of customers to low-cost suppliers by disrupting the disruptor may create competitive advantage and increase a firm's performance. Creating the right mix in a firm's portfolio, understanding interdependencies between business segments and identifying the right collaboration partners that enable fit between a firm's value creation activities as well as between a firm's interdependent businesses are challenges that chemical and pharmaceutical companies need to address more than ever. The management consultancy Accenture concludes in a recent chemicals report that "[a] changing, more complex world drives the need to rethink existing business and operating models that are key to delivering high performance" [30].

Based on these observations one can conclude that chemical and pharmaceutical companies need to consider business models to: (1) create value inside and outside their core area of operations, (2) to commercialize new technologies in new application fields, (3) to respond to shifts in competition and (4) to identify new ways of value creation and appropriation (Table 7.4).

7.4.1 Value Growth In- and Out-side the Core

Chemical and pharmaceutical companies have begun to consider business model transformation as a way to sustain competitive advantage in times of industry consolidation and restructuring. As an example, consider the following three cases:

- To create a more customer-focused and agile company, Dow Chemical transformed its business structure in 2012. Andrew Liveris, Dow's Chairman and CEO, emphasizes the need to constantly transform business models. He explains, "[w]e continue to adapt our business model to take advantage of the changing dynamics in the global marketplace" [31].

- The consumer goods company Unilever increased revenue while reducing its environmental impact by making sustainability part of its business model, that is, the company re-designed its supply chain activities. Paul Polman, CEO of Unilever, argues that it is "very clear for businesses that if they make [sustainability] their business model [...] and plan for this carefully, it is actually an accelerator of growth," and further explains, "to provide new products and services in that world, the need to continue to guarantee a stream of resources, thinking a little differently about the use of scarce resources in itself. All these things add up to a very valid business model" [32]. According to Polman, Unilever has saved €200 million annually by embracing sustainability in its business model, for example, by changing the use of scarce resources in its business model, while growing its core business at the same time.
- The business model of the chemical distributor Brenntag connects chemical manufacturers (its suppliers) and chemical users (its customers). Brenntag purchases industrial and specialty chemicals on a large scale, stores them, packs them into smaller quantities and delivers them (typically in less-than truckloads) to its customers – customers that are (because of their size) often not directly served by large chemical manufacturers. The company has adapted its business model over recent years by including new value-adding services such as product mixing, drum return handling, inventory management and laboratory services. These new service activities not only generate additional revenues but also bring Brenntag closer to its customers.

The re-organization of corporate structures, enhancing the sustainability of supply chains and combining product and service offerings are examples of how companies adapt their core businesses to grow value by means of business model design and re-design.

In response to industry dynamics and market change, chemical and pharmaceutical companies not only emphasize the transformation of existing business models but also the exploration and development of new business models. Several chemical and pharmaceutical companies have established units that specifically search for new business models. For example, DSM has established a unit called Ventures and Licensing. DSM's Chief Innovation Officer, Rob van Leen, explains the purpose of this unit as "to provide a window to the world for the DSM businesses and create strategic options such as access to new technologies, markets and business models" [33]. Similarly, Pfizer Venture Investment, the venture capital arm of the pharmaceutical company Pfizer, "has an interest in working with others to explore new business models that can create value for all players in the healthcare/life sciences ecosystem and ensure the continued development of therapeutics, technologies and services for all those whose medical needs are not being met" [34].

With established business models in transformation and new business models in development, more and more chemical and pharmaceutical companies also realize that they not only have to manage product or technology portfolios but also business model portfolios. For example, the petrochemical manufacturer SABIC aims to "build a portfolio of technology and business model options with long-term strategic business potential for current and possible new SBUs [strategic business units]" [35]. Managing business model portfolios is hence recognized by companies in order to achieve long-term success but also represents challenges. As Verity *et al.* note, chemical companies "will be compelled to operate many business models simultaneously to manage global and regional as well as commodity and specialty businesses" [36].

7.4.2 New Technologies – New Applications

Traditionally, chemical and pharmaceutical companies invest heavily in R&D and seek to gain deep customer insights to develop new products with better functionality and/or performance. This strategy has been and still is very successful. For example, BASF generated €8 billion sales in 2013 with product innovations that have been on the market for less than five years. Two examples illustrate this observation further. In 2007, BYK, a Germany-based supplier in the additives and instruments sector, introduced a new adhesion promoter, labelled BYK-4500, to the market for aqueous decorative and architectural paints. Traditionally painters needed to carefully sand and prepare surfaces with old paint on them prior to then coating with new paint. BYK's product innovation makes this preparatory work unnecessary. Accordingly, the innovation improves adhesion to substrates such as old alkyd resin coatings as well as to substrates that cannot be sanded for structural reasons. BYK targeted its existing customer base and used its established distribution channels and resources to deliver this new product. A different case is P&G's Swiffer, a lightweight stick with disposable (electrostatic) cleaning wipes. When P&G introduced the Swiffer in 1999, a market for electrostatic household cleaning wipes was literally non-existent, and P&G's product innovation competed against the established segment of mops and brooms, making it a radical product innovation. Moreover, P&G chose to give away the durable broomstick in order to lock consumers into purchasing the consumable cleaning wipes (razor-blade model). With this combination of technology and revenue model, P&G served customers in a fundamentally different way. Although the Swiffer technology was radically new to the market, for example, it changed the way consumers were cleaning, and its revenue model (razor-blade) was new to P&G, the company was able to use its existing resources, such as relationships with retailers, distribution and sales channels and competencies, and especially its marketing capabilities, to offer the product. Accordingly, P&G leveraged its core strengths in product commercialization to introduce the Swiffer. Overall, P&G succeeded in serving new customers in a fundamentally different (and very profitable) way but using its existing strengths in marketing and operations.

Though these two examples are distinct with respect to the nature of the customer (serving existing customers in traditional ways versus serving new customers in new ways), they have one thing in common. In both cases the nature of the opportunity had a good fit with the established organization, that is, the current business model. Even if the Swiffer may seem to be very different from P&G's core businesses, its existing business model was suitable to be able to commercialize this radical product innovation – P&G seized an adjacent area of its core business. However, business opportunities that emerge from new technologies do not always fit with a firm's established business model. That is especially the case when new technologies open up new application fields in which the focal firm has no experience, namely opportunities outside a firm's core business and beyond adjacencies. In other words, new technologies may create opportunities in areas "where, relatively speaking, assumptions are high and knowledge is low, the opposite of conditions in the company's core space" [8]. When new technologies do not fit with a firm's existing business model, the existing business model needs to be adapted or a new business model is needed. To illustrate this situation, consider the following two examples:

- A pioneer and today a market leader in liquid crystals (LCs) for displays, the Germany-based chemical company Merck has developed a strong application and production know-how in LC technologies. In the future display market, however, alternative technologies such as OLEDs could potentially threaten Merck's leadership position in the display technology market. Merck therefore invests in OLED technologies but not primarily because of a potential substitution for LC technologies but because OLEDs open up new application fields beyond LC displays. Merck estimates that OLEDs will enable new display features such as flexible displays, but estimates that non-flexible displays – the strength of LC technologies – will remain the dominating display format until 2025, for example because an effective technology to efficiently manufacture OLED TVs on a mass scale does not exist as yet [37]. It is very likely that Merck's existing market position and particularly its established customer relationships with display producers will allow Merck to integrate OLED display technologies into its existing business model. That is because even if prices for OLED TVs were to drop significantly, a mass-scale production technology for OLED TVs would become available, and OLEDs would become the dominant display format, then Merck could still benefit by means of its established business model. What makes OLEDs more interesting for Merck are new application fields beyond traditional displays. Accordingly, OLEDs have a high potential for differentiation in applications where LCs are not used. Flashing clothes, intelligent light guiding, transparent advertising on curved windows, glittering facades and luminous wallpapers are examples of applications beyond traditional displays where OLED technologies can be used. Since these application fields are not covered by Merck's current LC business, they offer new customer value propositions for new customer segments and require different resources and competencies, such as new distribution channels and production know-how; the company may thus have to build a new business model to commercialize OLED applications beyond LC displays.

- The Dutch life science company DSM is developing a new business model to capture business opportunities created by new bio-based product technologies. Specifically, DSM uses partnerships to create "sustainable business models" for the bio-based economy of the future. According to DSM this open business model is "either used to get access to intellectual property generated by third parties, what's known as in-licensing, in order to accelerate product development in [the partner] company; or it is used to create value by 'out-licensing' DSM technologies, either as part of a business model, or as value creation for idle IP" [38]. Some of the product technologies that DSM is co-developing with its partners replace existing products made from non-renewable sources, for example, second-generation biofuels, and hence target existing customers (by means of DSM's traditional business model) but other products and processes, such as cradle-to-cradle processes, require a fundamental re-design of established business models. As an example, in collaboration with the France-based biochemical company Roquette, DSM has developed a patented biotechnological process for producing succinic acid. Succinic acid is estimated to create a portfolio of products and applications that go beyond DSM's current area of operations, enabling the company to enter new markets and target new customers in different ways.

To commercialize new technologies in new application fields, and accordingly create value in future areas of growth, chemical and pharmaceutical companies need to consider the development of new business models to seize these opportunities. This may include the adaptation of the firm's present business model if the current model does not favour the commercialization of new technologies that originate from outside its organizational boundaries.

7.4.3 Shifts in Competition

When specialty chemical companies commercialize new molecules or formulations they typically compete on the basis of product performance or product functionality. The firms' technology and application know-how enables them to offer customers differentiated products with unique properties or features. Sometimes these specialty products are so unique that customers, although aware of the unique features, require dedicated support in using them. As an example, under its TEGO brand, Evonik Industries offers its coatings customers web-based seminars as well as product training, for example, for its hydrophobing agents, co-binders, nanoresins and heat-resistant silicone binders. Specialty manufacturers operate large R&D departments to generate the necessary expertise, build strong IP protection for their inventions, are innovation oriented, rely on skilled technical sales forces and bundle products and services, such as by offering on-site training for their customers. In turn, manufacturers not only create high customer value but also capture a premium of that value by charging high prices for their offerings. At this stage, the firm's business model is designed to deliver this unique product–service bundle to customers.

Over time, competition shifts from product performance to product reliability and process innovations become key to success. Specialty chemical companies use their strong technology and application know-how to improve product resistance or durability, and/or their own production processes and/or customer processes. In so doing, they can still realize high profit margins from their tailored bundle of products and services. The firm's business model becomes refined in accordance with process innovations, which may include a change in the importance of activities or the creation of new activities in its business model design.

For a while a specialty chemical company can use its innovation capabilities to fulfil customers' demand in a way that rivals cannot, but inevitably competition will finally shift to costs. Firstly, this is because high margins naturally attract new entrants, which will result in intensified competition (or in the worst case in price wars). Secondly, because over time, new entrants and competitors gradually improve their specialty chemistry capabilities, they thus begin to offer similar products and services. These products may not achieve the same performance as the incrementally improved product–service bundle of the established firm but they may be "good enough" to fulfil customers' requirements. Established firms normally respond to this threat in two ways: (1) they begin to offer existing customers additional features, for example, they try to foster the uniqueness of their offering (e.g. increase R&D expenditures), accepting lower margins and (2) they spend time and money to enforce their IP rights. Notwithstanding that both strategies may help to fend off imitators for a while, incumbents are trying to achieve something with these strategies that is impossible, namely to shift competition back to performance/functionality or reliability. Commoditization in the specialty chemical business is an inevitable part of the business model life cycle. Commoditization can for instance be observed with polycarbonates, where gross margins have dropped from 90% to less than 60%. This effect is particularly intense in the pharmaceutical industry where drug sales may drop up to 80% after generic copies have been introduced. Instead of focusing on providing additional and more advanced products/services (which customers often don't need), companies should understand that a competition shift towards costs is a business opportunity. However, as we observe, companies are in general very good at competing on the basis of performance and reliability but struggle in competing on the basis of costs. Competition shifts to costs can be seized by means of business model innovation, that is, rejuvenating the organization and extending the "life time" of the business model. Yet, innovating a business model that is under direct threat of disruption is exceptionally difficult. Studies in the chemical industry find that "[t]urning flawed business models around requires looking at the specialty chemical business with an entirely new lens in the process of analysing a company's operations objectively and making tough decisions that ultimately transform laggard organizations into those more suited for today's operational conditions in the industry" [39: 10]. This shift is particularly difficult because it requires focusing more on customer value proposition than on particular products/services or established norms of how to serve customer segments. Companies that proactively innovate their business model to compete on costs are indeed rare. More often chemical and pharmaceutical companies begin to search for a way out of the commodity trap after a company crisis has occurred.

7.4.4 New Ways of Value Creation

Chemical and pharmaceutical companies have begun to explore new ways of value creation and value capture. The capability of exploring new business models, and particularly of experimenting with new business model designs, enables business model innovation. In contrast to the previously described need to innovate an existing business model because of commoditization, the focus here is on the systematic search for business model innovations in order to reinvent an industry's dominant business model design. In so doing, companies can enter new markets that cannot be reached with traditional business models or by reorganizing existing markets. The latter is of particular relevance for the pharmaceutical industry where the traditional blockbuster model is no longer sustainable.

Industries typically have a dominant business model theme/pattern that has evolved over time and is often hard to change. For several years pharmaceutical companies have tried to transform their traditional blockbuster model but have so far not identified a new way to create and appropriate value. In this context, studies find that the degree of adopting new business models is negatively associated with the length of experience with an industry's dominant business model design. Companies therefore need to find a way to overcome this obstacle against business model innovation. In a first step, this requires that firms understand the current dominant business model design and how it has evolved over time. As an example, offering integrated solutions becomes more and more important in the chemical industry. Today, several chemical companies operate solution business models rather than product business models. Evonik Industries for instance notes that "customers expect tailored solutions" [40] and Dow Chemical is committed to maximizing customer value "by offering innovative, customized solutions that can help solve some of the market's most pressing challenges, such as maximizing supply, improving efficiencies and managing emissions" [41]. Solving strategically important customer problems allows Dow Chemical thus to enhance value creation. Strong orientations towards customers and collaboration with customers are hence important characteristics of business models in the chemical industry, or, as Evonik Industries explains, "customer proximity and solution partnerships are the key to success" [42]. Solution business models are hence dominant business model designs in the specialty chemicals segment. Besides understanding one's own industry's dominant business model design, companies should also consider analysing the business models of their customers or even those of their customers' customers. Battery electric vehicles for instance may not only compete with the established technology standard and infrastructure for internal combustion engines (which is already a challenge), but the automotive industry's dominant business model may also be incompatible with electric vehicle technologies and hence not capable of effectively commercializing them. Alternative business models such as Daimler's Car2Go concept may be important enablers of the future electric vehicle market that battery materials suppliers should consider in their analysis, for example to test more collaborative models with alternative positions along the value chain.

Several chemical and pharmaceutical companies have begun to systematically test and experiment with new business model designs, as the following two examples illustrate:

- At its location in Möndal, Sweden, the pharmaceutical company AstraZeneca is testing a "new pharma business model of adding value and reducing risks" [43] in the field of biotechnology. Through opening AstraZeneca's laboratories and offices to external partners, for example, academic groups, SMEs, biotechnology companies, venture capital funds, university incubators and consultants, the company aims to combine its corporate R&D infrastructure and its R&D competencies with external know-how in order to achieve higher capital efficiency and risk reduction. The company explains that this new business model is AstraZeneca's "contribution to a more dynamic and competitive life science ecosystem" [43].
- Infineum, a formulator, manufacturer and marketer of petroleum additives for lubricants and fuels, is engaged in business model experimentation to "leverage its product technology and know-how and create a list of profitable new opportunities that fit with its core competencies" [44]. With its dominant business model design, Infineum was for example not able to enter the

additive market for lubrication of high-precision instruments such as robots. For instance, Infineum was used to selling products in tons but the required amounts of lubricant in this market are much smaller. Testing new design elements and considering alternative views of Infineum's product offering along the value chain enabled the firm to develop an innovative business model to enter the market for high-precision instrument lubricants. Today, Infineum operates in this market in a position further forward in the value chain than it did traditionally [44].

Chemical and pharmaceutical companies are experimenting with different business model design elements in order to enter new markets, for example, by testing new positions along the value chain, and identifying new ways to create and capture value, for example, through new ways to collaborate.

7.5 Summary

- **A business model describes how companies create and capture value.** It consists of four *interlocking* elements, namely the customer value proposition, the profit formula, the key resources and the key processes. The internal consistency of choices among these elements determines the effectiveness of a business model.
- **Successful companies help their customers to get an important job done.** Successful business model innovators do not start with a product idea, they start with identifying how to help customers get an important job done and then develop a blueprint of how to fulfil that need profitably.
- **Business model, strategy and tactics are related but they are not the same.** The business model, as a system of choices and consequences, is a reflection of a firm's strategy but it is not the strategy. Strategy is about which business model a company should choose. The business model determines the tactics that are available to a firm.
- **Business model innovation is not about how a business looks in 20 years; it is about the changes a company needs to make today to still exist in 20 years.** Business model innovation is the introduction of a new business model in an existing market and/or a fundamental change in the design elements of an established business model. Business model innovators have an excellent understanding of their established business model, can sense the need to adapt it, integrate customers and suppliers during the development of a new business model, have the necessary strategic and operational flexibility to manage two business models simultaneously and have a strong strategic learning capability.
- **Do not overestimate business model innovation and never forget competitive strategy.** Companies should only engage in business model innovation if they are confident that the opportunity is great enough to

warrant the effort. Moreover, the best business model is worthless if the company has no competitive strategy, namely, a plan to create a unique position in a competitive marketplace.

References

1 Williams D. 2014. *The Steep Climb Back to Profit*. http://www.chemanager-online.com/en/topics/management/steep-climb-back-profit (accessed 20 August 2014).

2 Zachert M. 2014. Lanxess – Jefferies 10th Annual Industrials Conference 2014: Realignment program and Q2 2014 results. Available at: https://lanxess.de/uploads/tx_lanxessmatrix/q2_2014_jefferies_final_01.pdf (accessed 27 June 2017).

3 Magretta J. 2002. Why business models matter. *Harvard Business Review*, **80**(5): 86–92.

4 Birkinshaw J and Goddard J. 2009. What is your management model? *MIT Sloan Management Review*, **50**(1): 81–90.

5 Casadesus-Masanell R and Ricart JE. 2011. How to design a winning business model. *Harvard Business Review*, **89**(1-2): 100–107.

6 Johnson MW, Christensen CM, and Kagermann H. 2008. Reinventing your business model. *Harvard Business Review*, **86**(12): 57–68.

7 Keeney RL. 1999. The value of Internet commerce to the customer. *Management Science*, **45**(4): 533–542.

8 Johnson MW. 2010. *Seizing the White Space*. Harvard Business Press: Cambridge, MA.

9 Berggren E and Nacher T. 2001. Introducing new products can be hazardous to your company: Use the right new-solutions delivery tools. *Academy of Management Executive*, **15**(3): 92–101.

10 Kim WC and Mauborgne R. 2000. Knowing a winning business idea when you see one. *Harvard Business Review*, **78**(5): 129–138.

11 Johnson MW. 2010. *A New Framework for Business Models*. http://blogs.hbr.org/2010/01/is-your-business-model-a-myste-1/ (accessed 1 November 2014).

12 Chesbrough H. 2007. Why companies should have open business models. *MIT Sloan Management Review*, **48**(2): 22–28.

13 Chesbrough H. 2006. *Open Business Models*. Harvard Business Press: Cambridge, MA.

14 Mahadevan B. 2000. Business models for internet-based e-commerce: An anatomy. *California Management Review*, **42**(4): 55–69.

15 Feldmann J. 2006. *BASF Plastics: Business Models, Strategy and Growth Opportunities*. http://www.plasticsportal.net/wa/plasticsEU~en_GB/function/conversions:/publish/common/plasticsportal_news/docs/2007/07_262_Charts_e.pdf (accessed 25 July 2014).

16 Feldmann J. 2004. *Plastics – Materials of the 21st Century.* http://www.basf.de/basf2/html/plastics/images/presse/docs_e/261_feldmann.doc (accessed 25 July 2014).

17 BASF. 2013. *General Motors and BASF - a Long-term Partnership.* http://www.basf-coatings.com/global/ecweb/en/content/press/coatings-partner-magazine/archive/automotive-oem-coatings/gm-und-basf_eine-langlebige-partnerschaft (accessed 25 July 2014).

18 BASF Coatings. 2013. *Coatings Partner e-journal* 02/2013. http://www.basf-coatings.com/global/ecweb/de_DE/function/conversions:/publish/content/press/coatings-partner-magazine/pdf/E-Journal-2-2013/E-Journal_02-2013_EN.pdf (accessed 25 July 2014).

19 BASF Coatings. 2013. *Coatings Partner e-journal* 03/2013. http://www.basf-coatings.com/global/ecweb/en/function/conversions:/publish/content/press/coatings-partner-magazine/pdf/E-Journal-3-2013/E-Journal_03-2013_EN.pdf (accessed 25 July 2014).

20 Chesbrough H. 2010. Business model innovation: Opportunities and barriers. *Long Range Planning,* **43**(2): 354–363.

21 Huston L and Sakkab N. 2006. Connect and develop. *Harvard Business Review,* **84**(3): 58–66.

22 Lichtenthaler U and Ernst H. 2006. Attitudes to externally organising knowledge management tasks: A review, reconsideration and extension of the NIH syndrome. *R&D Management,* **36**(4): 367–386.

23 Ehret M, Kashyap V, and Wirtz J. 2013. Business models: Impact on business markets and opportunities for marketing research. *Industrial Marketing Management,* **42**(5): 649–655.

24 Frankenberger K, Weiblen T, and Gassmann O. 2014. The antecedents of open business models: An exploratory study of incumbent firms. *R&D Management,* **44**(2): 173–188.

25 P&G. 2014. *Strengths in Structure.* http://www.pg.com/en_US/company/global_structure_operations/corporate_structure.shtml (accessed 31 July 2014).

26 Giesen E, Berman SJ, Bell R, and Blitz A. 2007. Three ways to successfully innovate your business model. *Strategy & Leadership,* **35**(6): 27–33.

27 Govindarajan V and Trimble, C. 2011. The CEO's role in business model reinvention. *Harvard Business Review,* **89**(1–2): 108–114.

28 Markides C. 1998. Strategic innovation in established companies. *MIT Sloan Management Review,* **39**(3): 31–42.

29 Markides C and Charitou CD. 2004. Competing with dual business models: A contingency approach. *Academy of Management Executive,* **18**(3): 22–36.

30 Accenture. 2012. *Chemical Portfolio Players: Balancing the Needs of Today and Tomorrow in Difficult Times.* http://www.accenture.com/SiteCollectionDocuments/PDF/Accenture-Chemical-Portfolio-Players.pdf (accessed 25 July 2014).

31 Dow Chemical. 2012. *Dow Announces New Business Structure and Executive Leadership Appointments.* http://www.dow.com/news/press-releases/article/?id=6064 (accessed 25 July 2014).

32 Ignatius A. 2012. *Unilever's CEO on Making Responsible Business Work.* http://blogs.hbr.org/2012/05/unilevers-ceo-on-making-respon/ (accessed 27 July 2014).

33 Gaule A. 2012. *Gaule's Question Time: DSM Ventures and Licensing.* http://www.globalcorporateventuring.com/article.php/3576/gaules-question-time-dsm-ventures-and-licensing (accessed 27 July 2014).

34 Pfizer. 2013. *Partnering with Pfizer Worldwide R&D.* http://www.pfizer.com/files/partnering/partnering_rd_brochure.pdf (accessed 25 July 2014).

35 SABIC. 2010. *Report & Accounts* 2010. http://www.sabic.com/system/aspx/ImageHandler.aspx?file=/me/en/images/Annual_Report_2010_tcm14-762.pdf (accessed 4 August 2010).

36 Verity R, van den Heuvel R, Kubis N, and Meyer O. *Future of Chemicals Part IV: New Operating Models – Facing a Challenging Chemical Industry Landscape.* http://www.strategyand.pwc.com/media/file/Future_of_Chemicals_IV.pdf (accessed 20 August 2014).

37 Reckmann B. 2013. *Leadership Built on Unique Strengths.* http://www.merck.de/company.merck.de/de/images/LC_OLED_Day_2013_Reckmann_tcm1613_108818.pdf?Version= (accessed 27 July 2014).

38 DSM. 2014. *Open Innovation – Sharing the Brightest Ideas.* http://www.dsm.com/corporate/about/innovation-at-dsm/open-innovation.html (accessed 27 July 2014).

39 Morawietz M, Bäumler M, Caruso P, and Gotpaga J. 2010. *Future of Chemicals III: The Commoditization of Specialty Chemicals – Managing the Inevitable.* http://www.strategyand.pwc.com/media/uploads/Future_of_Chemicals_III.pdf (accessed 25 July 2014).

40 Evonik Industries. 2014. *Evonik's Feed Additives.* http://feed-additives.evonik.com/product/feed-additives/en/about/business-model/pages/default.aspx (accessed 20 August 2014).

41 Dow Chemical. 2014. *Oil & Gas.* http://www.dow.com/products/market/oil-and-gas/ (accessed 14 July 2014).

42 Evonik Industries. 2014. *Products & Solutions.* http://corporate.evonik.com/en/products/pages/default.aspx (accessed 20 August 2014).

43 AstraZeneca. 2014. *About the BioVentureHub.* http://www.astrazenecamolndal.com/bioventure.php (accessed 31 July 2014).

44 Sinfield JV, Calder E, McConnell B, and Colson S. 2011. How to identify new business models. *MIT Sloan Management Review*, **53**(2): 85–90.

8

External Integration: Why, When, and How to Integrate Suppliers and Customers

Carsten Gelhard[1] *and Irina Tiemann*[2]

[1] *University of Twente, Chair of Product-Market Relations*
[2] *University of Oldenburg, Chair of Innovation Management and Sustainability*

Your most unhappy customers are your greatest source of learning.
Bill Gates, Co-founder of Microsoft

This particular chapter discusses the principles, benefits, and challenges of integrating two of the most important stakeholders into a firm's value creation process: a firm's customers and its suppliers. You will learn why companies integrate external partners and why customers and suppliers are important knowledge resources for chemical companies (Section 8.1). In Section 8.2 you will be provided with more detailed information on customer integration, including a discussion of different degrees of collaborative activities with customers, their corresponding up- and down-sides, and existing typologies of customer co-creation. At the end of the chapter, we present a framework that is intended to particularly support you in designing and assessing collaborative activities with customers. In Section 8.3 we similarly provide fundamental insights on supplier integration. We discuss the emergence and importance of supplier-induced innovations, existing typologies of supplier integration, as well as thoroughly discussing the characteristics that determine suppliers' willingness to become involved in collaborative activities. The present chapter on customer and supplier integration eventually closes with an illustrative description of Beiersdorf's "Invisible for Black & White" as best practice for becoming engaged in collaborative efforts with both suppliers and customers.

Business Chemistry: How to Build and Sustain Thriving Businesses in the Chemical Industry, First Edition. Edited by Jens Leker, Carsten Gelhard, and Stephan von Delft.

8.1 Introduction

8.1.1 Why Do Companies Integrate External Partners?

Chemical firms are regarded as an important source of innovative products that often trigger the development of innovations in various downstream industries, such as construction, automobile, furniture, or the fast-moving consumer goods industry. In order to fulfill their customers' expectations of continuously developing innovative products, chemical firms are forced to search for new ways of improving their capability to innovate. In so doing, they increasingly build on close interactions with external partners. To be more precise, while focusing on their core competencies, many firms outsource these activities to external partners that are beyond their own sphere of expertise. This outsourcing approach becomes common practice even for the firm's innovation-related activities. Following such an Open Innovation strategy, firms specifically exploit external resources for innovations to complement limited internal resources. Owing to the increasing diversity and complexity of new technologies it is barely possible to combine all required competencies for the development of successful new products in one single company. Therefore, establishing research and development collaborations with different external partners is indispensable to staying competitive. In order to approach and collaborate with external partners, firms can rely on various initiatives, such as the consumer goods company Henkel, for instance, which successfully refers to "Open Innovation" intermediates (e.g., NineSigma or InnoCentive).[1] In this way, the company gains access to solutions to specific problems through collaborating with researchers, suppliers, or customers from all over the world [1, 2].

Along the value chain, customers and suppliers represent important stakeholder groups with whom chemical firms increasingly engage. For instance, by strengthening interactions with customers, these firms are in a better position to understand their customers' needs and are able to reduce market uncertainties. Initiatives aimed at gaining a better understanding of customer needs are particularly relevant, since chemical manufacturers often only seem to possess limited knowledge of their customers' needs. As a recent study among customers of chemical manufacturers shows, only half of the interviewed customers stated that chemical manufacturers have at least a good knowledge of their needs and only one out of five customers indicated that chemical manufacturers hold advanced knowledge of their needs. Furthermore, the fact that collaborations between chemical manufacturers and their customers are primarily triggered by customers underlines the emergence of empowered customers who increasingly seek to initiate close interactions with their suppliers [1, 3].

---

1 NineSigma and InnoCentive are consultancies that offer their services in the field of new product development. They provide a platform on which innovation searching companies publish, via the internet, their problems and can receive solutions from selected experts.

In addition to collaborations with their customers, chemical manufacturers also rely on their supply base for similar collaborations. Collaborations with suppliers have the greatest influence on the successful generation of innovations. The responsibilities of the suppliers extend from formal and informal consultation concerning technical issues, through to joint development projects, up to the complete takeover of development work. As the example of the consumer goods company Procter & Gamble (P&G), for instance, shows, 50% of the innovations of the global Baby Care business unit are developed by suppliers. Here, companies like BASF take over the development of super absorbents and support P&G in expanding its market position [2].

Taken together, the ongoing pressure to innovate forces companies in the chemical and pharmaceutical industry to integrate external partners more than ever before. The "voice of the customer" as well as the technological competences of suppliers are essential inputs for the successful development of innovations. Therefore, the present chapter sheds some more light on established strategies for both customer and supplier integration.

8.1.2 The Sources of Innovation

When categorizing various sources of innovation, one typically refers to two established and distinct approaches: market pull versus technology push. According to a market pull strategy, firms typically start with identifying unsatisfied needs of their customers or, in other words, firms begin with acquiring information on these needs. In a next step, firms acquire (either internally or externally) complementary solution information in order to adequately respond to the needs of their customers. With a technology push strategy, firms, on the other hand, typically start with acquiring solution information. Apart from relying on internally held solution information, firms can also refer to external partners (e.g., suppliers), which very often also provide a considerable amount of solution information. In a next step, firms aim at combining their newly acquired solution information with existing need information, or, put differently, firms seek appropriate markets comprising needs that will be satisfied by the newly developed solution. Since each of these two complementary business strategies solely focuses on one part of a greater whole, firms are recommended to consider a holistic approach to innovation. Hence, they might refer to a functional perspective of possible innovation sources. Functional innovation sources can be understood as being the explicit origin of an innovation, comprising the following categories: internal value chain,[2] external value chain (suppliers and customers), private and public universities and research institutions, as well as competitors and related industries [1, 2, 5].

2 This work differentiates, according to Porter, between the internal value chain of a company and the whole value chain, consisting of the vertical and the horizontal connections between internal value chains of single up- and down-stream companies (Porter, 2000 [4]).

Internal sources for innovation are diverse and not only derive from the new product development function, but also from the purchasing, manufacturing, or marketing and sales function. The resulting innovations, therefore, are not limited to the product domain, but also cover new product processes, new distribution channels, or new value-adding services. As an example, Air Liquide shifted its focus from technological product innovations to an innovative distribution concept – termed "local customer support" – in the field of industrial gas, which was suffering from declining margins. Searching for new distribution opportunities, Air Liquide established sites at its customers' locations and increasingly became active in the field of gas supply and safety management. To offer a customized set of services, a team of specialized employees worked on-site at the customer's plant [6].

Similar to internal sources of innovation, external sources are diverse and comprise various external partners such as customers, suppliers, competitors, and research institutes. The shifting paradigm of developing all innovations internally towards opening the boundaries of the firm is known as "Open Innovation". Describing the renunciation from the traditional innovation process, the fundamental principle refers to the disclosure of the previously outward closed innovation process and the exploitation of creative and innovative sources of innovation that are located outside the firm. Here, both up- (suppliers) and down-stream (customers) value chain partners represent very important sources of innovations. The benefits of absorbing external resources and knowledge or the outsourcing of various tasks to external partners with special competencies and expert knowledge are numerous. Apart from significantly shortening innovation cycles, firms can benefit in terms of cost and risk reductions of about 60 to 90%. As a consequence of these promising advantages, the Open Innovation approach has become established in various industries, including research intensive industries (e.g., Bayer, BASF, Evonik, and Merck Serono) as well as the consumer goods industry (e.g., 3M, Bosch, Henkel, Procter & Gamble, and Siemens) [7, 8].

Both the firm's customers and suppliers eventually represent two important sources of innovation. The integration of customers and suppliers for the purpose of innovation represents two distinct types of Open Innovation strategy. While customer integration reasonably refers to a market pull strategy, supplier integration refers to a technology push strategy. Suppliers often have better expertise and specialist knowledge about components and raw materials, which can be relevant for new product development of chemical manufacturers. Collaborating with suppliers provides the manufacturer with the suppliers' technology-related know-how, which might lead to substantial changes to the product. If customer integration – as a market pull strategy – however, is proposed, this results in the development of fairly incremental innovations. Apart from the different outcomes that might derive from customer and supplier integration, there are further differences with both integration strategies with regard to their driving factors as well as with the management tools required to transform the integration of the respective partner into beneficial outcomes. However, at this point it is important to mention that the primary focus of this chapter is not to decide which innovation source (i.e., customer or supplier) is more important, but rather to support managers in fully exploiting the benefits provided by both integration strategies. Hence, to support managers in developing and implementing more successful integration strategies, the present chapter seeks to answer the questions of why, when, and how to integrate partners and suppliers into a firm's creative process [1, 2].

8.2 Customer Integration

Firms operating in various industries have acknowledged that the customer – or rather the voice of the customer – plays an essential role within the firm's value creation. The firm's intention to consider the voice of the customer is a result of various uncertainties a firm faces, particularly while developing and commercializing new products. These uncertainties can be primarily categorized into technology- and market-related uncertainties. Since firms often lack a sufficient level of knowledge to cope with these uncertainties, they rely on external knowledge sources, including their customers. More specifically, in order to reduce these uncertainties and to successfully develop and commercialize new products, firms need to combine two essential types of information, namely, need information and solution information. Need information refers to information about problems and demands the firm faces in current or future markets and, therefore, reduces the risk of market failure. Solution information refers to information about the development of technologies and products in order to respond to customer needs. Typically, both types of information lie with different players. Whereas customers usually possess precise and detailed knowledge of their individual needs, firms usually possess precise and detailed knowledge of how to solve customer-related problems (see Figure 8.1).[3] Since a successful development and commercialization of new products requires both types of information, firms have to combine both solution and need information at one single locus [1, 10–12].

8.2.1 Degree of Collaborative Activities with Customers

8.2.1.1 Listening to the Voice of the Customer

To overcome the asymmetry between solution and need information with the purpose of reducing the risk of market failure, firms typically make use of

3 Similar to need information, solution information can also be located outside the firm and resides in suppliers, customers, competitors, or other external stakeholders (e.g., Piller, Ihl, and Vossen, 2011 [9]).

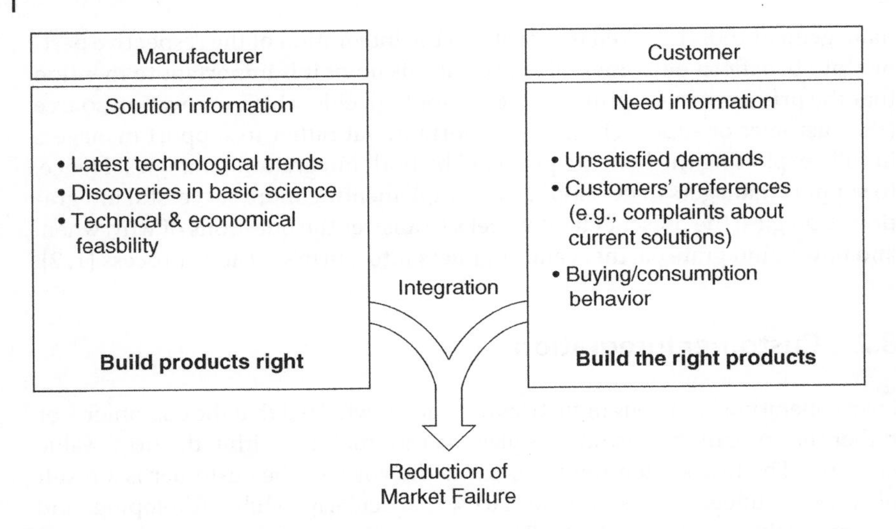

Figure 8.1 Need and solution information. Own representation

various traditional market research techniques [13]. Such traditional market research techniques are:

- interviews
- surveys
- focus groups
- sales reports analysis
- experimentation
- ethnography/netnography
- observations.

Although firms orient their activities towards their customers by employing these techniques, the customer is still regarded as a passive entity. It is the role of the firm to proactively select a group of customers and to interview, survey, or observe them. The interaction and flow of information between the firm and its customers is one-way, initiated by the firm with information flowing from the customer to the firm. As passive entities of the firm's value creation process, customers are dependent on the firm and its final market offering. Since the firm is exclusively responsible for designing, developing, manufacturing, and delivering the final product, customers can only indirectly influence the firm's market offering through participating in market research techniques such as surveys or interviews [1].

8.2.1.2 Customer Integration (outsourcing)

Apart from these traditional market research techniques, customers might also be more actively involved. Following an outside-in process of Open Innovation, firms can outsource certain tasks of their innovation process to

Box 8.1 Lead user approach [1, 17]

The best-known concept for customer integration is von Hippel's lead user approach – an approach being applied by many companies, including 3M, Johnson & Johnson Medical, and BASF. Lead users refer to a sub-group of customers that face specific needs earlier than the rest of the market and significantly benefit from solutions that address those needs. Apart from need-related knowledge, lead users are usually acquainted with specific technologies and thus possess valuable solution-related knowledge. Lead users represent a valuable source of novel ideas, provide valuable input for idea realization and product development, and sometimes even independently create prototypes. If a lead user independently creates a prototype and, in so doing, takes on the role of an external innovator, both need and solution information are integrated directly at the locus of the customer.

their customers and, by these means, benefit from the integration of external know-how and enhanced access to need-information. Customers shift their role from passive entities to more actively engaged partners within the firm's innovation process. Apart from reducing technology- and market-related uncertainties, customer integration can increase the firm's innovation performance in terms of quality as well as cost- and time-efficiency (for example, see Box 8.1) [1, 14–16].

8.2.1.3 Customer Co-creation

In recent years, the notion of customer integration has increasingly been replaced by the notion of "customer co-creation". Generally speaking, customer co-creation can be understood as the firm's interaction with customers to co-construct the value for the customer. Hence, compared with listening to the voice of the customer and customer integration, the concept of customer co-creation more holistically covers the collaborative activities between the firm and customers with the purpose of value creation. Although all three concepts are undoubtedly intertwined, they particularly differ in terms of:

- the degree of shared responsibility for value creation (i.e., power and control over the firm's value creation process) and
- the degree of firm–customer interactions
  - frequency of interaction
  - intensity of interaction and
  - direction of interaction (one-way versus two-way) [1].

As shown in Figures 8.2 and 8.3, customer co-creation is characterized by a higher degree of firm–customer interactions as well as a higher degree of shared responsibility for value creation. Co-creative practices, however, might vary with regard to the extent to which firm–customer interactions occur

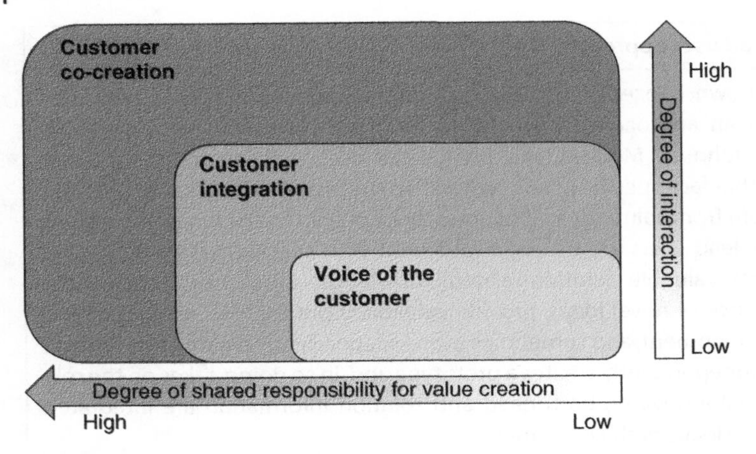

Figure 8.2 Relationship between the voice of the customer, customer integration, and customer co-creation [1]

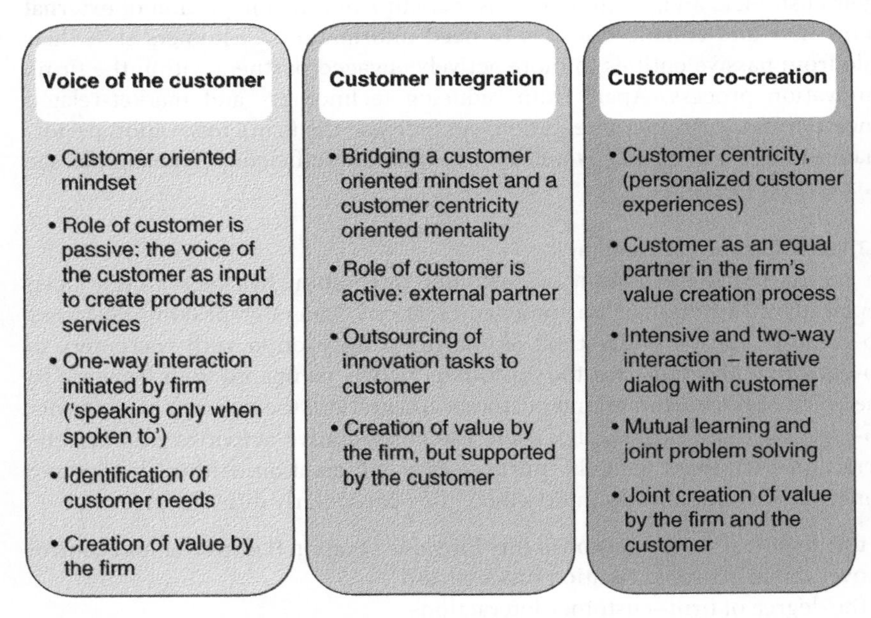

Figure 8.3 Comparison of the voice of the customer, customer integration, and customer co-creation [1]

across various stages of the firm's value creation process (e.g., the front end of innovation, idea realization and development, commercialization, product design, production, and marketing/sales) as well as during the customer's process of usage (e.g., through after sales services). The most comprehensive approach to customer co-creation implies interactions that span the firm's entire value creation process and also incorporates the customer's process of usage [1].

8.2.2 Up- and Down-sides of Collaborative Activities with Customers

Collaborative activities with customers are associated with various advantages for the focal firm, including the access to new ideas, identification of market trends, increased customer satisfaction, and accelerated time-to-market. However, each type of inter-firm relationship has to be considered as an investment that – depending on the intensity of the collaborative activity – demands different levels of financial-related (e.g., infrastructure, incentives, hiring of new personnel) and time-related (e.g., coordination activities, relationship development and maintenance, idea assessment, provision of feedback) resource investments. In addition, collaborative activities with customers do not automatically and with certainty result in the successful introduction of new products or the significant improvement of existing products; a strong focus on the customer base, for instance, bears the risk of only *exploring the already explored*. However, on the other hand, neglecting the voice of the customer when developing new solutions bears the risk of *over-exploring the undesirable*. Thus, before deciding to invest in any collaborative activity with customers, decision makers have to seriously estimate whether the expected benefits outweigh the potential risks and costs associated with the desired activity. Table 8.1 provides an overview of the potential benefits and costs of being engaged in collaborative activities with customers. We will elaborate on these subsequently.

8.2.2.1 Mutual Learning and Trial and Error

Close collaborations with customers and the establishment of an ongoing dialog foster a process of mutual learning. Through engaging in iterative interactions, the firm and its customers can exchange solution and need information, which are then jointly interpreted and integrated into the final solution. The occurrence of a mutual learning process is fostered by direct and two-way communication channels between the firm and its customers. The closer and the more holistically the firm and customer processes are connected, the more likely and effectively can both parties engage in a mutual learning process. Whereas traditional market research techniques (e.g., sales reports analysis, observations) are primarily directed towards the collection of need information, closer and more intensive interactions with customers (e.g., co-creation sessions) also allow interpretation of such information jointly with the customer. Since learning and the subsequent generation of knowledge do not automatically follow from the exchange of information, firms are asked to

Table 8.1 Benefits and costs of collaborative activities with customers [1, 18].

Benefits	Risks and costs
• increased innovativeness • access to valuable solution- and need-related information • reduced risk of market failure • improved solution–customer need fit • improved identification of customer needs • occurrence of trial and error process • engagement in mutual learning process (growing knowledge base) • creation of personalized, close, and intense relationships with customers • greater customer loyalty • positive word-of-mouth • improved profitability of market offerings • reduced transaction costs	• limitation to exploitative innovation • loss of know-how • dependence on customers' demands, views, and input • risk of serving a niche market (participating customer is not representative of the majority of customers) • negative word-of-mouth triggered by perceived unfairness and dissatisfaction • difficulty in predicting customers' intended action (uncertainty) • required costs for appropriate infrastructure • coordination and maintenance costs • offering monetary incentives (e.g., financial rewards)

closely interact with their customers and jointly reflect on the newly gathered information. The mutual learning process is further accompanied by the occurrence of a trial and error process: in line with the "fail fast, fail often" mantra, failures (or alternatively termed "the rejection of hypotheses") during the development process is endured and even endorsed by many firms – provided those errors are identified early on in the development process. Collaborative activities throughout different stages of the firm's value creation process enable firms to timely and incrementally adjust their products to their customers' needs. Through these means, firms can reduce the risk of costs associated with post-launch modifications.

To let these close and intense interactions with customers become a common practice, decision makers eventually have to revise their mindsets that customers not only represent a pure source of information, but also are active contributors with valuable knowledge and capabilities [1, 13].

8.2.2.2 Innovativeness

The question of whether listening to customers might significantly increase the innovativeness of newly developed products or merely lead to the development of incremental innovations has still not been answered clearly. Efforts made to shed some more light on this important research question produced contradictory findings. On the one hand, it is widely argued that actively involved customers propose ideas that comprise a higher degree of innovativeness compared with ideas that are generated by means of traditional market research techniques. Since customers' ideas derive from outside the firm (i.e., from a context in which usage takes place and customer needs are from personal experience), they might be highly valuable and new to the firm. Hence, it can be reasonably argued that involving customers in a joint problem-solving process induces employees to think outside the "box" and, in this way, enhances new product innovativeness [11, 19].

On the other hand, it can also be argued that listening to customers can carry various risks, such as the mere development of exploitative innovations, the development of products that simply serve a niche market, or the risk of losing knowledge. The fundamental problem refers to the latent needs of the customers. Latent needs are not in the consciousness of the customer and, for this reason, these needs are difficult if not impossible for customers to articulate. Considering these contradictory arguments and findings, the question of whether collaborative activities with customers might lead to the development of exploitative innovation and/or the development of exploratory innovation, cannot generally be answered. The type of customers involved (e.g., lead users versus traditional users), type of innovation project, openness of employees involved, employees' ability to interpret and handle customers' suggestions, or the customers' creativity and ability to articulate, certainly represent only a few of the various factors that determine the degree to which firms benefit from the active involvement of customers in terms of increased innovativeness. Thus, decision makers should always consider the holistic picture when developing a strategy for customer collaborations and avoid trusting in any generally valid statements, such as "listening to customers always fosters/hampers the firm's innovation efforts" [1, 18, 20].

8.2.2.3 Reduction of Market Failure

When using traditional market research techniques (e.g., surveys, interviews, observation), employees have to independently translate customer needs into the final products. Since customer needs are frequently very complex and difficult to articulate, very often they are actully transferred insufficiently and incorrectly into the firm's value creation process. The active involvement of customers, on the other hand, ensures the direct embeddedness of the voice of the customer into the firm's value creation process, and, thus, supersedes the translation process – a success-determining step that is prone to failures. The joint translation of customer needs into product requirements eventually reduces the risks of misinterpretations. This eventually results in the development of products with an improved solution–customer need fit. Since the resulting products are a better match to the needs of the customers, firms benefit in terms of improved customer satisfaction and increased likelihood of adoption. Iterative feedback loops, reflections, practical involvement, and the use of diagnosis tools applied by the firm further support firms in stimulating their customers to holistically express their needs, similarly increasing the likelihood that the firm's products are successfully adopted in the marketplace. Furthermore, once firms pursue a holistic approach to joint value creation with customers, they can adapt their products to customer demands at each stage of the value creation process, for example, by implementing product modifications during the production stage or by choosing adequate and customer-specific distribution channels. By these means, firms ensure that the final market offering (i.e., the core product as well as various additional value components such as packaging or delivery) fits holistically with their customers' needs and expectations [1, 12, 21].

8.2.2.4 Customer Relationship Management

Collaborative activities with customers go beyond the joint creation of value and also support firms in improving their process of value capturing. Driven by the co-creative approach to value creation, the view of the market as an aggregation of customers is replaced by a more contemporary view that specifically stresses the need to consider the personalized experiences of each individual customer. Through individual customer interactions during the firm's process of value creation (e.g., product design, production, marketing) as well as during the customer's process of usage (e.g., after sales services), firms are in a better position to respond to individual customer needs and preferences. This, in turn, leads to improved customer satisfaction, greater customer loyalty, increased willingness to pay, positive word-of-mouth, as well as a higher degree of perceived customer centricity. The occurrence of these marketing-related outcomes is equally driven by the empowerment of the customers themselves: giving customers a bigger say in constructing the market offering leads to "more favorable corporate attitudes and more favorable behavioral intentions (purchase, loyalty, positive word of mouth, corporate commitment)" [15].

The pursuit of joint goals as well as the mutual investment of resources (e.g., physical resources, knowledge resources, or time) strengthens the level of trust between the firm and its customer, with the reward being greater customer loyalty. However, apart from having an effect on the behavioral intentions of current, actively involved customers, co-creative efforts might also improve a firm's reputation among customers that do not actively participate in the firm's value creation activities or even among the firm's prospective customers. By offering customers the opportunity to become engaged in joint value creation activities, firms signal their customer centricity as part of their business mentality.

Taken together, co-creative activities with customers also fulfill various marketing-related tasks (e.g., relationship building, customer acquisition, promotional activities) and, in this way, support firms in efficiently and effectively delivering and capturing the co-created value [1, 15, 22–25].

8.2.2.5 Increased Dependency and Uncertainty

While collaborative activities with customers, on the one hand, reduce technology- and market-related uncertainties, on the other hand, they add a new variable into the uncertainty of the firm's value creation process: the customer itself. Since firms become more dependent on their customers, joint value creation activities challenge the firm's strategic planning efforts. For instance, it is difficult to predict how customers will engage in co-creation practices, and also what types of ideas (e.g., radical, incremental, technically unfeasible, non-profitable) will be suggested. Another challenge refers to the management collaborative activities with customers who are willing to or are actually engaged in co-creation practices, but lack the required skills to contribute to the firm's value creation process. If firms do not consider the input (e.g., ideas, suggestions, or complaints) received from participating customers, these customers might perceive unfairness and dissatisfaction. To avoid the risk of protest and negative word-of-mouth, firms are advised to treat the input from each customer seriously. The threat of negative word-of-mouth is even larger when customers are embedded and interconnected within a customer community. Customer communities provide dissatisfied customers with a platform to attract attention from other customers with the purpose of mobilizing them as supporters for any type of protest.

Since firms either interact with the entire customer community, a selected group of customers, or each individual customer, the selection process of the respective customer(s) as collaborative partner(s) implies another source of dependency and risk. If the firm interacts solely with a small sub-group of customers, the firm needs to ensure that the customers of this sub-group can be regarded as representative of the majority of its customers. Otherwise, the firm may face the risk that the members of this sub-group will be the only ones that are interested in and willing to pay for the new market offering [1, 18, 23, 24, 26].

8.2.2.6 Associated Costs

In order to establish and maintain collaborative activities with customers, firms have to invest in an adequate infrastructure and face various costs that are associated with the overall coordination of their co-creative efforts. Independently of whether firms intend to interact with an entire customer community (e.g., via an internet-based platform) or individually with each customer (e.g., via customerization techniques, face-to-face activities), they have to provide an adequate infrastructure that enables an efficient and effective exchange with customers. The costs of coordinating and maintaining collaborative activities with customers increase with the number of customers the firm intends to interact with, as well as with the level of intensity of the intended co-creation effort (e.g., online versus face-to-face). Moreover, since not all customers are intrinsically motived to engage in co-creation activities, firms may also have to bear the costs in order to motivate their customers to participate in often time-consuming activities. In general, customers are motivated to engage in customer co-creation practices for several reasons, including the opportunity to take a more active role in the value creation process, to achieve a sense of self-efficacy and reputation, to engage in a mutual learning process leading to the generation of technology- and product-related knowledge, to express themselves, to receive market offerings that respond to their individual needs, as well as the opportunity to gain increased transparency of the value creation process. Apart from these non-monetary incentives, firms are often forced to additionally provide monetary incentives (e.g., financial rewards) in order to mobilize their customers [12, 22, 24, 25, 27–30].

The conditions for and the concrete outcomes of collaborative activities with customers vary between different projects. Since no generally valid advice is available on whether to invest in co-creative practices or not, decision makers have to sincerely estimate the associated benefits and cost before allocating their resources to individual co-creative practices (see Box 8.2).

8.2.3 Typologies of Customer Co-creation

Apart from deciding whether to collaborate with customers or not with the purpose of joint value creation, firms have to decide at which stage of the value creation process these collaborative efforts with customers should be directed. The most comprehensive approach to customer co-creation spans the firm's entire value creation process (e.g., the front end of innovation, idea realization and development, commercialization, product design, production, and marketing/sales) and also incorporates the customer's process of usage (e.g., after sales services). Depending on the specific stage of the value creation process, the concept of customer co-creation can be classified into seven subcategories: customer co-ideation, customer co-development, customer co-launch, customer co-design, customer co-production, customer co-marketing, and customer co-usage (see Figure 8.4).

8.2.3.1 Co-ideation

The front end of innovation comprises the following two activities: idea generation and idea selection. By making use of various co-creative practices, firms can open up the front end of innovation and eventually release control over the generation and/or the selection of new ideas and concepts to their customers. These co-creation practices mainly differ with regard to the following two characteristics [1, 9, 12, 15].

Box 8.2 Roster for evaluating the appropriateness of investing in collaborative activities with customers

While deciding whether to invest in co-creative practices or not, decision makers are advised to consider the nature of the product (or service) intended to be co-created, their previous experiences with co-creative practices, as well as prior knowledge of the intended target customer group. The following roster supports their decision making process by systematically assessing the appropriateness of collaborative activities with customers with regard to different projects.

- Our customers have well-developed solution-related capabilities
- The intended product is important to the quality of the customer's products
- The function of the intended product is highly complex
- Our customers are easily accessible
- Our customers are generally willing and able to provide us with feedback on the performance of our products
- We know the target customer group for the intended product
- Customer feedback is a valuable resource for us
- The intended product implicates significant changes for the usage/consumption behavior of our customers

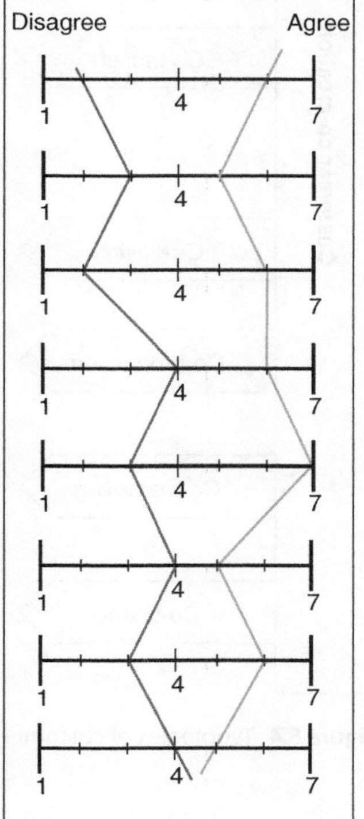

—— Right-hand line: Invest in collaborative activities with customers

—— Left-hand line: Do not invest in collaborative activities with customers

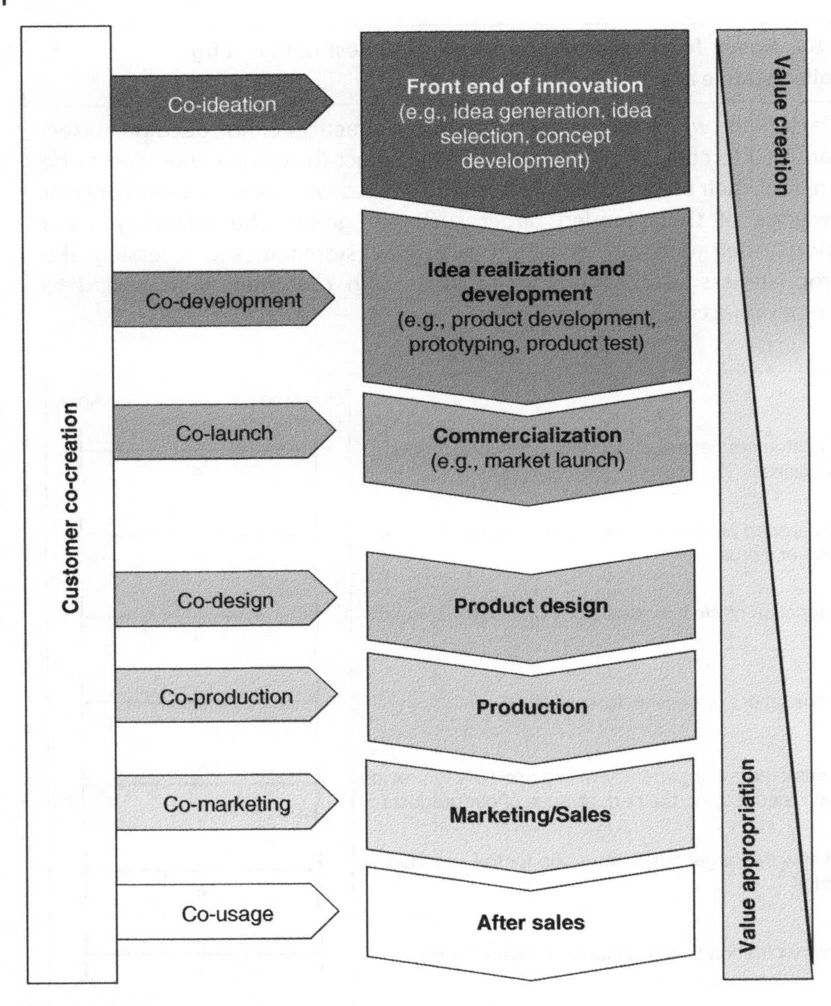

Figure 8.4 Typologies of customer co-creation [1]

i) Degree to which customers are empowered to generate and select new ideas and concepts: while customers can become empowered to submit their ideas to a broadly defined and creative task and to select (or vote for) any idea they prefer, the empowerment of customers can also be restricted to the submission of ideas to narrow tasks as well as to the selection of predefined ideas and concepts.
ii) Extent of simultaneously occurring firm–customer interactions: firms can either engage in dyadic firm–customer relationships (i.e., interact with one

customer at a time) or they can interact with a group or network of customers, who additionally may interact among themselves (e.g., online communities).

Various common practices for co-ideation can be summarized as follows.

Co-creative Sessions Co-creative sessions are typically based on the selection of a small group of customers (lead users or ordinary customers) for the purpose of idea generation and selection (Box 8.3). In contrast to traditional focus groups, which refer to a market research technique that "involves convening a group of respondents, usually eight–10, for a more or less open-ended discussion about a product", one-way interviews are replaced by in-depth face-to-face discussion sessions on new ideas or concepts, implying the practical involvement of customers [31: 353]. The degree to which customers are empowered to propose and select new product ideas depends on the overall nature of the project. Since firms that collaborate with lead users usually aim at developing exploratory innovations, namely, new product offerings that comprise a high degree of innovativeness and require a high degree of creativity, lead users typically receive (almost) full authority to generate their new ideas and concepts and become fully engaged in the selection process. Co-creative sessions are further characterized by the use of face-to-face communication channels, which are appropriate for interacting with individual customers or a small group of customers. Face-to-face communication channels enable the exchange of tacit and in-depth knowledge, which is typically difficult to articulate and to codify [1, 21, 32].

Crowdsourcing The main limitation of co-creative sessions is with respect to the maximum number of customers (usually up to ten) that can be approached simultaneously. To overcome this downside, many companies have started to rely on crowdsourcing as an alternative approach to co-creation. Crowdsourcing refers to a "mode of openness where firms broadcast innovation challenges in the form of open calls to undefined (and generally large) groups of external contributors" [39: 344]. Thus, by means of crowdsourcing practices firms can interact with various individuals and/or groups of individuals simultaneously (i.e., the "crowd"). Typically, these crowdsourcing practices are based on a self-selection process among customers that are willing and able to contribute to a specific innovation challenge. In contrast to co-creative sessions, crowdsourcing practices typically make use of the internet as the interaction channel [1].

In general, two different types of crowdsourcing practices exist: tournament-based crowdsourcing and collaborating-based crowdsourcing practices. In tournament-based crowdsourcing practices (e.g., idea contests,[4] idea

4 See also Chapter 6, Box 6.1.

Box 8.3 Overview of creativity techniques for co-creative sessions with customers [33–38]

Creativity techniques, such as brainstorming, brainwriting, synectics, morphological analysis, or SCAMPER, are indispensable tools for successful idea generation. They can be applied to stimulate both employees and customers (ordinary or lead users) to generate new ideas during co-creative sessions. In the following, we outline and briefly describe the most commonly practiced creativity techniques. While some of these techniques can only be used by groups (e.g., 635 method), others can also be applied by individuals (e.g., morphological analysis).

Creativity Technique	Group Size	Procedure	Duration (minutes)
Brainstorming	5–12	1) Definition of problem 2) Participants receive required information 3) Co-creative session with no structured process 4) Screening of ideas participants, analysis by experts	20–60
Morphological Analysis	≥1	1) Definition of problem 2) Parameter of problems determined 3) Morphological box designed, in which each parameter is classified in a number of different dimensions 4) All dimensions of all parameters are systematically combined 5) Results of each alternative combination forecast 6) Selection of best solution	30–240
Progressive Abstraction	≥1	1) Definition of problem 2) Developing alternative problem definition at a higher level of problem abstraction (What do we actually need? What are the underlying conditions?) 3) Developing solution to alternative problem definition (How can we solve the problem?) 4) Developing alternative problem definition at the next higher level of problem abstraction (What do we actually need? What are the underlying conditions?) 5) ...	30–240

Creativity Technique	Group Size	Procedure	Duration (minutes)
Fishbone Diagram (Cause and Effect Diagram)	≥1	1) Definition of problem 2) Definition of main categories of causes (people, methods, machines, materials, measurements, environment) 3) Analysis of causes; causes present sources of variation (combination with other creativity techniques) 4) Completeness check 5) Prioritization of identified causes	30–240
Mind Mapping	≥1	1) Definition of problem (central concept at center of page) 2) Major associations are directly connected to central concept 3) Creation of tree-like map of ideas: problem = stem, 1. generation of solutions = major branches, higher generation of solutions = minor branches 4) Connections used for structuring and visualization, interconnections taken into account 5) Elimination of minor aspects, adding new points, showing new connections	20–60
635 Method	6	1) Definition of problem 2) Each participant writes down three ideas on a form (5 minutes) 3) Forms are passed around, every participant adds ideas to the ideas of his neighbor 4) Forms passed once again 5–7) Forms passed three more times 8) Systematic evaluation	30–40
Osborn Checklist	≥1	1) Definition of problem 2) Use of checklist to modify existing ideas into new ideas; checklist provides a catalog of questions, aimed at facilitating idea generation 3) Participants answer questions (free associations)	15–240

(*Continued*)

(Continued)

Creativity Technique	Group Size	Procedure	Duration (minutes)
SCAMPER	≥1	Derivation of Osborn's checklist S = Substitute (Which part/material/component can be substituted?) C = Combine (What parts/ideas can be combined?) A = Adapt (What might be changed or used in a different way?) M = Modify (What can be added? What might be made larger/smaller?) P = Put (Put to another use?) E = Eliminate (What could be omitted?) R = Reverse (What other arrangements might be better?)	15–240
Synectics	5–7	1) Definition of problem 2) Spontaneous solution singled out and/or eliminated 3) Distancing from the problem, redefinition of problem, alienation, analogies[1] 4) Creation of new connections, associations, and analogies[1] 5) "Force-fit": associations and analogies woven together with the problem to be solved 6) Development of concrete solutions	120–240
Bionics	≥1	1) Definition of problem 2) Systematic search for analogous solutions to problems in nature 3) The evolution process in nature theoretically explained 4) Transferability tested (material, function, structure, organization) 5) The evolution process systematically comprehended	120–240
Attribute Listing	≥1	1) Definition of problem 2) Predefining core features (shape, function, material …) 3) Creating a list with major attributes of an idea (product, device, service ...) 4) Changing and modifying all listed attributes	20–60

[1] Analogies can be personal, direct, or symbolic

(Continued)

Creativity Technique	Group Size	Procedure	Duration (minutes)
TRIZ Problem Solving	≥1	1) Definition of concrete problem 2) Transferring concrete problem to an abstract level 3) Finding abstract solution (relation to abstract problem via contradiction matrix) 4) Transferring abstract solution principles back to concrete problem Tools for TRIZ *Systematics*: Problem and contradiction definition (innovation checklist, problem formulation, ideality) *Knowledge:* Use of available knowledge (contradictory analysis, effects database, material-field-analysis) *Analogies:* Analogies to past problems (40 principles, separation principles, 76 standard solutions) *Vision:* Definition of further development of product (S-curve, laws of evolution)	120–240

screening) firms interact with one customer at a time (see also Box 8.4). Customers self-select to work on their own solution, that is, customers submit and/or evaluate their ideas and concepts without interacting with other customers. The best solution is selected as the winning solution. In collaborating-based crowdsourcing practices (e.g., discussion forums), on the other hand, firms interact with a group of individuals, the so-called virtual customer community. Since members of the community interact amongst themselves and work together on one specific problem, firms only receive one collaborative solution from the crowd. In contrast to tournament-based crowdsourcing practices, firms can also leverage the social dimension of customer knowledge, namely, knowledge that solely becomes apparent when shared among a group of customers that share a common interest [1, 9, 39, 40].

8.2.3.2 Co-development

Collaborative activities that occur during the realization and development stage (e.g., product design, prototyping, and product testing) demand customer input that is more concrete and elaborated during the fuzzy front end (see

Box 8.4 Idea contests [1, 9, 42]

In idea contests, a group of independent and competing customers are asked to submit their ideas and potential solutions to a given task within a defined timeframe. Idea contests make use of competition conditions in order to encourage customers to engage in the firm's process of problem solving, to increase the quality of submissions, as well as to reduce the time for receiving new ideas and concepts. While some idea contests build on non-monetary incentives (e.g., peer or brand recognition, reputation, self-efficacy) in order to motivate customers, others are additionally based on monetary awards (e.g., cash rewards, licensing contracts). Idea contests are usually performed by using internet-based innovation platforms. While some of these internet-based innovation platforms are organized and operated by the seeking firm itself (e.g., P&G's Co-Creation Channel), others are operated by third parties that function as virtual knowledge brokers (e.g., InnoCentive).

Apart from empowering customers to submit ideas, firms can also involve them in the process of idea evaluation and screening. Instead of releasing control to a committee that consists of representatives of the seeking firm and/or external experts, customers are empowered to evaluate submitted ideas and to vote on which ideas or concepts should actually be continued. By these means, customers can actively select (or at least vote on) those ideas that they find most valuable. While evaluation committees typically base their decisions on several, pre-defined criteria (e.g., technological feasibility, strategic fit, financial risk/reward, or market potential), evaluations deriving from customers are based more on soft factors (e.g., expected usability, overall appearance, personal preferences) and are supposed to represent the actual demand for the respective solutions.

Section 8.2.3.1, "Co-ideation"). Through customer co-development practices, the firm and its customers jointly engage in a trial and error process that allows firms to test solutions early, often, and continuously in close coordination with customers. By means of this "as-you-go" process, firms can test their preliminary solutions in terms of both technical usability and customer acceptance. Furthermore, by providing customers with the opportunity to experience how a new product might be used, they are better enabled to express and define their needs. P&G, for instance, interacts intensively with a small group of customers in terms of product testing activities and, thereby, continuously receives feedback on how to improve its products. By participating in P&G's Baby Discovery Centers, customers can choose from a variety of studies (e.g., on-site product testing, group discussions, or home-use studies) and become empowered to actively influence the development process of new products. While such interactions with a rather small group of customers are typically based on face-to-face interactions, customer co-development practices with a larger number of customers typically make use of toolkits. Following von Hippel and Katz (2002), "[t]oolkits for user innovation are coordinated sets of 'user-friendly' design tools that enable users to develop new product innovations for themselves" [43: 821]. Toolkits for user innovation enable customers to (i) create a preliminary solution, (ii) simulate or prototype it, (iii) evaluate its performance in a user-specific context, and (iv) allow the customer to continuously improve it [1, 41, 43–45].

8.2.3.3 Co-launch

Activities that are conducted jointly between the firm and its customers during the commercialization process can be referred to as customer co-launch practices. Customers that are engaged as co-creators can, for instance, function as promoters and testers during the commercialization of new products or services. In the case where customers are actively engaged as testers, the focal firm can primarily benefit from (almost) real-time and hands-on customer feedback, which eventually supports a successful monitoring and controlling of the market launch. In this way, the focal firm is also in a better position to adjust its products or services in a timely manner in the event that these should not holistically satisfy their customers' demands or even disgruntle their customers. In addition, customers can function as promoters. Considering that actively engaged customers tend to speak positively about the firm and its market offerings – assuming the absence of any negative experience with the focal firm – customers that take part in for the firm's commercialization process might influence other customers and, thereby, encourage them to also buy or support the firm's new product. Here, primarily the lead user can be regarded as a trend setter that might influence the purchasing behavior of the larger community [1, 21, 27, 46].

8.2.3.4 Co-design

Firms can also involve their customers during the design stage of the value creation process. Thus, they can primarily benefit from product customization, which enables firms to respond to the individual needs of various customer groups or even the needs of each individual customer. In line with customer co-development activities, firms can make use of toolkits for actively engaging their customers in the design process. However, in comparison with toolkits that are used for the development of entirely new solutions (customer co-development), toolkits for customer co-design provide a smaller solution space and, thus, result in the development of new product variations rather than entirely new products (innovations). Typically, the focal firm provides its customers with a set of predefined building blocks (e.g., modules or components) that allows customers to configure a product that matches their individual needs (e.g., composition, functionality) [1, 9, 26, 29, 47].

8.2.3.5 Co-production

Customer co-production, also referred to as customer co-manufacturing, describes joint value creation activities that occur during the process of transforming raw materials into products. In customer co-production, customers work alongside the firm's production process. While firms gain first-hand and real-time information on their customers' requirements (e.g., product quality, demand, changing requirements in terms of product features), customers are empowered to actively monitor and control the production process. By performing on-site quality checks (e.g., analysis of the level of purity) customers can ensure that product batches can actually be used and do not have to be reclaimed. Customer co-production practices also enable both parties to align their production and demand planning systems. By continuously adjusting and aligning delivery and demand information, firms are in a better position to serve their customers on demand and in time [1, 26].

8.2.3.6 Co-marketing

Customer co-marketing implies the shift of more power and control over marketing activities, such as distribution-related activities, towards the customers. Instead of placing predefined and fixed offerings in the market place, firms can 3 also offer their customers some degree of flexibility by empowering them to adjust the firm's offerings to their individual needs. Thus, firms, for instance, can offer their customers the opportunity (e.g., by means of toolkits) to configure customized product bundles as well as to select customized packaging and delivery services. Furthermore, firms can seek to benefit from the involvement of their customers with regard to their promotion-related marketing activities. For instance, to foster the emergence of positive word of mouth triggered by satisfied customers, firms are recommended to provide their customers with any type of platform (e.g., internet-based platform) in which their customers can interact between each other and share experiences gained with the firm and the firm's products and services (e.g., P&G's Pampers Village). Here, customers can even become co-creators of the firm's brand identity, namely, customers become actively involved in shaping and communicating the firm's brand[5] [1, 48, 49].

8.2.3.7 Co-usage

Joint value creation activities between the firm and its customers might also occur during the customers' process of usage and are termed customer co-usage. In contrast to the previous sub-typologies of customer co-creation, customer co-usage implies the extension of the firm's traditional role: the firm enters the customer's process of usage (i.e., the customer's process of value

---

5 Co-branding, i.e., shaping and communicating the firm's brand, might be especially apparent in consumer groups such as brand communities (Payne *et al.*, 2009 [48]).

creation or consumption). Customer co-usage practices, which are typically classified as after sales services, mainly fulfill the following two functions:

i) getting access to direct customer feedback (e.g., complaints) and
ii) supporting customers in gaining more value [1, 50].

Instead of offering (standardized) products, firms provide their customers with tailored and comprehensive solutions (i.e., product–service combinations). In general, firms can contribute to the customer's process of usage by:

i) decreasing costs or
ii) increasing and prolonging benefits.

The list of after sales services firms can offer is extensive, including call center services, delivery services, technical services, repair services, installation services, maintenance services, recovery services, or R&D services (e.g., feasibility studies, prototype design, product tailoring, and manufacturability analysis). The interactions implied by these services can be either face-to-face (e.g., on-site technical support) or based on digital communication tools (e.g., online forums, feedback channels, contact sheets, live chat). The provision of on-site technical support is primarily important in the B2B sector. As shown by prior studies, new product performance within the chemical industry depends on the firm's capabilities in offering technical support and customer service, which might be explained by the high degree of complexity of the solutions offered by chemical firms. Customer co-usage practices further provide firms with access to valuable first-hand information on how customers use a certain product. By these means, firms face new opportunities to modify and further improve their offers and to support their customers in gaining superior experiences with the total market offering [1, 26, 50–52].

8.2.4 Designing and Assessing Customer Co-creation Practices

When designing concrete practices for customer co-creation, firms have to make several choices. Figure 8.5 provides an overview of all decisions that have to be made with regard to the architectural design of customer co-creation practices. The first decision refers to the intended range of customer co-creation. As outlined in the previous section, co-creative activities with customers can either take place at one specific stage of the value creation process or – following the most comprehensive approach of customer co-creation – across the entire value creation process (front of innovation, idea realization and development, commercialization, product design, production, and marketing/sales) reaching into the customer's process of usage. To cause these collaborative activities with a customer to occur, firms further have to provide an adequate

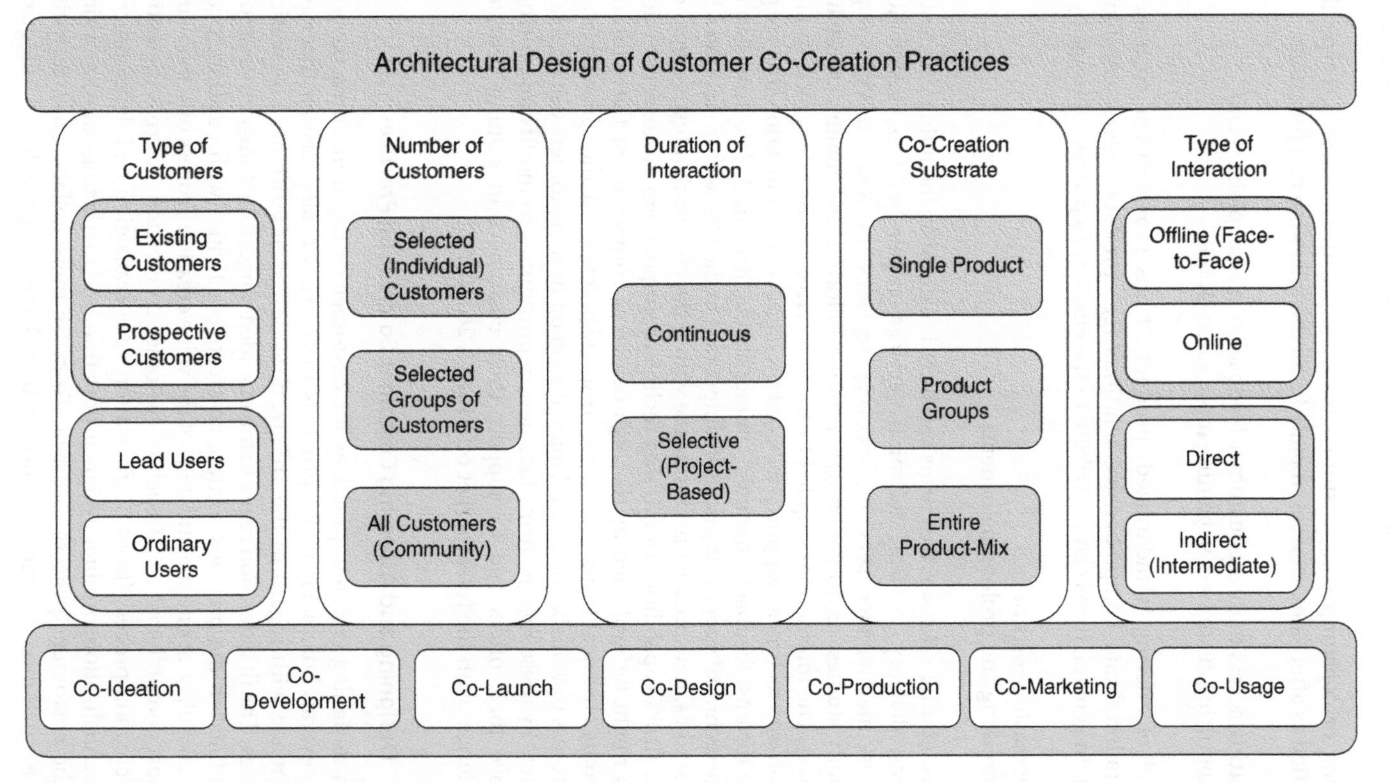

Figure 8.5 Architectural design of customer co-creation practices. Own representation

infrastructure that enables and fosters two-way communications with their customers. In general, the type of interaction between firms and their customers can comprise the following four characteristics: offline or online as well as direct or indirect. Offline (or face-to-face) communication channels, such as personal meetings, interviews, and workshops, are basically appropriate for reaching a selected (small) group of customers and enable the exchange of tacit and in-depth knowledge, which is difficult to articulate and to codify. Online communication channels, on the other hand, are beneficial for addressing a large group of customers at low costs. In contrast to offline communication channels, they, however, merely allow the exchange of feedback information, suggestions, or complaints. Independent of the concrete type of communication channel, the communication infrastructure has to be centered on the customer, with the purpose of encouraging the customers to actively participate in the firm's value creation process. The direction of the communication should be bilateral, that is, the communication channels should not only foster the inflow of information to the firm, but should also enable firms to assist their customers in better utilizing their own as well as the resources provided by the firm. In this way, firms can actively support their customers' processes usage.

Moreover, firms can either directly interact with their customer base through customer co-creation practices, or indirectly by engaging an independent third party. One of these examples refers to InnoCentive, which can be described as an independent service provider that connects companies facing a specific problem with a global community of scientists, who are able to provide solutions to these specific R&D challenges. Put differently, InnoCentive represents a virtual knowledge broker between two or more parties. Customers of InnoCentive, for instance, are BASF, Novozymes, Dow Chemical, DuPont, Henkel, P&G, Syngenta, Nestlé PURINA, and Evonik Industries. The solver community of InnoCentive consists of more than 300 000 individuals from nearly 200 countries, who have submitted more than 40 000 solutions. For firms that face a specific problem ("seekers"), InnoCentive is a cost- and time-efficient mechanism for tapping into scientific knowledge distributed across the world as well as for connecting with various sources of innovation, including customers, suppliers, and scientists from diverse disciplines [1, 53–56].

In addition, firms may have to define the extent to which customer co-creation practices become an integral part of their overall business practices. This decision mainly includes the intended duration of customer co-creation practices (continuous versus selective) as well as the underlying co-creation substrate (single product versus product group versus entire product mix). In stable markets with a relatively low degree of changing customer needs, a selective pursuit of customer co-creation practices on a project basis might be sufficient to require customer feedback on the performance of existing products as well as the emergence of new customer needs. In dynamic markets with volatile customer needs, firms benefit more from customer co-creation practices that are continuously implemented in the firm's value creation activities. In this way, firms can ensure that they steadily receive direct customer feedback when developing new products and improving their existing product portfolio. Apart from the duration of their customer co-creation practices, firms also have to decide for which of their products they might make use of customer co-creation practices. This decision is eventually driven by various factors (e.g., access to customers, degree of complexity, importance, urgency). As a basis for decision making, managers can refer to the various criteria outlined in Box 8.2, "Roster for evaluating the appropriateness of investing in collaborative activities with customers".

Another degree of freedom refers to the number of customers the firm intends to co-create value with. Thus, firms might decide to either interact with the entire customer community, a selected group of customers, or each individual customer. In order to holistically benefit from customer co-creation practices, firms need to engage in personalized interactions with all of their customers and replace the view of the market as an aggregation of customers. Through individually interacting with customers during the firm's process of value creation (e.g., product design, production, marketing/sales) as well as during the customer's process of usage (e.g., after sales services), firms are in a better position to consider the needs and expectations of each customer. Customerization techniques (in both B2B and B2C markets) provide firms with new opportunities to actively involve customers in the design, production, as well as marketing of individualized and differentiated products. Instead of interacting with one customer at a time, firms can further interact with a group or network of customers (e.g., online communities, focus groups). Hence, firms not only benefit from individual customer insights but – considering that customers also interact amongst themselves – can also tap into the social dimension of customer knowledge, that is, knowledge that becomes apparent only when shared among a group of customers with shared interests. In all of these cases, firms, however, have to install an adequate infrastructure that enables the efficient and effective engagement in close interactions with customers. The larger the number of customers the firm interacts with, the higher the costs for appropriately coordinating and maintaining these firm–customer relationships [1, 22, 24, 29, 30, 41, 57].

Another core decision refers to the type of customer that becomes involved in customer co-creation practices. Thus, firms firstly have to decide whether to interact with existing or prospective customers. This decision is primarily based on the firm's overall (growth) strategy: while firms that seek to either penetrate an existing market (i.e., increasing their market share by incremental product modifications) or to extend their product range by developing new products for an existing market benefit from the collaboration with existing customers, firms that seek to diversify by expanding into new markets benefit more from the incorporation of prospective customers into their value creating activities. Related to these decisions, firms have to decide to include either lead users or ordinary users. The decision might depend on the intended degree of newness of the new products or services to be developed. While lead users are generally preferred when aiming at the development of exploratory innovations, ordinary users should be chosen when the overall purpose of the customer co-creation practice is to incrementally improve the existing offerings.

In order to assess whether the implemented customer co-creation practices fulfill the firm's targets, firms should monitor the performance of their customer co-creation practices on a regular basis. Figure 8.6 provides an overview of various indicators firms can rely on when assessing the performance impact of their customer co-creation practices. While performance indicators might differ with regards to the different stages at which customers are involved in the value creation process (reaching from the front end of innovation to the consumption/usage phase), performance indicators also differ with regard to the extent to which the performance implications can be assessed by means of objective measures. Whereas objective measures include the number of generated product ideas, the number of process/product innovations, or the increase in market share, subjective measures include perceived innovativeness, perceived usefulness, or positive word of mouth streaming from the firm's customer co-creation practices (see Figure 8.6).

8.2.5 BASF as Best Practice for Providing Customized Solutions

The following case of BASF is to emphasize in particular the relevance of customer co-creation practices in a B2B setting such as the chemicals industry. BASF – the largest chemical company in the world – generally shows a strong customer-orientation and is seriously committed to serving its customers with superior value through close interactions with them. According to Kurt Bock, CEO and chairman of the executive board at BASF, in 2013 "[BASF] sold more, worked more closely together with [its] [...] customers and enhanced [its] [...] portfolio" [59: 20]. BASF's declared goal is to align its businesses even more closely with its customers' needs. Through close collaborations, BASF develops customized products as well as functional materials and system solutions jointly with its customers. Customers, however, have various needs and expectations with regard to the way they interact with BASF as suppliers. A proper approach to dealing with those diverse expectations is the definition of various need-based customer segments and the subsequent derivation of various Customer Interaction Models (CIMs). By means of these CIMs, BASF seeks to ensure

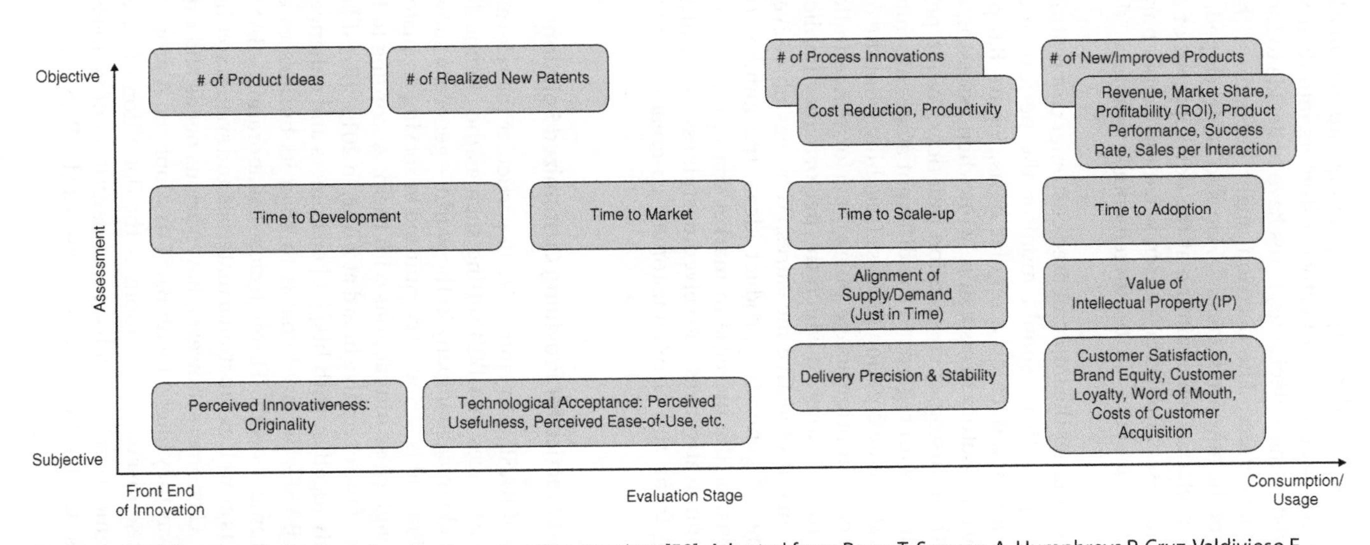

Figure 8.6 Performance indicators of customer co-creation practices [58]. Adapted from: Roser T, Samson A, Humphreys P, Cruz-Valdivieso E. 2009. *Co-creation: New Pathways to Value: An Overview*. Promise & LSE Enterprise

that customers' needs and expectations are effectively addressed in terms of both the form and level of firm–customer interactions. These six CIMs are:

i) trader/transactional supplier (numerous anonymous or shallow buyer–supplier relationships);
ii) lean/reliable basics supplier (supply reliability is an important buying factor for customers);
iii) product/process innovator (customers are interested in superior performance products/services);
iv) value chain integrator (customers open to shifting parts of their own value chain);
v) standard package provider (customers configure their own packages);
vi) customized solution provider (customers are willing to partner with supplier to jointly develop solutions) [1, 59–61].

The CIM to be chosen depends on the specific business. In the classical chemical business, BASF primarily offers commodities, namely, basic products from the "Chemicals" segment (e.g., sulfuric acid, plasticizers, caprolactam, propylene). In this cost-driven business segment, BASF primarily aims at supplying its customers reliably and cost-effectively. Accordingly, interactions between BASF and its customers are limited to the sole exchange of (standardized) products. Following the CIM "trader/transactional supplier", BASF may not offer any after sales services nor engage in personalized interactions during sales. In the "Performance Products" segment (e.g., personal care ingredients, food additives, home and personal care items), BASF, on the other hand, offers a wide range of customized products. Here, BASF interacts more closely with its customers and, for instance, also collaborates with its customers in joint projects. Employees of BASF "work closely together with customers from an early stage in order to develop new products or formulations for a specific industry" [59: 50]. Moreover, in the "Functional Materials and Solutions" and "Agricultural Solutions" segments, BASF also offers functionalized materials and solutions (e.g., engineering plastics, concrete additives, automobile and industrial coatings). Since the primary target of these businesses refers to the supply of tailor-made intelligent solutions that help customers to achieve more value, businesses within these segments require close interactions between BASF and its customers. In these businesses, BASF may follow the CIM "customized solution provider". Thus, BASF engages in customer-specific collaborations to fulfill the unmet needs of its customer, who "fairly shares value jointly created" (BASF, 2011, p. 35) [1, 59–61].

The supply of tailor-made intelligent solutions very often implies that engineers of BASF work alongside its customers and provide on-site support, for example, during the introduction of a new production system and/or during an ongoing production process. We can refer to BASF Coatings as a best practice, where its customers in the automobile industry are provided with extensive on-site support. Instead of selling paints to its customers, BASF Coatings follows a "cost per unit approach", implying that customers are charged per unit of painted car instead of per ton of paint. As a consequence, engineers at BASF Coatings are active at the customer's plant, working jointly with the engineers of the automobile manufacturer. In this way, BASF Coatings actively supports the value creation process of its customer (customer co-usage, Figure 8.4). As part of these co-creation practices, BASF Coatings supports its customers with regard to managing the material flows as well as the coating process, provides technical and analytical services, conducts laboratory analyses, detects and remedies defects, and is responsible for documentation of material consumption and quality data. Thus, BASF Coatings can directly and immediately interact with its customer as soon as any questions or problems arise during the coating process. For example, Achim Harms, former head of the BASF Coatings service team, stated that "[w]hen VW built their new paint shop, we were right at their side with plenty of valuable advice. And in the implementation phase as well, we checked the seals and rubber parts that were used in the new line" (SpecialChem, 2004) [1, 62, 63].

Another example is demonstrated by the collaborative activities between the BMW Group and BASF Coatings during the development of the new coating technology "Integrated Process II" (an example of customer co-development and customer co-launch, see Figure 8.4). To successfully complete the joint project at the Mini plant in Oxford, employees of both partners worked closely together. Wolfgang Duschek, head of the joint project between BASF Coatings and BMW Group, stated that "[we] worked with the BMW Group crew for many months, including many weekend and night shifts, in order to make this project successful. It attests to the fact that with the right ideas, perfect teamwork, and great partnership, great success can follow" [64]. The new coating technology had several considerable advantages for the BMW Group, including an increase in production from 200 000 vehicles to 240 000 vehicles per year. Owing to the development of a basecoat that integrates the positive properties of the primer (e.g., stone chip protection, ultraviolet resistance), both partners were able to cut out an entire coating step (primer application and the primer oven) without suffering any loss in terms of quality. The new technology – the result of a joint value creation project between BASF Coatings and one of its customers – eventually led to significant savings in terms of time, materials, as well as energy and labor costs [64–66].

While the previous two examples emphasize the benefits of co-locating engineers at customers' plants, BASF also offers manufacturing firms or other industrial operation providers with various opportunities to co-locate their businesses at one of BASF's plants. By these means, these firms can directly profit from shared infrastructure, shared utilities, and shared services. Through offering intelligent system solutions and high-quality products on-site, BASF actively and directly supports its suppliers, customers, and other co-location partners in successfully doing business [67].

8.3 Supplier Integration

8.3.1 Emergence of Chemical Supplier-induced Innovations

Research on supplier integration in the new product development processes disclosed meaningful advantages for manufacturers in terms of reduced development costs, increased time-to-market, and improved product qualities. In particular, the development of products that comprise a high degree of complexity requires a more intensive integration of suppliers [68–70]. Hence, suppliers of research intensive industries, such as the chemical industry, play an essential role. The relevance of suppliers in inducing innovations can be ideally shown by various examples in the field of detergents. In a case-based analysis, the following three classifications of causes that induced changes in detergent formulation were identified: economic, ecological, and technological causes [2].

Economic-driven changes refer to adaptions in formulations that are supposed to improve the firm's competitive situation within a specific business. To this end, the newly applied raw materials, and subsequently the adapted end-product, need to address an additional customer need or preference. The communication of the new value proposition through, for instance, advertisements, should provide justification for the high prices charged and increase the overall sales. One example is the product Persil that contains an integrated softener effect. The "2-in-1" product offers two functions (cleaning and softening), which makes the additional use of a softener obsolete and, hence, provides the customer with an added value (i.e., high-performance products and simple handling) compared with standard products in this category. The development of the "2-in-1" product was eventually a result of Henkel's close collaborations with their raw materials and fragrance suppliers [2, 71].

Another economic-related cause for changes (more specifically substitutions of raw materials) in formulations refers to economic changes within the raw material market. While demand-driven shortages in the supply of raw materials can significantly impact the raw material prices, an increase in raw material prices can also stem from structural changes within the industry, such as the reduction of the total number of suppliers associated with the formation of oligopolies or – in an extreme case – the formation of monopolies. This, for instance, could be observed with regard to citric acid producers in the year 2008. Triggered by price dumping of Chinese suppliers, which started with the massive exports of citric acid in the EU in the 1990s, many European suppliers were not able to withstand the strong price pressure from China and, thus, had to discontinue their production in Europe (e.g., Tate & Lyle). The few citric acid producers that survived this price war formed an oligopoly and eventually initiated an investigation at the European Commission into suspicions of dumping. As a result, the European Commission imposed an antidumping duty on imports of citric acid from China (depending on the product of between 42.2 and 49.3%). As a consequence, the prices for citric acid rapidly rose (of both Chinese as well as the European suppliers' oligopoly), combined with a shortage of the raw material in the European markets. In general, increasing raw material prices, which cannot or only with difficulty can be passed on to the end customer, significantly reduce the margins of the manufacturer. Hence, in many cases the substitution of certain raw materials may remain the only option to offer the product at a competitive price while simultaneously exploiting adequate profit margins. In the search for substitutes, manufacturers might collaborate closely with raw material suppliers, which often provide a consultancy service (e.g., demonstration of alternative raw materials) [2, 72, 73].

Apart from economically driven causes, changes in formulations might be further derived from the manufacturer's increasing need to respond to ecological and eco-toxic problems (i.e., "ecology push"[6]). The pressure to respond to such problems might result either from voluntary agreements and self-commitments (e.g., abandonment of the use of phosphates in detergents) or from legal regulations (detergent regulation, wash and cleaning agents act, REACH[7]). Furthermore, the increasing awareness of environmentally friendly products among customers forces manufacturers to develop solutions that contribute to or – at least – do not harm the environment ("ecology pull"[8]). In addition to these ecological factors, social aspects also represent important decision criteria for the choice of raw materials, such as the selection of certified raw materials (e.g., certified palm kernel oil). In these decisions, manufacturers significantly depend on the knowledge and technological competences of their suppliers. The case of green surfactants, for instance, illustrates that suppliers to the chemical industry contribute significantly to the development of sustainable products. Hence, ecological factors have triggered the technological development of new products, which – as they met unsatisfied customer demands – were successfully introduced into the market [2].

Finally, raw material changes might derive from technological causes that are either initiated by the continuous improvement of existing products or through the development of novel products ("technology push"). Improvements to existing products in the field of detergents are normally related to the primary (removal of dirt) and secondary (prevention of re-contamination with previously removed dirt) washing effect of detergents. These innovations can be either achieved by substituting old components with new, more efficient

---

6 An "ecology push" represents a special type of "technology push" and refers to a regulatory pressure of the state as well as a societal pressure (Fichter and Behrendt, 2007 [74]).

7 Registration, Evaluation, Authorisation and Restriction of Chemicals.

8 An "ecology pull" represents a special characteristic of "market pull" and refers to a market demand associated with ecologically compatible products (Ehrenfeld, 1997 [75]).

components or by adding new components, such as enzymes or color transfer inhibitors. These developments are usually associated with increasing the overall user friendliness of a certain product. One example refers to the hygiene rinser, which substitutes a softener in the last rinse cycle and, thereby, ensures a sterile result for the laundry at temperatures from 15 °C. When developing new formulations, suppliers often significantly contribute to the overall success of the manufacturers' innovative efforts. For instance, in the case of Perwoll Sport (a mild detergent for special applications such as sport and functional textiles (sportswear)), Henkel developed a new formulation in close collaboration with the fragrance supplier Givaudan and the specialty chemical supplier Evonik, which provided zinc ricinoleate – a substance with a unique odor absorbent effect [2, 76].

Taken together, causes for raw material changes in formulations can be dedicated to the areas of "economy", "ecology", and "technology". If "technology push"/"ecology push" and "market pull"/"ecology pull" are successfully combined, these raw material changes in formulations may eventually result in innovations that might be successfully introduced and adopted in the market (Figure 8.7).

Box 8.5 outlines some of the benefits of supplier integration.

8.3.2 Typologies of Supplier Integration and Roles

The supplier's contribution to the value creating activities of a manufacturer depends on the extent to which the supplier is integrated into the practices and processes of the manufacturer. For less innovative collaborative projects (e.g., a simple product modification), the supplier will only be weakly integrated into the new product development process. Hence, the supplier only provides independently developed components and, thus, it is not actively integrated into the manufacturer's development process. When the primary aim of the manufacturer is to develop entirely new products, suppliers should be more intensively involved in the development process in order to actively benefit from the supplier's technological competence. In general, the following three types of supplier integration exist: "white box", "grey box", and "black box" (see Figure 8.8) [77–79].

White-box integration refers to the form of supplier integration with the lowest level of supplier responsibility. Suppliers are somewhat regarded as external consultants to the firm. Although suppliers participate in discussions about certain specifications and requirements, it is still the firm that is entirely responsible for making the final decision on the specifications and for accomplishing the final development process. Hence, although the firm considers the suppliers' know-how and expertise on certain solution-related topics, the final solutions are developed entirely in-house without any active contribution from the suppliers [2, 69].

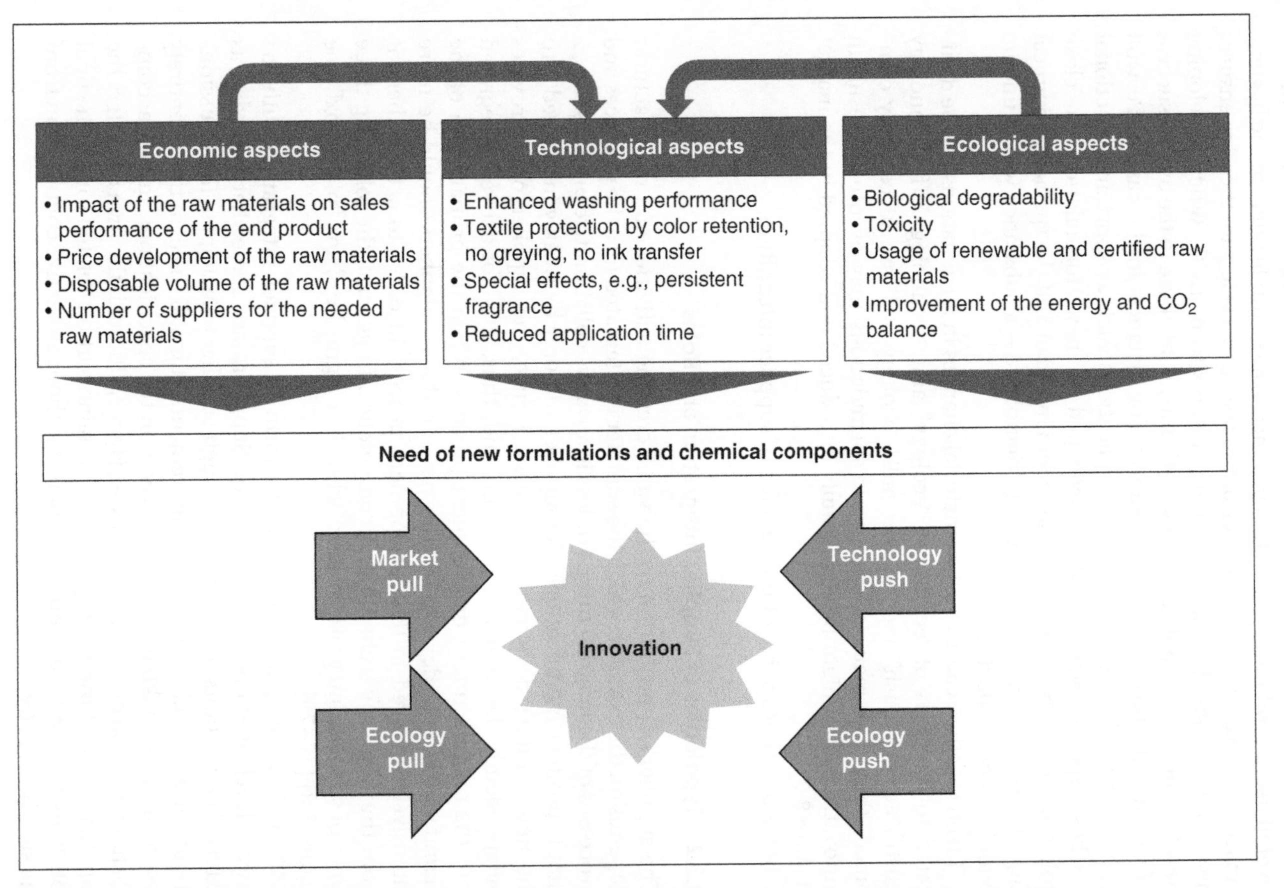

Figure 8.7 Three causes for raw material changes in formulations and the subsequent induction of innovations [2]

Box 8.5 Benefits of supplier integration [70]

Ragatz, Handfield, and Petersen (2002: 392) outline the following beneficial outcomes of supplier integration:

- "including suppliers on project teams adds information and expertise regarding new ideas and technology […] and helps to identify potential problems so they can be resolved up front […]. This improves the quality of the final product, and ensures that it meets or exceeds the final customers' expectations […]
- supplier integration provides outsourcing and external acquisition possibilities that reduce the internal complexity of projects […], and provides extra personnel to shorten the critical path for new product development projects […]
- [supplier integration] helps to better coordinate communication and information exchanges […], which further reduces delays and ensures the project is completed on time […]
- [supplier integration] broadens the scope of tasks and issues because accessibility of parts can be considered early on, thus eliminating rework and reducing costs […]"

White-box supplier integration
Discussions between firm and supplier about certain specifications and requirements.
Supplier as external consultant.
Grey-box supplier integration
Supplier works alongside the firm.
Joint development activities between firm and supplier.
Black-box supplier integration
Outsourcing of development activities to the supplier.
Supplier independently develops new parts.
Firm
Supplier
Responsibility for product development activities

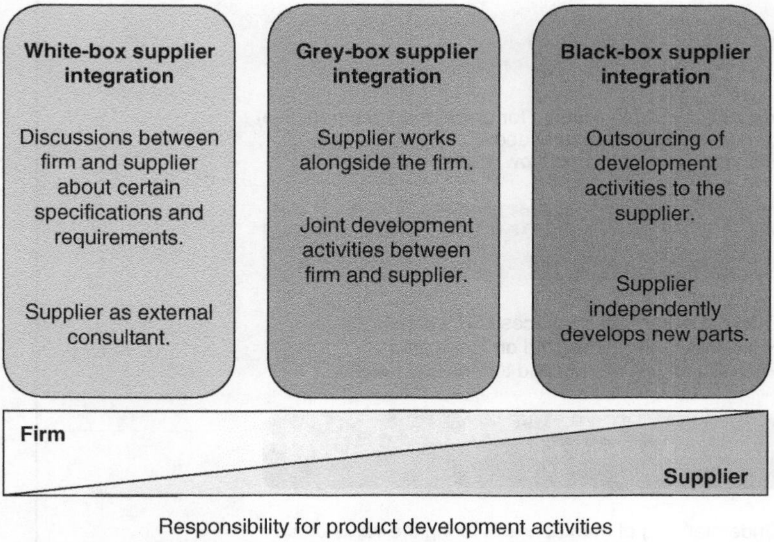

Figure 8.8 Spectrum of supplier integration [1, 77]. Adapted from: Petersen KJ, Handfield RB, and Ragatz GL. 2005. Supplier integration into new product development: Coordinating product, process and supply chain design. *Journal of Operations Management*, **23**(3): 371–388

Black-box and grey-box supplier integrations constitute supplier integration from a product and process point of view, respectively. Whereas black-box supplier integration refers to the integration of holistic solutions, such as externally developed parts, components, or subassemblies, into the manufacturer's new product development process, grey-box supplier integration refers to collaborative R&D activities in which the supplier's knowledge is directly incorporated into the manufacturer's new product development process. In black-box supplier integration "suppliers carry out product engineering activities on behalf of their customers and even develop components or entire subassemblies" [16: 102]. In grey-box supplier integration the "supplier's engineers work alongside the customer's engineers to jointly design the product so the supplier's process can be effectively integrated with the design" (Koufteros, Vonderembse, and Jayaram, 2005, p. 102) [1, 2, 16, 77, 80].

Instead of referring to the degree of responsibility for certain development activities that are transferred to suppliers, suppliers might also be classified by means of the overall contribution they make to the manufacturer's value creating activities. Here, the following three different classifications are made: "solutions provider", "technology provider", and "problem finder & solution provider" (see Figure 8.9). While the "solution provider" solely responds to the request from the manufacturer, the "technology provider" proactively offers

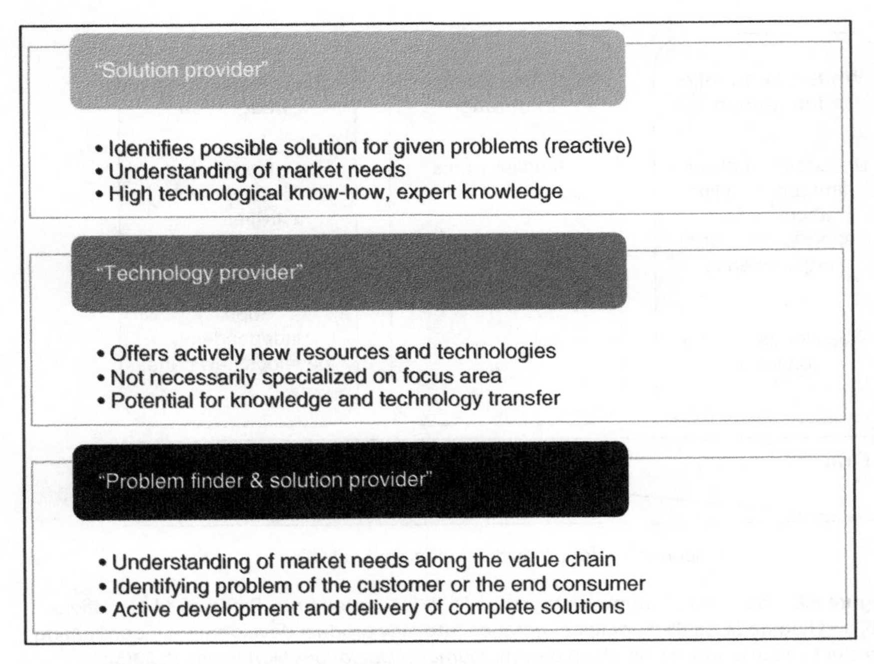

Figure 8.9 Typologies of suppliers according to their contribution [2]

new technologies. The "problem finder & solution provider" takes a step even further and offers proactively complete solutions for new products of the manufacturer. The last is particularly oriented towards the needs of the end-consumer [2].

A supplier that acts as a "solution provider" identifies solutions to problems of the manufacturer. Here, the manufacturer takes over the initiating of innovation activities by identifying the needs or rather the problems of consumers through various market research techniques, and subsequently transfers the problem of the consumer into a technological problem. If the manufacturer is not able to solve this technological problem on its own, its supplier supports the manufacturer in developing an adequate solution. Depending on the availability of the required technology for the required solution, the application of existing technologies (e.g., through purchase of a new raw material ("from the shelf")) or the development of new technologies in the context of "black-box" or "grey-box" supplier integration takes place. Depending on the degree of novelty, incremental or radical innovations might be the result of this type of collaboration activity. The supplier, however, needs to possess a specific technological knowledge and a certain understanding of the market. The problems have to be understand and suitable customer-tailored solutions have to be identified. Taken together, the overall contribution of the supplier in the new product development process can be described as being reactive rather than proactive [2, 81].

As a "technology provider", the supplier initiates innovations at the manufacturer's side by engaging intensively in its own research and development activities. Based on the developments of technologies or a raw material, the supplier actively searches for suitable fields of application. In so doing, the new technology is presented to manufacturers with the purpose of initiating downstream innovations. The manufacturer, who typically holds extensive knowledge of the consumer needs, steadily seeks to identify existing or new problems of the consumer and, when developing potential solutions to these problems, considers in particular the newly developed technologies by its suppliers. Thus, an adaption of the suppliers' newly developed technology might be required, which eventually can be conducted by means of "grey-box" supplier integration, implying an intensive exchange of technological knowledge by both partners. This type of collaboration, which is characterized by an active commitment and contribution of the supplier in the manufacturer's value creating activities, might eventually result in the development and introduction of either incremental or radical innovations [2, 81].

Suppliers that act as a "problem finder & solution provider" usually take over a large part of the manufacturer's innovation activities, such as ideation and development activities. The supplier simultaneously senses recent developments within the primary market (downstream manufacturers) as well as within the secondary market (consumer market). By identifying and combining the needs of both the manufacturer and the consumers, suppliers are in a better position to develop ideas and concepts for new products that are of superior value to the manufacturer. By combining the concepts of "market pull" and "technology push", suppliers particularly seek to combine both superior technology- and solution-related knowledge in supporting the manufacturer with valuable input for new products. The manufacturer, thus, profits from the supplier's efforts in new product development and the provision of complete solutions for products that address unsatisfied needs of consumers. The "problem finder & solution provider" follows a multi-stage marketing approach, which encompasses all kinds of marketing activities that are directed towards its customers' customers [2].[9]

8.3.3 Supplier Willingness to Be Involved in the New Product Development

In many cases manufacturers are presented as the most powerful organ in any buyer–supplier relationship, while suppliers only take on a subordinate role and are just expected to follow the manufacturers' strict guidelines. However, the tendency of shifting value-added activities from manufacturers towards suppliers demonstrates the increasingly important role suppliers have for the manufacturers' value creating activities. Suppliers can be regarded as an important source of competitive advantage, considering that they are able to generate new solutions and initiate valuable innovations for the manufacturers. While in the past contracting with multiple suppliers had been argued as the opportunity for the manufacturer to become independent of certain sources as well as to reduce the potential risk of shortages in supply, recent decades have shown that firms have increasingly reduced the number of suppliers they are contracting with and, rather, deal with a limited number of qualified suppliers. Inherent with this development, buyer–supplier relationships have become closer and more long-term oriented. Thus, rather than considering buyer–supplier relationships as being adversarial and solely focusing on the process of exchange, manufacturers increasingly establish a partnership mentality with their suppliers, implying that manufacturers also seek to tap into their suppliers' intangible resources (e.g., ideas, expertise, and solution-related knowledge). Thus, since this development might also imply the emergence of a strong demand for the strategically important suppliers, some suppliers may be in the position to more confidently select their collaboration partners [1, 77, 82–84].

8.3.4 Value Creation and Supplier Relationship

A collaborative manufacturer–supplier relationship might only be established and sustained if the (expected) outcome is beneficial for both parties, that is, if the (expected) outcome is of higher value than the value generation with an

---

9 For a detailed description of the multi-stage marketing cf. Backhaus and Voeth (2007) [81].

alternative relationship ("the whole is greater than the sum of its parts"). Thus, to be able to make adequate decisions about which relationships to invest in, firms have to understand and evaluate the value generated within different business relationships. The concept of value creation within business relations is an established concept in the marketing literature and can also be transferred to the supply chain context. Considering the increasing relevance of suppliers as a strategic resource for achieving competitive advantage, profound knowledge about the factors that lead to close relationships with certain suppliers is indeed very important. In so doing, value generation has to be considered from a supplier's perspective in order to offer suppliers adequate incentives to contribute to the overall success of the relationship [2, 85–87].

Based on social exchange theory, the value of a relationship can be defined by the difference between the benefits and the costs [88]. Here, different facets of value generation have to be taken into account. While most of the existing literature primarily considers economic advantages (e.g., reasonable prices or capacity utilization), other, less obvious, drivers of value generation are often neglected though they are also of great importance (e.g., strategic information exchange or access to important network). Following a supplier's perspective on value generation within manufacturer–supplier relationships, the functions of business relations are considered as activities and resources provided by the manufacturer. These activities and resources are performed and applied in the context of the specific manufacturer–supplier relationship and are supposed to produce an advantage from the supplier's perspective [87, 89–92].

A difference can be made between direct and indirect functions of business relations. While direct functions are defined by their immediate effect on the value generation through the manufacturer, indirect functions either have a future impact or have an effect across other partners within the network. Direct functions of the manufacturer–supplier relationship comprise activities and resources of the manufacturers that directly create an advantage for the supplier. These are subdivided into [90, 93, 94]:

- sales function – manufacturers purchase large quantities and thereby ensure the capacity utilization of the supplier;
- profit function – manufacturers generate a value for the supplier by securing the profitability of the relationship through reasonable prices;
- safeguard function – manufacturers also buy in unstable market situations and during crises and, thereby, ensure cost efficiency for suppliers.

Indirect functions, on the other hand, generate a value within a manufacturer–supplier relationship by offering a more long-term orientated and, thus, strategic advantage for the supplier. They are subdivided into:

- innovation function – manufacturers contribute to the new product development of suppliers;

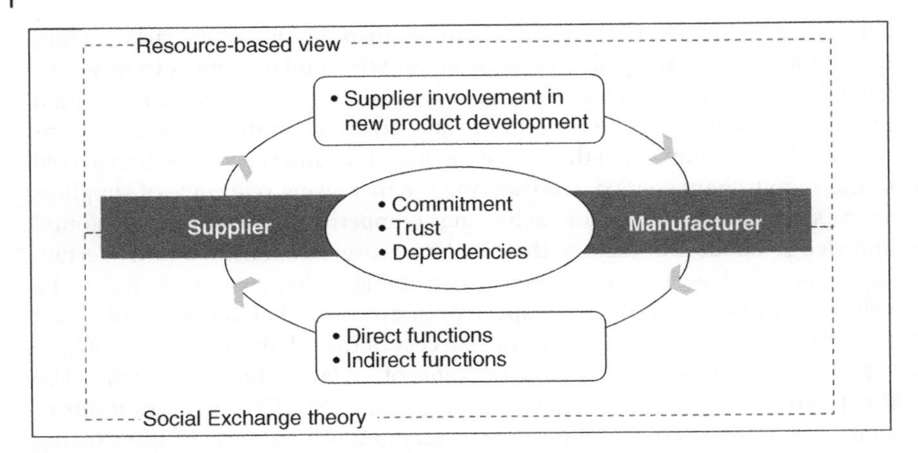

Figure 8.10 Illustration of manufacturer–supplier relationship [2]

- reference function – manufacturers serve as positive reference and make the supplier credible to other potential manufacturers;
- information function – manufacturers convey important market and competition information, which is not accessible to the suppliers;
- network function – manufacturers provide access to important organizations and partners from the relevant network.

While the economic cost-benefit focus might certainly explain to some extent how value is generated in manufacturer–buyer relationships, relational factors such as trust and commitment are of additional importance to ensure a holistic consideration of value generation in business relationships. Successful manufacturer–supplier relationships typically derive from trusting, long-term oriented associations that are characterized by immediate and close interactions between both parties. As a consequence, many manufacturers reduce their supply base while simultaneously binding strategically important suppliers more strongly. In so doing, evaluating the quality of these relationships (in terms of trust and long-term orientation) from a supplier's perspective is of great significance (see Figure 8.10) [95–97].

8.3.5 How Do You Attract the Most Innovative Chemical Suppliers?

To answer this question adequately an empirical study with 94 participants was conducted. The participants were all employees of suppliers for the processing chemical industry in Germany. The suppliers, producing goods for the chemical processing industry, were chosen from the company list of the "Vereinigung der chemischen Industrie" (VCI). The interviewed participants were mainly from the marketing, R&D or sales departments and had maintained contact for

Table 8.2 Important steps in attracting the most innovative chemical suppliers [2].

(1)	The constant development of internal technological competences cannot be neglected by the manufacturer. It increases the attractiveness to serve as a collaboration partner in new product development, enables the advancement of knowledge, and reduces the technological dependence on suppliers.
(2)	Future earnings of the new product development collaboration have to be guaranteed for the supplier. In so doing, fair and collaborative conditions have to be established.
(3)	The collaboration has to lead to long-term advantages such as exchange of strategic information and access to a strategic relevant network.
(4)	Manufacturers have to ensure a foundation of trust for desired collaborations with suppliers by avoiding opportunistic behavior (exploitation of power, tough price negotiations) as well as through the establishment of mutual targets and measures that are beneficial for both partners.
(5)	Making use of any dependencies is inappropriate in encouraging the suppliers to invest in the manufacturer–supplier relationship in the long-term. Manufacturers should avoid the opportunistic exploitation of their advantages in power.
(6)	Ensuring the supplier's commitment represents the key to the supplier's decision to engage in collaborative new product development with the manufacturer. The supplier's commitment might derive from ensuring clear advantages of the relationship, such as future earnings, access to the relevant network, or the exchange of strategic information.

at least one year with their B2B contacts within the processing chemical industry, which ensured that the person had sufficient information to participate in the online survey.

Based on this study, the following factors influencing the willingness of a supplier to engage in new product development of a manufacturer are identified. They build important incentives (see Table 8.2), which should be guaranteed by manufacturers to ensure the suppliers' willingness to participate [2].

1) The engagement in the product development activities of the particular partner is often based on the mutual competences that both partners possess with regard to their relevant technology sectors. Hence, high-performance partners often come together to establish a collaboration with the purpose of strengthening and expanding their knowledge through an intensive exchange of their experiences. For the manufacturer, this approach implies that – contrary to the strong tendency of outsourcing certain value creating activities to external partners – the manufacturer's expertise in strategic relevant technology sectors remains within the company and can be further developed. Advanced research and development competences in relevant technology sectors increase the attractiveness of the manufacturer for high-performance, innovation- and technology-orientated suppliers

and thus can positively influence the decision of the supplier to collaborate in the new product development. Furthermore, the development of their own technological competences enables the manufacturer to better absorb knowledge and also to better utilize insights gained from the collaboration. In addition, manufacturers can reduce the technological dependence on their suppliers. For these reasons, the constant enhancement of their own technological competences cannot be neglected [2, 98].

2) Furthermore, the supplier's evaluation of the benefits of the entire manufacturer relationship based on economic as well as noneconomic criteria can considerably influence the supplier's decision to collaborate. Thus, direct functions, that is, direct economic contributions by the manufacturer, can have a positive influence on the supplier's general commitment and increase the supplier's long-term commitment to maintain the relationship. The direct functions, however, reveal no influence on the supplier's engagement in new product development activities. This interesting finding might be explained by the following assumption. The integration of suppliers into the manufacturer's new product development process might be required for new businesses rather than for established businesses. Since in new businesses the focus is more on potential future earnings, the manufacturer's current economic contributions might not necessarily have an influence on the supplier's willingness to engage in the new product development process of the manufacturer. Thus, manufacturers should offer suppliers an incentive that is related to the supplier's future earnings that will derive from the collaborative development. In so doing, fair and collaborative conditions have to be created [2].
3) With regard to indirect functions (i.e., long-term orientated advantages of a manufacturer–supplier relationship), the same study discloses a positive influence of manufacturer–supplier relationships on the supplier's general commitment as well as on the supplier's engagement in the new product development. A slight influence on the supplier engagement is conveyed by the supplier's commitment. Thus, when deciding about their engagement in the new product development, suppliers consider the long-term advantages of the relationship, which, for instance, might result from an open exchange of information with the manufacturer. In accordance with the "Open Innovation" paradigm (see Chapter 5), suppliers seek to achieve competitive advantage through a continuous and open exchange of information and the subsequent acquisition of valuable knowledge. Further, the manufacturer can act as a network promoter, becoming responsible for the connection with the supplier within the relevant network. Here, increasing competition among entire networks replaces the predominant competition among single companies. Thus, getting access to strong competitive networks is becoming an important factor for success in the chemical industry. To increase their own attractiveness and, through these means, the supplier's

willingness to become integrated into new product development, manufacturers need to possess strategic information as well as access to a strategic relevant network. Manufacturers subsequently have to share this information with suppliers and offer them the opportunity to participate in the manufacturer's network [2, 99].

4) The empirical analysis further confirms a strong influence of the trust factor that the supplier shows towards the manufacturer. From the manufacturers' perspective, these findings imply that the supplier's long-term interest to maintain the manufacturer–supplier relationship can be influenced through the design of a trusting collaborative culture. However, particularly in relationships between manufacturers and suppliers, the trusting collaboration is often threatened by price negotiations and opportunistic behavior of the partners. Thus, to foster long-term collaboration, both partners should define mutual targets that represent satisfactory results for both partners. To foster the emergence of trust, manufacturers not only have to positively reward the trust granted by the supplier, but in return they also have to proactively offer a significant level of trust to their suppliers. Further, the compliance with agreements, an open treatment of problems, as well as the strict consideration of the interests of the partner have to be ensured to strengthen a trusting climate between both partners. Moreover, inter-organizational trust is always based on the personal relationships between the employees of the companies. Thus, employees, who are involved in close interactions with the representatives of the supplier company and take over the responsibility for managing the relationship, play an essential part in building and shaping strong relationships with suppliers. They act as a relationship promoter and can strongly influence the quality of the relationship. From the manufacturers' perspective, it is preferable to position qualified employees in this area of responsibility. They should function as direct contact partners and take over the maintenance of the relationship with a certain continuity. Furthermore, empirical evidence is provided that commitment completely mediates the relationship between trust and supplier engagement. This finding illustrates the central role of commitment in manufacturer–supplier relationships. Only if suppliers show a certain degree of commitment will a trusting relationship affect the supplier's engagement in the manufacturer's new product development activities [2, 100].
5) The influence of the suppliers' dependency can differ significantly regarding the specific dependency situation and its causes. On the one hand, a specific dependency situation can enforce the formation of a collaborative relationship. On the other hand, the dependency and the resulting disadvantages can harm a collaboration through opportunistic and non-collaborative behavior of the partner, and thus lead to the termination of the relationship. In general, if manufacturers are in a more powerful position

than their suppliers, they should seek to avoid any opportunistic behavior, since such a behavior might reduce the level of trust and harm the durability of the manufacturer–supplier relationship [2, 101–103].

6) The commitment of the supplier represents a central influential factor on the supplier's engagement in the manufacturer's new product development activities. It not only has a direct influence, but further strengthens the effects of other factors. Thus, commitment represents the key factor for successful supplier integration and can be considered as a control mechanism: if the supplier's commitment is guaranteed, suppliers might be more willing to engage in the manufacturers' new product development activities that are characterized by high risk. In addition, a supplier's commitment can complement the control mechanisms that are established in a relationship or can even partially replace inefficient and bureaucratic control processes implied by contractual regulations. Hence, the supplier's commitment is based on the recognized advantages of the relationship and the resulting intention of the manufacturer to retain the manufacturer–supplier relationship. This positive attitude forms the basis for successful collaborations between manufacturers and suppliers [2].

8.4 Invisible for Black & White – A Best Practice for Collaborating with Both Suppliers and Customers

To increase its innovation capacity, the German personal care company Beiersdorf relies on various co-creation practices. In 2011, Beiersdorf, for instance, launched the internet-based platform "Pearlfinder". By means of this innovation platform, Beiersdorf aims at "actively integrating external ideas and solutions into the development of new products – i.e. collaborating as equals with external partners" (Klaus-Peter Wittern, Head of R&D at Beiersdorf) [104]. Through Pearlfinder, Beiersdorf's external partners (e.g., customers, suppliers, scientists, universities, start-ups, etc.) can share their ideas, concepts, and inventions with Beiersdorf within a secure and confidential network [1, 104–106].

In addition, Beiersdorf also established the "Project House" initiative and the so-called "Incubation Labs". Both initiatives aim at intensifying the collaborations between Beiersdorf and specifically selected partners. Since scientists at Beiersdorf as well as scientists from its partners work constantly on-site and as equal members of the project team, they co-develop new products through actual face-to-face interactions. According to Beiersdorf, the early involvement of such partners in the innovation process promotes communication and facilitates the transfer of valuable knowledge [1, 104].

Apart from these continuous initiatives, Beiersdorf also runs a project-based customer co-creation initiative for its well-known skin care brand Nivea. More specifically, Beiersdorf's Nivea Deodorant and Antiperspirant Division followed a co-creation approach with its customers at the front end of innovation. This customer co-creation project can be basically divided into two parts: (i) a netnography method (i.e., analysis of virtual customer communities) for the identification and basic understanding of the voice of the customer, and (ii) a co-ideation method (i.e., idea generation and screening) for the active integration of the voice of the customer [1, 107].

In order to identify and develop a basic understanding of the customers' needs and expectations, Beiersdorf initially applied a netnography method. This market research technique – in general – implies the passive observation of online communities, forums, and other content published in the social media. The analysis focused on search fields, such as cosmetics, health, beauty, lifestyle, fashion, and sports. In total, Beiersdorf screened more than 200 online sources in different languages. After this relatively broad analysis, Beiersdorf analyzed the most meaningful online communities in greater detail, including a platform maintained by lead users: The Undershirt Guy. The lead users of this platform disclosed valuable need- and solution-related information, which they, for instance, generated through self-conducted experiments with deodorant stains. By referring to this product-related discussion forum – and other online sources – Beiersdorf could identify the following topics of interest [1, 107, 108]:

i) **Types of stain**. Users frequently complained about various types of stains associated with use of deodorants.
ii) **Causes of stain.** Users also proposed various factors that might cause these stains.
iii) **Stain removal.** Finally, users also suggested various methods in order to remove these stains from clothes or how to prevent the occurrence of stains. For instance, a user suggested the following: "I know that this is going to sound weird but you could try mouthwash. Soak a cotton ball with sugarless, alcohol-based mouthwash and swab your armpits. This will kill the odor causing bacteria. Let it dry and then apply your normal deodorant" [107: 38].

Building on these customer insights, Beiersdorf's R&D department came up with several new product ideas, which formed the basis for the subsequent customer co-creation practice. To evaluate and enrich the initial stock of product ideas, Beiersdorf conducted a follow-up online co-creation study and involved its customers in the process of idea evaluation and screening. Thus, customers were empowered to evaluate and to vote on product ideas that had been developed by its internal R&D department. In order to offer a compelling co-creation experience for participants, the online co-creation practice contained various drag and drop tools. Furthermore, participants were able to enrich and improve specific ideas, comment, ask questions, and suggest fields of applications [1, 107].

Among the top three ideas was the idea for the subsequently introduced product "Invisible for Black & White" deodorant. The concept of the new "Invisible for Black & White" deodorant addressed two main concerns that had been identified within the co-creation project: the protection of black clothing from stains and the reduction of yellow stains on white clothing. The entire success of Beiersdorf initiative becomes apparent from the following: Nivea's new Invisible for Black & White was Beiersdorf's most successful international deodorant launch. After 9 months, Beiersdorf's new product had reached more customers than its nearest competitor product in an entire year [1, 104, 107].

Apart from the close collaborations with Beiersdorf's customer base, the success of the "Invisible for Black & White" project may have also been derived from the establishment of a specific "Project House" aimed at the adequate and timely development of the required chemical agents. Here, Beiersdorf actively involved selected partners of its supplier network (e.g., BASF Personal Care, Evonik Industries) in the development project. The "Project House" initiative eventually implied that the collaborative development activities did not take place in the partners' spatially separated laboratories. During the entire project, scientists of Beiersdorf as well as scientists of its suppliers worked alongside one another in a joint laboratory at one of Beiersdorf's subsidiaries [1, 109].

8.5 Summary

- **Customers and suppliers represent two important stakeholder groups with whom chemical firms continuously collaborate.** The integration of customers and suppliers has gained considerable attention among practitioners as both supply chain partners provide valuable input into a firm's value creation process. Customer and supplier integrations for the purpose of triggering innovations represent two distinct types of Open Innovation strategy: while customer integration refers to a market pull strategy, supplier integration refers to a technology push strategy. Suppliers often have a better expertise and special knowledge about components and raw materials, which might lead to substantial changes to the product. Customer integration – as a market pull strategy – though is suggested as resulting in the development of fairly incremental innovations.
- **The primary aim of customer integration is to reduce an asymmetric need and solution information for the purpose of reducing the level of uncertainty.** To reduce various uncertainties during the development of new products (i.e., mainly technology- and market-related uncertainties), firms have to combine two essential types of information: need information and solution information. Whereas customers usually possess precise and detailed knowledge of their individual needs (i.e., need information), firms usually possess precise and detailed knowledge of how to solve

customer-related problems (i.e., solution information). The primary purpose of customer integration (and similarly to traditional market research techniques) is to overcome this asymmetry between solution and need information. Depending on the degree of shared responsibility for value creation and the degree of firm–customer interactions, we primarily distinguish between (i) listening to the voice of the customer, (ii) customer integration (outsourcing), and (iii) customer co-creation.

- **Firms have to decide at which stage of the value creation process their collaborative efforts with customers should be directed.** The most comprehensive approach of customer co-creation spans the firm's entire value creation process (e.g., front end of innovation, idea realization and development, commercialization, product design, production, and marketing/sales) and also incorporates the customer's process of usage (e.g., after sales services). Depending on the specific stage of the value creation process, customer co-creation can be classified into seven subcategories: customer co-ideation, customer co-development, customer co-launch, customer co-design, customer co-production, customer co-marketing, and customer co-usage.
- **Designing customer co-creation practices.** Since a generally applicable best practice for customer co-creation does not exist, firms have to consider their personal circumstances (available resources, priority of product to be developed/improved, etc.) and make several choices on how to design their individual customer co-creation practices. These choices basically refer to the intended range of customer co-creations, type of customers to be involved, number of customers to be involved, duration of interaction, underlying object of co-creation effort, as well as the concrete type of interaction.
- **Constant development of internal technological competences is a precondition for the integration of the technology-oriented partners.** Despite the ongoing pressure to innovate, companies should not blindly follow the outsourcing approach and outsource all innovation-related activities. The constant development of internal technological competences in relevant fields should not be neglected by the manufacturer. It increases the attractiveness of serving as a collaboration partner in new product development, enables the absorption of knowledge, and reduces the technological dependence on suppliers.
- **Differentiation between direct and indirect value functions of manufacturer–supplier relationship.** Considering the increasing relevance of suppliers as a strategic resource for achieving competitive advantage, profound knowledge about the factors that lead to close relationships with certain suppliers is indeed very important. Particularly, value generation has to be considered from a supplier's perspective in order to offer suppliers adequate incentives to contribute to the overall success of the relationship. A difference should be made between direct and indirect functions of business relationships. While direct functions are defined by their immediate effect on the value

generation through the manufacturer, indirect functions either have a future impact or affect other partners within the network. In particular, the indirect functions have a positive influence on the supplier's general commitment as well as on the supplier's engagement in the new product development.

- **The commitment of the supplier represents a central influence factor on the supplier's willingness to engagement in the manufacturer's new product development activities.** Ensuring the supplier's commitment represents the key to the supplier's decision to engage in collaborative new product development with the manufacturer. The supplier's commitment might derive from ensuring clear advantages of the relationship, such as future earnings, access to the relevant network, or the exchange of strategic information.

Note: The main parts of this chapter are adapted from [1] and [2].

References

1 Gelhard C. 2014. A Backward and Forward Consideration of the External Value Co-creation Process: Performance Implications in the Short-and Long-term. Doctoral dissertations, University of Münster.

2 Tiemann I. 2011. *Lieferantenengagement in der Neuproduktentwicklung*. Sierke Verlag: Münster.

3 Kearney A.T. 2012. *Collaboration: A New Mantra for Chemical Industry Growth*. http://www.atkearney.com/chemicals/ideas-insights/article/asset_publisher/ LCcgOeS4t85g/content/collaboration-a-new-mantra-for-chemical-industry-growth /10192 (accessed 10 April 2014); Kearney A.T. 2013. *C3X: Looking Back at Five Years in the Chemicals Industry*. http://www.atkearney.com/chemicals/ideas-insights/article/asset_publisher/LCcgOeS4t85g/content/c3x-looking-back-at-five-years-in-the-chemicals-industry/10192 (accessed 10 April 2014).

4 Porter ME. 2000. Wettbewerbsvorteile: Spitzenleistungen erreichen und behaupten. Frankfurt am Main (u. a.).

5 Afuah A. 2003. *Innovation Management: Strategies, Implementation and Profits*. Oxford University Press: New York.

6 Jasper J. 2006. Eine Sorge weniger: Anforderungen an die Gaseversorgung von Produktionsstätten. *P&A-Fachmagazin*, December 2006: 28–29.

7 Chesbrough H. 2003. The logic of open innovation: Managing intellectual property. *California Management Review*, **45**(3): 33–58.

8 Gassmann O and Enkel E. 2006. Open innovation. *Zeitschrift Führung + Organisation*, **75**(3): 132–138.

9 Piller F, Ihl C, and Vossen A. 2011. Customer co-creation: Open innovation with customers. *Wittke V./Hanekop H*, 31–63.

10 Thomke S and Von Hippel E. 2002. Innovators. *Harvard Business Review*, **80**(4): 74–81.
11 Lau AKW, Tang E, and Yam RCM. 2010. Effects of supplier and customer integration on product innovation and performance: Empirical evidence in Hong Kong manufacturers. *Journal of Product Innovation Management*, **27**(5): 761–777.
12 O'Hern M and Rindfleisch A. 2010. Customer co-creation. *Review of Marketing Research*, **6**: 84–106.
13 Witell L, Kristensson P, Gustafsson A, and Löfgren M. 2011. Idea generation: Customer co-creation versus traditional market research techniques. *Journal of Service Management*, **22**(2): 140–159.
14 Bendapudi N and Leone RP. 2003. Psychological implications of customer participation in co-production. *Journal of Marketing*, **67**(1): 14–28.
15 Fuchs C and Schreier M. 2011. Customer empowerment in new product development. *Journal of Product Innovation Management*, **28**(1): 17–32.
16 Koufteros X, Vonderembse M, and Jayaram J. 2005. Internal and external integration for product development: The contingency effects of uncertainty, equivocality, and platform strategy. *Decision Sciences*, **36**(1): 97–133.
17 Lüthje C and Herstatt C. 2004. The lead user method: An outline of empirical findings and issues for future research. *R&D Management*, **34**(5): 553–568.
18 Enkel E, Kausch C, and Gassmann O. 2005. Managing the risk of customer integration. *European Management Journal*, **23**(2): 203–213.
19 Fang E. 2008. Customer participation and the trade-off between new product innovativeness and speed to market. *Journal of Marketing*, **74**(4): 90–104.
20 Narver JC, Slater SF, and MacLachlan DL. 2004. Responsive and proactive market orientation and new-product success. *Journal of Product Innovation Management*, **21**(5): 334–347.
21 Mahr D, Lievens A, and Blazevic V. 2014. The value of customer cocreated knowledge during the innovation process. *Journal of Product Innovation Management*, **31**(3): 599–615.
22 Prahalad CK and Ramaswamy V. 2004. Co-creation experiences: The next practice in value creation. *Journal of Interactive Marketing*, **18**(3): 5–14.
23 Gebauer J, Füller J, and Pezzei R. 2013. The dark and the bright side of co-creation: Triggers of member behavior in online innovation communities. *Journal of Business Research*, **66**(9): 1516–1527.
24 Hoyer WD, Chandy R, Dorotic M, Krafft M, and Singh SS. 2010. Consumer cocreation in new product development. *Journal of Service Research*, **13**(3): 283–296.
25 Stump RL, Athaide GA, and Joshi AW. 2002. Managing seller-buyer new product development relationships for customized products: A contingency model based on transaction cost analysis and empirical test. *Journal of Product Innovation Management*, **19**(6): 439–454.

26 Etgar M. 2008. A descriptive model of the consumer co-production process. *Journal of the Academy of Marketing Science*, **36**(1): 97–108.

27 Nambisan S and Baron RA. 2009. Virtual customer environments: Testing a model of voluntary participation in value co-creation activities. *Journal of Product Innovation Management*, **26**(4): 388–406.

28 Zhao X, Huo B, Flynn BB, and Yeung JHY. 2008. The impact of power and relationship commitment on the integration between manufacturers and customers in a supply chain. *Journal of Operations Management*, **26**(3): 368–388.

29 Payne AF, Storbacka K, and Frow P. 2008. Managing the co-creation of value. *Journal of the Academy of Marketing Science*, **36**(1): 83–96.

30 Zhang X and Chen R. 2008. Examining the mechanism of the value co-creation with customers. *International Journal of Production Economics*, **116**(2:) 242–250.

31 Calder BJ. 1977. Focus groups and the nature of qualitative marketing research. *Journal of Marketing Research*, 353–364.

32 Lettl C. 2007. User involvement competence for radical innovation. *Journal of Engineering and Technology Management*, **24**(1): 53–75.

33 Gaubinger K, Rabl M, Swan S, and Werani T. 2015. *Innovation and Product Management*. Springer: Berlin, Heidelberg.

34 Hauschildt J. 1996. Innovation, creativity and information behaviour. *Creativity and Innovation Management*, **5**(3): 169–178.

35 Williams B and Hummelbrunner R. 2010. *Systems Concepts in Action: A Practitioner's Toolkit*. Stanford University Press: Stanford, CA.

36 Schlicksupp H. 1999. *Innovation, Kreativität und Ideenfindung*. Vogel: Würzburg.

37 Yan HS. 1998. *Creative Design of Mechanical Devices*. Springer Science & Business Media: Berlin, Heidelberg.

38 Robson C. 2002. *Real World Research*, 2nd edn. Blackwell: Oxford.

39 Frey K, Lüthje C, and Haag S. 2011. Whom should firms attract to open innovation platforms? The role of knowledge diversity and motivation. *Long Range Planning*, **44**(5): 397–420.

40 Afuah A and Tucci CL. 2012. Crowdsourcing as a solution to distant search. *Academy of Management Review*, **37**(3): 355–375.

41 Sawhney M, Verona G, and Prandelli E. 2005. Collaborating to create: The Internet as a platform for customer engagement in product innovation. *Journal of Interactive Marketing*, **19**(4): 4–17.

42 Piller FT and Walcher D. 2006. Toolkits for idea competitions: A novel method to integrate users in new product development. *R&D Management*, **36**(3): 307–318.

43 Von Hippel E and Katz R. 2002. Shifting innovation to users via toolkits. *Management Science*, **48**(7): 821–833.

44 Pampers. 2013. *Discovery Center*. http://www.pampers.com/pg-baby-discovery-center (accessed 11 October 2013).

45 Neale MR and Corkindale DR. 1998. Co-developing products: Involving customers earlier and more deeply. *Long Range Planning*, **31**(3): 418–425.

46 Enkel E, Perez-Freije J, and Gassmann O. 2005. Minimizing market risks through customer integration in new product development: Learning from bad practice. *Creativity and Innovation Management*, **14**(4): 425–437.

47 Franke N and Piller F. 2004. Value creation by toolkits for user innovation and design: The case of the watch market. *Journal of Product Innovation Management*, **21**(6): 401–415.

48 Payne A, Storbacka K, Frow P, and Knox S. 2009. Co-creating brands: Diagnosing and designing the relationship experience. *Journal of Business Research*, **62**(3): 379–389.

49 Frohlich MT, Westbrook R. 2001. Arcs of integration: An international study of supply chain strategies. *Journal of Operations Management*, **19**(2): 185–200.

50 Wikström S. 1996. The customer as co-producer. *European Journal of Marketing*, **30**(4): 6–19.

51 Cooper RG, Kleinschmidt EJ. 1993. Major new products: What distinguishes the winners in the chemical industry? *Journal of Product Innovation Management*, **10**(2): 90–111.

52 Grönroos C. 2011. Value co-creation in service logic: A critical analysis. *Marketing Theory*, **11**(3): 279–301.

53 Johnson EJ, Moe WW, Fader PS, Bellman S, and Lohse GL. 2004. On the depth and dynamics of online search behavior. *Management Science*, **50**(3): 299–308.

54 InnoCentive, Inc. 2014. *Facts About InnoCentive*. https://www.innocentive.com/about-innocentive/facts-stats (accessed 15 April 2014).

55 InnoCentive, Inc. 2014. *Frequently Asked Questions*. https://www.innocentive.com/fa q/Solver#26n1254 (accessed 15 April 2014).

56 Sawhney M, Prandelli E, and Verona G. 2003. The power of innomediation. *MIT Sloan Management Review*, **44**(2): 77.

57 Wind J and Rangaswamy A. 2001. Customerization: The next revolution in mass customization. *Journal of Interactive Marketing*, **15**(1): 13–32.

58 Roser T, Samson A, Humphreys P, and Cruz-Valdivieso E. 2009. *Co-creation: New Pathways to Value: An Overview*. A report by Promise & LSE Enterprise.

59 BASF. 2014. *BASF Report 2013. Economic, Environmental and Social Performance*. Available at: http://www.basf.com/group/corporate/en/function/conversions:/ publish/content/about-basf/facts-reports/reports/2013/BASF_Report_2013.pdf (accessed 15 June 2014).

60 BASF. 2011. *BASF Factbook 2011. We Create Chemistry*. Available at: http://www.basf. com/group/corporate/en/function/conversions:/publish/content/investorrelations/news-publications/presentations/images/BASF_Factbook_2011.pdf (accessed 7 May 2014).

61 Schmidt R and Lange O. 2014. Account Management 2.0: From silo thinking to integrated account development. *Journal of Business Chemistry*, **11**(2), 77–86.

62 Leker J and Herzog DKP. 2004. *Marketing in der chemischen Industrie, in Handbuch Industriegütermarketing*, Gabler Verlag: Wiesbaden, pp. 1171–1193.
63 SpecialChem. 2004. *BASF Coatings Service Teams Stay On-site and Close to Customers*. http://www.specialchem4coatings.com/news-trends/displaynews.aspx?id=1942 (accessed 12 June 2014).
64 International Sheet Metal Review. 2007. *New Coating Innovation from BASF*. http://publishing.yudu.com/Abo6/ismrmarapril07/resources/10.htm (accessed 15 April 2014).
65 BASF. 2014. *Innovative Application Processes*. http://www.basf-coatings.com/global /ecweb/en/content/products_industries/automotive-oem-oatings/processes/index (accessed 15 June 2014).
66 Webwire. 2007. *BASF Coatings' World Innovation Successfully Integrated in Automotive OEM Coating*. http://www.webwire.com/ViewPressRel.asp?aId=27666#.U8yp DLHNziA (accessed 7 May 2014).
67 BASF. 2014. *Welcome to the BASF Co-location Web Site*. http://www2.basf.us/colocation/ (accessed 15 June 2014).
68 Johnsen TE. 2009. Supplier involvement in new product development and innovation: Taking stock and looking to the future. *Journal of Purchasing and Supply Management*, **15**(3): 187–197.
69 Clark KB and Fujimoto T. 1991. *Product Development Performance: Strategy, Organization, and Management in the World Auto Industry*. Harvard Business Press: Cambridge, MA.
70 Ragatz GL, Handfield RB, and Petersen KJ. 2002. Benefits associated with supplier integration into new product development under conditions of technology uncertainty. *Journal of Business Research*, **55**(5): 389–400.
71 Wagner SM. 2010. Supplier traits for better customer firm innovation performance. *Industrial Marketing Management*, **39**(7): 1139–1149.
72 Mandelson P. 2008. Verordnung (EG) Nr. 488/2008 der Kommission vom 2. Juni 2008 zur Ein-führung eines vorläufigen Antidumpingzolls auf die Einfuhren von Zitronensäure mit Ursprung in der Volksrepublik China. *Amtsblatt der Europäischen Union*, Brussels.
73 Milmo S. 2008. Good for the wallet, good for the planet. *Chemical Engineer*, **807**: 40–41.
74 Fichter K and Behrendt S. 2007. Grundlagen einer interaktiven Innovationstheorie. Beschreibungs-und Erklärungsmodelle als Basis für die empirische Untersuchung von nachhaltigkeitsrelevanten Innovationsprozessen in der Displayindustrie, in *Innovationsforschung. Ansätze Methoden Grenzen und Perspektiven, (Innovationsforschung, 1)*, (eds H Hof and U Wengenroth). LIT-Verlag: Hamburg, pp. 211–226.
75 Ehrenfeld JR. 1997. Industrial ecology: A framework for product and process design. *Journal of Cleaner Production*, **5**(1–2), 87–95.

76 Henkel AG & Co. KGaA. 2010. *Persil Hygiene Spüler*. http://www.henkel.de/de/content_data/PersilHygieneSpueler_Fakten2010.pdf. (accessed 18 April 2011).

77 Petersen KJ, Handfield RB, and Ragatz GL. 2005. Supplier integration into new product development: Coordinating product, process and supply chain design. *Journal of Operations Management*, **23**(3): 371–388.

78 Handfield RB and Nichols EL. 1999. *Introduction to Supply Chain Management*. Vol. 1. Prentice Hall: Upper Saddle River, NJ.

79 Wagner SM and Hoegl M. 2006. Involving suppliers in product development: Insights from R&D directors and project managers. *Industrial Marketing Management*, **35**(8): 936–943.

80 Koufteros XA, Cheng TE, and Lai KH. 2007. "Black-box" and "gray-box" supplier integration in product development: Antecedents, consequences and the moderating role of firm size. *Journal of Operations Management*, **25**(4): 847–870.

81 Backhaus K and Voeth M. 2007. *Industriegütermarketing*. Verlag Franz Vahlen: Munich.

82 Hartley JL, Zirger BJ, and Kamath RR. 1997. Managing the buyer-supplier interface for on-time performance in product development. *Journal of Operations Management*, **15**(1): 57–70.

83 Shin H, Collier DA, and Wilson DD. 2000. Supply management orientation and supplier/buyer performance. *Journal of Operations Management*, **18**(3): 317–333.

84 Chen IJ and Paulraj A. 2004. Towards a theory of supply chain management: The constructs and measurements. *Journal of Operations Management*, **22**(2): 119–150.

85 Sydow J. 2001. Management von Unternehmungsnetzwerken – Auf dem Weg zu einer reflexiven Netzwerkentwicklung? in *Kooperationsverbünde und regionale Modernisierung*. Gabler Verlag: Wiesbaden, pp. 79–101.

86 Thibaut JW and Kelley HH. 1986. *The Social Psychology of Groups*. John Wiley and Sons, Inc.: New York.

87 Wilson DT. 1995. An integrated model of buyer-seller relationships. *Journal of the Academy of Marketing Science*, **23**(4): 335–345.

88 Homans GC. 1961. *Social Behavior: Its Elementary Forms*. Routledge K. Paul: London.

89 Anderson JC. 1995. Relationships in business markets: Exchange episodes, value creation, and their empirical assessment. *Journal of the Academy of Marketing Science*, **23**(4): 346–350.

90 Walter A, Ritter T, and Gemünden HG. 2001. Value creation in buyer–seller relationships: Theoretical considerations and empirical results from a supplier's perspective. *Industrial Marketing Management*, **30**(4): 365–377.

91 Hogan JE. 2001. Expected relationship value: A construct, a methodology for measurement, and a modeling technique. *Industrial Marketing Management*, **30**(4): 339–351.

92 Lindgreen A and Wynstra F. 2005. Value in business markets: What do we know? Where are we going? *Industrial Marketing Management*, **34**(7): 732–748.

93 Buckley W. 1967. *Sociology and Modern Systems Theory*. Prentice Hall: Englewood Cliffs, NJ.

94 Dixon DF and Wilkinson IF. 1989. An alternative paradigm for marketing theory. *European Journal of Marketing*, **23**(8): 59–69.

95 Sheth JN and Sharma A. 1997. Supplier relationships: Emerging issues and challenges. *Industrial Marketing Management*, **26**(2): 91–100.

96 Kalwani MU and Narayandas N. 1995. Long-term manufacturer-supplier relationships: Do they pay off for supplier firms? *The Journal of Marketing*, 1–16.

97 Walter A. 2003. Relationship-specific factors influencing supplier involvement in customer new product development. *Journal of Business Research*, **56**(9): 721–733.

98 Cohen WM and Levinthal DA. 1990. Absorptive capacity: A new perspective on learning and innovation. *Administrative Science Quarterly*, 128–152.

99 Gomes-Casseres B. 1994. Group versus group: How alliance networks compete. *Harvard Business Review*, **72**(4): 62–66.

100 LaBahn DW and Krapfel R. 2000. Early supplier involvement in customer new product development: A contingency model of component supplier intentions. *Journal of Business Research*, **47**(3): 173–190.

101 Dwyer FR, Schurr PH, and Oh S. 1987. Developing buyer-seller relationships. *The Journal of Marketing*, 11–27.

102 Andaleeb SS. 1996. An experimental investigation of satisfaction and commitment in marketing channels: The role of trust and dependence. *Journal of Retailing*, **72**(1): 77–93.

103 Anderson E and Weitz B. 1989. Determinants of continuity in conventional industrial channel dyads. *Marketing Science*, **8**(4): 310–323.

104 Beiersdorf. 2014. *Open Innovation*. http://www.beiersdorf.com/innovation/open-innovation (accessed 2 July 2014).

105 Beiersdorf. 2014. *About Pearlfinder. Here's How Pearlfinder Works*. http://pearlfinder. beiersdorf.com/about-pearlfinder/how-pearlfinder-works (accessed 2 July 2014).

106 Beiersdorf. 2014. *About Pearlfinder. Request for Proposals*. http://pearlfinder.beiersdorf.com/about-pearlfinder/request-for-proposals (accessed 2 July 2014).

107 Bilgram V, Bartl M, and Biel S. 2011. Getting closer to the consumer – how Nivea co-creates new products. *Marketing Review St. Gallen*, **28**(1): 34–40.

108 Undershirt Guy Blog. 2014. *The Undershirt Guy. Ask Me Anything!* http://www.undershirtguy.com (accessed 26 May 2014).

109 Bilgram V. 2013. Eine Allianz gegen Flecken. *Harvard Business Manager*, March, 62–68.

Index

Business Chemistry: How to Build and Sustain Thriving Businesses in the Chemical Industry,
First Edition. Edited by Jens Leker, Stephan von Delft, and Carsten Gelhard.

Printed and bound by CPI Group (UK) Ltd, Croydon, CR0 4YY
16/04/2025
14658544-0003